Light and photosynthesis in aquatic ecosystems

# Light and photosynthesis in aquatic ecosystems

## JOHN T. O. KIRK

*Division of Plant Industry, Commonwealth Scientific and Industrial Research Organization, Canberra, Australia*

CAMBRIDGE UNIVERSITY PRESS

*Cambridge*

*London   New York   New Rochelle*

*Melbourne   Sydney*

Published by the Press Syndicate of the University of Cambridge
The Pitt Building, Trumpington Street, Cambridge CB2 1RP
32 East 57th Street, New York, NY 10022, USA
296 Beaconsfield Parade, Middle Park, Melbourne 3206, Australia

First published 1983

Printed in Great Britain by the University Press, Cambridge

Library of Congress catalogue card number: 83-5189

*British Library cataloguing in publication data*
Kirk, John T. O.
Light and photosynthesis in aquatic ecosystems.
1. Marine ecology 2. Light – Physiological effect
I. Title
574.5′2636   QH651

ISBN 0 521 24450 1

UP

# CONTENTS

..........................................................................................

# PREFACE

......................................................................................................

Four things are required for plant growth: energy in the form of solar radiation; inorganic carbon in the form of carbon dioxide or bicarbonate ions; mineral nutrients; and water. Those plants which, in the course of evolution, have remained in, or have returned to, the aquatic environment have one major advantage over their terrestrial counterparts: namely, that water – lack of which so often limits productivity in the terrestrial biosphere – is for them present in abundance; but for this a price must be paid. The medium – air – in which terrestrial plants carry on photosynthesis offers, within the sort of depth that plant canopies occupy, no significant obstacle to the penetration of light. The medium – water – in which aquatic plants occur, in contrast, both absorbs and scatters light. For the phytoplankton and the macrophytes in lakes and rivers, coastal and oceanic waters, both the intensity and the spectral quality of the light vary markedly with depth. In all but the shallowest waters, light availability is a limiting factor for primary production by the aquatic ecosystem. The aquatic plants must compete for solar radiation not only with each other (as terrestrial plants must also do), but also with all the other light-absorbing components of the aquatic medium. This has led, in the course of evolution, to the acquisition by each of the major groups of algae of characteristic arrays of light-harvesting pigments which are of great biochemical interest, and also of major significance for an understanding both of the adaptation of the algae to their ecological niche and of the phylogeny and taxonomy of the different algal groups. Nevertheless, in spite of the evolution of specialized light-harvesting systems, the aquatic medium removes so much of the incident light that aquatic ecosystems are, broadly speaking, less productive than terrestrial ones.

Thus, the nature of the light climate is a major difference between the terrestrial and the aquatic regions of the biosphere. Within the underwater environment, light availability is of major importance in determining how much plant growth there is, which kinds of plant predominate and, indeed, which kinds of plants have evolved. It is not the whole story – biotic factors, availability of inorganic carbon and mineral nutrients, temperature, all make their contribution – but it is a large part of that story. This book is a study of light in the underwater environment from the point of view of photosynthesis. It sets out to bring together the physics of light transmission through the medium and capture by the plants, the biochemistry of photosynthetic light-harvesting systems, the physiology of the photosynthetic response of aquatic plants to different kinds of light field, and the ecological relationships which depend on the light climate. The book does not attempt to provide as complete an account of the physical aspects of underwater light as the major works by Jerlov (1976) and Preisendorfer (1976); it is aimed at the limnologist and marine biologist rather than the physicist, although physical oceanographers should find it of interest. Its intention is to communicate a broad understanding of the significance of light as a major factor determining the operation and biological composition of aquatic ecosystems. It is hoped that it will be of value to practising aquatic scientists, to university teachers who give courses in limnology or marine science, and to postgraduate and honours students in these and related disciplines.

Certain features of the organization of the book merit comment. Although in some cases authors and dates are referred to explicitly, to minimize interruptions to the text, references to published work are in most cases indicated by the corresponding numbers in the complete alphabetical reference list at the end of the book. Accompanying each entry in the reference list is (are) the page number(s) where that paper or book is referred to in the text. Although coverage of the field is, I believe, representative, it is not intended to be encyclopaedic. The papers referred to have been selected, not only on the grounds of their scientific importance, but in large part on the basis of their usefulness as illustrative examples for particular points that need to be made. Inevitably, therefore, many equally important and relevant papers have had to be omitted from consideration, especially in the very broad field of aquatic ecology. I have therefore, when necessary, referred the reader to more specialized works in which more comprehensive treatments of particular topics can be found. Because its contribution to total aquatic primary production is usually small I have not attempted to deal with bacterial photosynthesis, complex and fascinating though it is.

I would like to thank Professor D. Branton, Dr M. Bristow, Mr S.

Craig, Dr W. A. Hovis, Dr S. Jeffrey, Dr D. Kiefer, Professor V. Klemas, Professor L. Legendre, Dr Y. Lipkin, Professor W. Nultsch, Mr D. Price, Professor R. C. Smith, Dr M. Vesk, and Biospherical Instruments Inc., who have provided original copies of figures for reproduction in this work, and Mr F. X. Dunin and Dr P. A. Tyler for unpublished data.

John Kirk

*Canberra*
*November 1982*

# PART I

THE UNDERWATER LIGHT FIELD

# 1

## Concepts of hydrological optics

### 1.1    Introduction

The purpose of the first part of this book is to describe and explain the behaviour of light in natural waters. The word 'light' in common parlance refers to radiation in that segment of the electromagnetic spectrum – about 400 to 700 nm – to which the human eye is sensitive. Our prime concern is not with vision but with photosynthesis. Nevertheless, by a convenient coincidence, the waveband within which plants can photosynthesize corresponds approximately to that of human vision and so we may legitimately refer to the particular kind of solar radiation with which we are concerned simply as 'light'.

Optics is that part of Physics which deals with light. Since the behaviour

Fig. 1.1. The relation between hydrological optics and other branches of optics (after Preisendorfer, 1976).

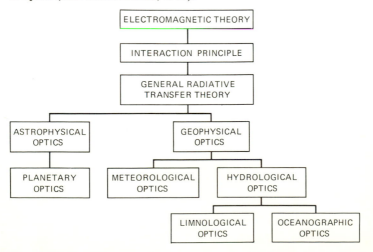

of light is greatly affected by the nature of the medium through which it is passing, there are different branches of optics dealing with different kinds of physical systems. The relations between the different branches of the subject and of optics to fundamental physical theory, are outlined diagrammatically in Fig. 1.1. Hydrological optics is concerned with the behaviour of light in aquatic media. It can be subdivided into limnological and oceanographic optics according to whether fresh, inland or salty, marine waters are under consideration. Hydrological optics has, however, up to now been mainly oceanographic in its orientation.

## 1.2    The nature of light

Electromagnetic energy occurs in indivisible units referred to as *quanta* or *photons*. Thus a beam of sunlight in air consists of a continual stream of photons travelling at $3 \times 10^8$ m s$^{-1}$. The actual numbers of quanta involved are very large. In full summer sunlight, for example, 1 m$^2$ of horizontal surface receives about $10^{21}$ quanta s$^{-1}$ of visible light.

Electromagnetic radiation, despite its particulate nature, behaves in some circumstances as though it has a wave nature. Every photon has a wavelength, $\lambda$, and a frequency, $v$. These are related in accordance with

$$\lambda = c/v \tag{1.1}$$

where $c$ is the speed of light. Since $c$ is constant in a given medium, the greater the wavelength, the lower the frequency. If $c$ is expressed in m s$^{-1}$, and $v$ in cycles s$^{-1}$, then the wavelength, $\lambda$, is expressed in m. For convenience, however, wavelength is more commonly expressed in nanometres, a namometre (nm) being equal to $10^{-9}$ m.

The energy, $\varepsilon$, in a photon varies with the frequency, and therefore inversely with the wavelength, the relation being

$$\varepsilon = hv = hc/\lambda \tag{1.2}$$

where $h$ is Planck's constant and has the value $6.63 \times 10^{-34}$ J s. Thus, a photon of wavelength 700 nm, from the red end of the photosynthetic spectrum, contains only 57% as much energy as a photon of wavelength 400 nm, from the blue end of the spectrum. The actual energy in a photon of wavelength $\lambda$ nm is given by the relation

$$\varepsilon = (1988/\lambda) \times 10^{-19} \text{ J} \tag{1.3}$$

A monochromatic radiation flux expressed in quanta s$^{-1}$ can thus readily be converted to J s$^{-1}$, i.e. to watts (W). Conversely, a radiation flux, $\Phi$, expressed in W, can be converted to quanta s$^{-1}$ using the relation

$$\text{quanta s}^{-1} = 5.03 \; \Phi\lambda \times 10^{15} \tag{1.4}$$

In the case of radiation covering a broad spectral band, such as for example the photosynthetic waveband, a simple conversion from quanta s$^{-1}$ to W, or *vice versa*, cannot be carried out accurately since the value of $\lambda$ varies across the spectral band. If the distribution of quanta or energy across the spectrum is known, then conversion can be carried out for a series of relatively narrow wavebands covering the spectral region of interest and the results summed for the whole waveband. Alternatively, an approximate conversion factor which takes into account the spectral distribution of energy that is likely to occur may be used. For solar radiation in the 400–700 nm band above the water surface, Morel & Smith (1974) found that the factor ($Q:W$) required to convert W to quanta s$^{-1}$ was $2.77 \times 10^{18}$ quanta s$^{-1}$ W$^{-1}$ to an accuracy of plus or minus a few per cent, regardless of the meteorological conditions.

As we shall discuss at length in a later section (§6.2) the spectral distribution of solar radiation under water changes markedly with depth. Nevertheless Morel & Smith found that for a wide range of marine waters the value of $Q:W$ varied by no more than $\pm 10\%$ from a mean of $2.5 \times 10^{18}$ quanta s$^{-1}$ W$^{-1}$. As expected from eqn 1.4, the greater the proportion of long-wavelength (red) light present, the greater the value of $Q:W$. For yellow inland waters with more of the underwater light in the 550–700 nm region (see §6.2), by extrapolating the data of Morel & Smith we arrive at a value of approximately 2.9 for the value of $Q:W$.

In any medium, light travels more slowly than it does in a vacuum. The velocity of light in a medium is equal to the velocity of light in a vacuum, divided by the refractive index of the medium. The refractive index of air is 1.00028, which for our purposes is not significantly different from that of a vacuum (exactly 1.0, by definition), and so we may take the velocity of light in air to be equal to that in a vacuum. The refractive index of water, although it varies somewhat with temperature, salt concentration and wavelength of light, may with sufficient accuracy be regarded as equal to 1.33 for all natural waters. Assuming that the velocity of light in a vacuum is $3 \times 10^8$ m s$^{-1}$, the velocity in water is therefore about $2.25 \times 10^8$ m s$^{-1}$. The frequency of the radiation remains the same in water but the wavelength diminishes in proportion to the decrease in velocity. When referring to monochromatic radiation, the wavelength we shall attribute to it is that which it has in a vacuum. Because $c$ and $\lambda$ change in parallel, eqns 1.2, 1.3 and 1.4 are as true in water as they are in a vacuum: furthermore, when using eqns 1.3 and 1.4, it is the value of the wavelength in a vacuum which is applicable, even when the calculation is carried out for underwater light.

### 1.3     The properties defining the radiation field

If we are to understand the ways in which the prevailing light field changes with depth in a water body, then we must first consider what are the essential attributes of a light field, in which changes might be anticipated. The definitions of these attributes, in part, follow the report of the Working Group set up by the International Association for the Physical Sciences of the Ocean (1979), but are also influenced by the more fundamental analyses given by Preisendorfer (1976).

We shall generally express direction within the light field in terms of the *zenith angle*, $\theta$ (the angle between a given light pencil, i.e. a thin parallel beam, and the upward vertical), and the *azimuth angle*, $\phi$ (the angle between the vertical plane incorporating the light pencil and some other specified vertical plane such as the vertical plane of the sun). In the case of the upwelling light stream it will sometimes be convenient to express a direction in terms of the *nadir angle*, $\theta_n$ (the angle between a given light pencil and the downward vertical). These angular relations are illustrated in Fig. 1.2.

*Radiant flux*, $\Phi$, is the time rate of flow of radiant energy. It may be expressed in W (J s$^{-1}$) or quanta s$^{-1}$.

Fig. 1.2. The angles defining direction within a light field. The figure shows a downward and an upward pencil of light, both, for simplicity, in the same vertical plane. The downward pencil has zenith angle $\theta$; the upward pencil has nadir angle $\theta_n$, which is equivalent to a zenith angle of $(180° - \theta_n)$. Assuming the $xy$ plane is the vertical plane of the Sun, or other reference vertical plane, then $\phi$ is the azimuth angle for both light pencils.

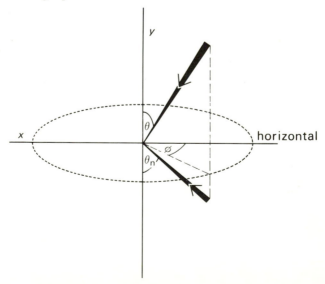

*Radiant intensity*, $I$, is a measure of the radiant flux per unit solid angle in a specified direction. The radiant intensity of a source in a given direction is the radiant flux emitted by a point source, or by an element of an extended source, in an infinitesimal cone containing the given direction, divided by that element of solid angle. We can also speak of radiant intensity at a point in space. This, the *field* radiant intensity, is the radiant flux at that point in a specified direction in an infinitesimal cone containing the given direction, divided by that element of solid angle. $I$ has the units W (or quanta s⁻¹) steradian⁻¹.

$$I = d\Phi/d\omega$$

If we consider the radiant flux not only per unit solid angle but also per unit area of a plane at right angles to the direction of flow, then we arrive at the even more useful concept of radiance, $L$. Radiance at a point in space is the radiant flux at that point in a given direction per unit solid angle

Fig. 1.3. Definition of radiance. (*a*) Field radiance at a point in space. The field radiance at P in the direction D is the radiant flux in the small solid angle surrounding D, passing through the infinitesimal element of area d$A$ at right angles to D, divided by the element of solid angle and the element of area. (*b*) Field radiance at a point in a surface. It is often necessary to consider radiance at a point on a surface, from a specified direction relative to that surface. d$S$ is the area of a small element of surface. $L(\theta, \phi)$ is the radiance incident on d$S$ at zenith angle $\theta$ (relative to the normal to the surface) and azimuth angle $\phi$: its value is determined by the radiant flux directed at d$S$ within the small solid angle, d$\omega$, centred on the line defined by $\theta$ and $\phi$. The flux passes perpendicularly across the area d$S \cos \theta$, which is the projected area of the element of surface, d$S$, seen from the direction $\theta, \phi$. Thus the radiance on a point in a surface, from a given direction, is the radiant flux in the specified direction per unit solid angle per unit projected area of the surface. (*c*) Surface radiance. In the case of a surface which emits radiation the intensity of the flux leaving the surface in a specified direction is expressed in terms of the surface radiance which is defined in the same way as the field radiance at a point in a surface except that the radiation is considered to flow away from, rather than on to, the surface.

(*a*)

$$L = \frac{d^2\Phi}{dA\, d\omega}$$

(*b*)

$$L(\theta,\phi) = \frac{d^2\Phi}{dS\cos\theta\, d\omega}$$

(*c*)

per unit area at right angles to the direction of propagation. The meaning of this *field* radiance is illustrated in Figs 1.3*a* and *b*. There is also *surface* radiance which is the radiant flux emitted in a given direction per unit solid angle per unit projected area (apparent unit area, seen from the viewing direction) of a surface: this is illustrated in Fig. 1.3*c*.

To indicate that it is a function of direction, i.e. of both zenith and azimuth angle, radiance is commonly written as $L(\theta, \phi)$. The angular structure of a light field is expressed in terms of the variation of radiance with $\theta$ and $\phi$. Radiance has the units W (or quanta s$^{-1}$) m$^{-2}$ steradian$^{-1}$.

$$L(\theta, \phi) = d^2\Phi/dS \cos\theta \, d\omega$$

*Irradiance* (at a point of a surface), $E$, is the radiant flux incident on an infinitesimal element of a surface, containing the point under consideration, divided by the area of that element. Less rigorously, it may be defined as the radiant flux per unit area of a surface. It has the units W (or quanta s$^{-1}$) m$^{-2}$.

$$E = d\Phi/dS$$

*Downward irradiance*, $E_d$, and *upward irradiance*, $E_u$, are the values of the irradiance on the upper and the lower faces, respectively, of a horizontal surface. Thus, $E_d$ is the irradiance due to the downwelling light stream and $E_u$ is that due to the upwelling light stream.

The relation between irradiance and radiance can be understood with the help of Fig. 1.3*b*. The radiance in the direction defined by $\theta$ and $\phi$ is $L(\theta, \phi)$ W (or quanta s$^{-1}$) per unit projected area per steradian (sr). The projected area of the element of surface is $dS \cos\theta$ and the corresponding element of solid angle is $d\omega$. Therefore the radiant flux on the element of surface within the solid angle $d\omega$ is $L(\theta, \phi) dS \cos\theta \, d\omega$. The area of the element of surface is $dS$ and so the irradiance at that point in the surface where the element is located, due to radiant flux within $d\omega$, is $L(\theta, \phi) \cos\theta \, d\omega$. The total downward irradiance at that point in the surface is obtained by integrating with respect to solid angle over the whole upper hemisphere

$$E_d = \int_{2\pi} L(\theta, \phi) \cos\theta \, d\omega \qquad (1.5)$$

The total upward irradiance is related to radiance in a similar manner except that allowance must be made for the fact that $\cos\theta$ is negative for values of $\theta$ between 90 and 180°

$$E_u = -\int_{-2\pi} L(\theta, \phi) \cos\theta \, d\omega \qquad (1.6)$$

Alternatively the cosine of the nadir angle, $\theta_n$ (see Fig. 1.2), rather than of the zenith angle, may be used

$$E_u = \int_{-2\pi} L(\theta_n, \phi) \cos \theta_n \, d\omega \qquad (1.7)$$

The $-2\pi$ subscript is simply to indicate that the integration is carried out over the $2\pi$ sr solid angle in the *lower* hemisphere.

The *net downward irradiance*, $\vec{E}$, is the difference between the downward and the upward irradiance

$$\vec{E} = E_d - E_u \qquad (1.8)$$

It is related to radiance by the eqn

$$\vec{E} = \int_{4\pi} L(\theta, \phi) \cos \theta \, d\omega \qquad (1.9)$$

which integrates the product of radiance and $\cos \theta$ over all directions: the fact that $\cos \theta$ is negative between 90 and 180° ensures that the contribution of upward irradiance is negative in accordance with eqn 1.8. The net downward irradiance is a measure of the net rate of transfer of energy downwards at that point in the medium, and as we shall see later is a concept which can be used to arrive at some valuable conclusions.

The *scalar irradiance*, $E_0$, is the integral of the radiance distribution at a point over all directions about the point

$$E_0 = \int_{4\pi} L(\theta, \phi) \, d\omega \qquad (1.10)$$

Scalar irradiance is thus a measure of the radiant intensity at a point, which treats radiation from all directions equally. In the case of irradiance, on the other hand, the contribution of the radiation flux at different angles varies in proportion to the cosine of the zenith angle of incidence of the radiation: a phenomenon based on purely geometrical relations (Fig. 1.2, eqn 1.5), and sometimes referred to as the Cosine Law. It is useful to divide the scalar irradiance into a downward and an upward component. The *downward scalar irradiance*, $E_{0d}$, is the integral of the radiance distribution over the upper hemisphere

$$E_{0d} = \int_{2\pi} L(\theta, \phi) \, d\omega \qquad (1.11)$$

The *upward scalar irradiance*, $E_{0u}$, is defined in a similar manner for the lower hemisphere

$$E_{0u} = \int_{-2\pi} L(\theta, \phi) \, d\omega \qquad (1.12)$$

Scalar irradiance (total, upward, downward) has the units W (or quanta $s^{-1}$) $m^{-2}$.

It is always the case in real-life radiation fields that irradiance and scalar irradiance vary markedly with wavelength across the photosynthetic range. This variation has a considerable bearing on the extent to which the radiation field can be used for photosynthesis. It is expressed in terms of the variation of irradiance or scalar irradiance per unit spectral distance (in units of wavelength or frequency, as appropriate) across the spectrum. Typical units would be W (or quanta $s^{-1}$) $m^{-2}$ $nm^{-1}$.

If we know the radiance distribution over all angles at a particular point in a medium then we have a complete description of the angular structure of the light field. A complete radiance distribution, however, covering all zenith and azimuth angles, at reasonably narrow intervals, represents a large amount of data: with 5° angular intervals, for example, the distribution will consist of 1369 separate radiance values. A simpler, but still very useful, way of specifying the angular structure of a light field is in the form of the three average cosines – for downwelling, upwelling and total light – and the irradiance reflectance.

The average cosine for downwelling light, $\bar{\mu}_d$, at a particular point in the radiation field, may be regarded as the average value, in an infinitesimally small volume element at that point in the field, of the cosine of the zenith angle of all the downwelling photons in the volume element. It can be calculated by summing (i.e. integrating) for all elements of solid angle (d$\omega$) comprising the upper hemisphere, the product of the radiance in that element of solid angle and the value of $\cos\theta$ (i.e. $L(\theta,\phi)\cos\theta$), and then dividing by the total radiance originating in that hemisphere. By inspection of eqns 1.5 and 1.11 it can be seen that

$$\bar{\mu}_d = E_d/E_{0d} \qquad (1.13)$$

i.e. the average cosine for downwelling light is equal to the downward irradiance divided by the downward scalar irradiance.

The average cosine for upwelling light, $\bar{\mu}_u$, may be regarded as the average value of the cosine of the nadir angle of all the upwelling photons at a particular point in the field. By a similar chain of reasoning to the above, we conclude that $\bar{\mu}_u$ is equal to the upward irradiance divided by the upward scalar irradiance

$$\bar{\mu}_u = E_u/E_{0u} \qquad (1.14)$$

The average cosine, $\bar{\mu}$, for the total light at a particular point in the field may be regarded as the average value, in an infinitesimally small volume element at that point in the field, of the cosine of the zenith angle of all the photons in the volume element. It may be evaluated by integrating the product of radiance and $\cos\theta$ over all directions and dividing by the total radiance from all directions. By inspection of eqns 1.8, 1.9 and 1.10, it can be seen that the average cosine for the total light is equal to the *net* downward irradiance divided by the scalar irradiance

$$\bar{\mu} = \frac{\vec{E}}{E_0} = \frac{E_d - E_u}{E_0} \tag{1.15}$$

That $E_d - E_u$ should be involved (rather than, say, $E_d + E_u$) follows from the fact that the cosine of the zenith angle is negative for all the upwelling photons ($90° < \theta < 180°$). Thus a radiation field consisting of equal numbers of downwelling photons at $\theta = 45°$ and upwelling photons at $\theta = 135°$ would have $\bar{\mu} = 0$.

The remaining parameter which provides information about the angular structure of the light field is the *irradiance reflectance* (sometimes called the *irradiance ratio*), $R$. It is the ratio of the upward to the downward irradiance at a given point in the field

$$R = E_u/E_d \tag{1.16}$$

In any absorbing and scattering medium, such as sea or inland water, all these properties of the light field change in value with depth (for which we use the symbol $z$): the change might typically be a decrease, as in the case of irradiance, or an increase, as in the case of reflectance. It is sometimes useful to have a measure of the rate of change of any given property with depth. All the properties with which we have dealt that have the dimensions of radiant flux per unit area, diminish in value, as we shall see later, in an approximately exponential manner with depth. It is convenient with these properties to specify the rate of change of the logarithm of the value with depth since this will be approximately the same at all depths. In this way we may define the *vertical attenuation coefficient* for

downward irradiance

$$K_d = -\frac{d\ln E_d}{dz} = -\frac{1}{E_b}\frac{dE_d}{dz} \tag{1.17}$$

upward irradiance

$$K_u = -\frac{d\ln E_u}{dz} = -\frac{1}{E_u}\frac{dE_u}{dz} \tag{1.18}$$

net downward irradiance

$$K_E = -\frac{d\ln(E_d - E_u)}{dz} = -\frac{1}{(E_d - E_u)}\frac{d(E_d - E_u)}{dz} \tag{1.19}$$

scalar irradiance

$$K_0 = -\frac{d \ln E_0}{dz} = -\frac{1}{E_0} \frac{dE_0}{dz} \qquad (1.20)$$

radiance

$$K(\theta, \phi) = -\frac{d \ln L(\theta, \phi)}{dz} = -\frac{1}{L(\theta, \phi)} \frac{dL(\theta, \phi)}{dz} \qquad (1.21)$$

In recognition of the fact that the values of these vertical attenuation coefficients are to some extent a function of depth they may sometimes be written in the form $K(z)$.

## 1.4    The inherent optical properties

There are only two things that can happen to photons within water: they can be absorbed or they can be scattered. Thus if we are to understand what happens to solar radiation as it passes into any given water body, we need some measure of the extent to which that water absorbs and scatters light. The absorption and scattering properties of the aquatic medium for light of any given wavelength are specified in terms of the absorption coefficient, the scattering coefficient and the volume scattering function. These have been referred to by Preisendorfer (1961) as *inherent* optical properties, because their magnitudes depend only on the substances comprising the aquatic medium and not on the geometric structure of the light fields that may pervade it.

They are defined with the help of an imaginary, infinitesimally thin, plane parallel layer of medium, illuminated at right angles by an infinitesimally narrow, parallel beam of monochromatic light (Fig. 1.4). Some of the incident light is absorbed by the thin layer. Some is scattered – that is, caused to diverge from its original path. The fraction of the incident flux which is absorbed, divided by the thickness of the layer, is the *absorption coefficient, a*. The fraction of the incident flux which is scattered, divided by the thickness of the layer, is the *scattering coefficient, b*.

To express the definitions quantitatively we make use of the quantities *absorptance, A*, and *scatterance, B*. If $\Phi_0$ is the radiant flux (energy or quanta per unit time) incident in the form of a parallel beam on some physical system, $\Phi_a$ is the radiant flux absorbed by the system, and $\Phi_b$ is the radiant flux scattered by the system, then

$$A = \Phi_a/\Phi_0 \qquad (1.22)$$

and

$$B = \Phi_b/\Phi_0 \qquad (1.23)$$

i.e. absorptance and scatterance are the fractions of the radiant flux lost from the incident beam, by absorption and scattering, respectively. The

sum of absorptance and scatterance is referred to as *attenuance*, *C*: it is
the fraction of the radiant flux lost from the incident beam by absorption
and scattering combined. In the case of the infinitesimally thin layer,
thickness $\Delta r$, we represent the very small fractions of the incident flux which
are lost by absorption and scattering as $\Delta A$ and $\Delta B$, respectively. Then

$$a = \Delta A / \Delta r \qquad\qquad (1.24)$$

and

$$b = \Delta B / \Delta r \qquad\qquad (1.25)$$

An additional inherent optical property which we may now define is the
*beam attenuation coefficient, c*. It is given by

$$c = a + b \qquad\qquad (1.26)$$

and is the fraction of the incident flux which is absorbed and scattered,
divided by the thickness of the layer. If the very small fraction of the
incident flux which is lost by absorption and scattering combined is given
the symbol $\Delta C$ (where $\Delta C = \Delta A + \Delta B$) then

$$c = \Delta C / \Delta r \qquad\qquad (1.27)$$

The absorption, scattering and beam attenuation coefficients all have units
of 1/length, and are normally expressed in $m^{-1}$.

In the real world we cannot carry out measurements on infinitesimally
thin layers, and so if we are to determine the values of *a*, *b* and *c* we need

Fig. 1.4. Interaction of a beam of light with a thin layer of aquatic
medium. Of the light that is not absorbed, most is transmitted without
deviation from its original path: some light is scattered, mainly in a
forward direction.

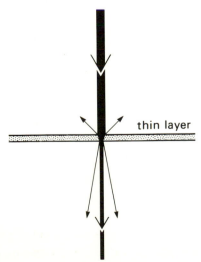

expressions which relate these coefficients to the absorptance, scatterance and beam attenuance of layers of finite thickness. Consider a medium illuminated perpendicularly with a thin parallel beam of radiant flux, $\Phi_0$. As the beam passes through it loses intensity by absorption and scattering. Consider now an infinitesimally thin layer, thickness $\Delta r$, within the medium at a depth, $r$, where the radiant flux in the beam has diminished to $\Phi$. The change in radiant flux in passing through $\Delta r$ is $\Delta\Phi$. The attenuance of the thin layer is

$$\Delta C = -\Delta\Phi/\Phi$$

(the negative sign is necessary since $\Delta\Phi$ must be negative)

$$\Delta\Phi/\Phi = -c\Delta r$$

Integrating between 0 and $r$ we obtain

$$\ln\frac{\Phi}{\Phi_0} = -cr \tag{1.28}$$

or

$$\Phi = \Phi_0 e^{-cr} \tag{1.29}$$

indicating that the radiant flux diminishes exponentially with distance along the path of the beam. Eqn 1.28 may be rewritten

$$c = \frac{1}{r}\ln\frac{\Phi_0}{\Phi} \tag{1.30}$$

or

$$c = -\frac{1}{r}\ln(1-C) \tag{1.31}$$

The value of the beam attenuation coefficient, $c$, can therefore, using eqn 1.30 or 1.31, be obtained from measurements of the diminution in intensity of a parallel beam passing through a known pathlength of medium, $r$.

The theoretical basis for the measurement of the absorption and scattering coefficients is less simple. In a medium with absorption but negligible scattering, the relation

$$a = -\frac{1}{r}\ln(1-A) \tag{1.32}$$

holds, and in a medium with scattering but negligible absorption, the relation

$$b = -\frac{1}{r}\ln(1-B) \tag{1.33}$$

holds, but in any medium which both absorbs and scatters light to a significant extent, neither relation is true. This can readily be seen by considering the application of these equations to such a medium.

In the case of eqn 1.33 some of the measuring beam will be removed by absorption within the pathlength $r$ before it has had the opportunity to be scattered, and so the amount of light scattered, $B$, will be lower than that required to satisfy the equation. Similarly, $A$ will have a value lower than that required to satisfy eqn 1.32 since some of the light will be removed from the measuring beam by scattering before it has had the chance to be absorbed.

In order to actually measure $a$ or $b$ these problems must be circumvented. In the case of the absorption coefficient, it is possible to arrange that most of the light scattered from the measuring beam still passes through approximately the same pathlength of medium and is collected by the detection system. Thus the contribution of scattering to total attenuation is made very small and eqn 1.32 may be used. In the case of the scattering coefficient there is no instrumental way of avoiding the losses due to absorption and so the absorption must be determined separately and appropriate corrections made to the scattering data. We shall consider ways of measuring $a$ and $b$ in more detail later (§§3.2 and 4.2).

The way in which scattering affects the penetration of light into the medium depends not only on the value of the scattering coefficient but also on the angular distribution of the scattered flux resulting from the primary scattering process. This angular distribution has a characteristic shape for any given medium and is specified in terms of the *volume scattering function*, $\beta(\theta)$. This is defined as the radiant intensity in a given direction from a volume element, $dV$, illuminated by a parallel beam of light, per unit of irradiance on the cross-section of the volume, and per unit volume (Fig. 1.5a). The definition is usually expressed mathematically in the form

$$\beta(\theta) = dI(\theta)/EdV \tag{1.34}$$

Since, from the definitions in §1.3

$$dI(\theta) = d\Phi(\theta)/d\omega$$

and

$$E = \Phi_0/dS$$

where $d\Phi(\theta)$ is the radiant flux in the element of solid angle $d\omega$ oriented at angle $\theta$ to the beam, and $\Phi_0$ is the flux incident on the cross-sectional area, $dS$, and since

$$dV = dS \cdot dr$$

where $dr$ is the thickness of the volume element, then we may write

$$\beta(\theta) = \frac{d\Phi(\theta)}{\Phi_0} \cdot \frac{1}{d\omega dr} \tag{1.35}$$

The volume scattering function has the units $m^{-1} sr^{-1}$.

Light scattering from a parallel light beam passing through a thin layer

Fig. 1.5. The geometrical relation underlying the volume scattering function. (*a*) A parallel light beam of irradiance *E* and cross-sectional area d*A* passes through a thin layer of medium, thickness d*r*. The illuminated element of volume is d*V*. d*I(θ)* is the radiant intensity due to light scattered at angle *θ*. (*b*) The point at which the light beam passes through the thin layer of medium can be imagined as being at the centre of a sphere of unit radius. The light scattered between *θ* and *θ*+Δ*θ* illuminates a circular strip, radius sin *θ* and width Δ*θ*, around the surface of the sphere. The area of the strip is $2\pi \sin \theta \, \Delta\theta$ which is equivalent to the solid angle (in steradians) corresponding to the angular interval Δ*θ*.

of medium is radially symmetrical around the direction of the beam. Thus, the light scattered at angle $\theta$ should be thought of as a cone with half-angle $\theta$, rather than as a pencil of light (Fig. 1.5*b*).

From eqn 1.35 we see that $\beta(\theta)$ is the radiant flux per unit solid angle scattered in the direction $\theta$, per unit pathlength in the medium, expressed as a proportion of the incident flux. The angular interval $\theta$ to $\theta + \Delta\theta$ corresponds to an element of solid angle, $d\omega$, equal to $2\pi \sin\theta\Delta\theta$ (Fig. 1.5*b*) and so the proportion of the incident radiant flux scattered (per unit pathlength) in this angular interval is $\beta(\theta)\,2\pi \sin\theta\Delta\theta$. To obtain the proportion of the incident flux which is scattered in all directions per unit pathlength – by definition, equal to the scattering coefficient – we must integrate over the angular range $\theta = 0°$ to $\theta = 180°$

$$b = 2\pi \int_0^\pi \beta(\theta) \sin\theta\,d\theta = \int_{4\pi} \beta(\theta)d\omega \tag{1.36}$$

Thus, an alternative definition of the scattering coefficient is the integral of the volume scattering function over all directions.

It is frequently useful to distinguish between scattering in a forward direction and that in a backward direction. We therefore partition the total scattering coefficient, $b$, into a *forward scattering coefficient*, $b_f$, relating to light scattered from the beam in a forward direction, and a *backward scattering coefficient* (or simply, *backscattering coefficient*) $b_b$, relating to light scattered from the beam in a backward direction

$$b = b_b + b_f \tag{1.37}$$

We may also write

$$b_f = 2\pi \int_0^{\pi/2} \beta(\theta) \sin\theta\,d\theta \tag{1.38}$$

$$b_b = 2\pi \int_{\pi/2}^{\pi} \beta(\theta) \sin\theta\,d\theta \tag{1.39}$$

The variation of $\beta(\theta)$ with $\theta$ tells us the absolute amount of scattering at different angles, per unit pathlength in a given medium. If we wish to compare the *shape* of the angular distribution of scattering in different media separately from the absolute amount of scattering that occurs, then it is convenient to use the *normalized volume scattering function*, $\tilde{\beta}(\theta)$, which is that dimensionless function obtained by dividing the volume scattering function by the total scattering coefficient

$$\tilde{\beta}(\theta) = \beta(\theta)/b \tag{1.40}$$

The integral of $\tilde{\beta}(\theta)$ over all solid angles is equal to 1. The integral of $\tilde{\beta}(\theta)$ up to any given value of $\theta$ is the proportion of the total scattering which occurs in the angular interval between $0°$ and that value of $\theta$. We can also define normalized forward scattering and backward scattering coefficients, $\tilde{b}_f$ and $\tilde{b}_b$, as the proportions of the total scattering in forwards and backwards directions, respectively

$$\tilde{b}_f = b_f/b \tag{1.41}$$

$$\tilde{b}_b = b_b/b \tag{1.42}$$

## 1.5    Apparent and quasi-inherent optical properties

The vertical attenuation coefficients for radiance, irradiance and scalar irradiance are, strictly speaking, properties of the radiation field since, by definition, each of them is the logarithmic derivative with respect to depth of the radiometric quantity in question. Nevertheless experience has shown that their values are largely determined by the inherent optical properties of the aquatic medium and are not very much altered by changes in the incident radiation field such as a change in solar elevation.[24] For example, if a particular water body is found to have a high value of $K_d$ then we expect it to have approximately the same high $K_d$ tomorrow, or next week, or at any time of day, so long as *the composition of the water remains about the same.*

Vertical attenuation coefficients, such as $K_d$, are thus commonly used, and thought of, by oceanographers and limnologists as though they are optical properties belonging to the water, properties which are a direct measure of the ability of that water to bring about a diminution in the appropriate radiometric quantity with depth. Furthermore they have the same units ($m^{-1}$) as the inherent optical properties $a$, $b$ and $c$. In recognition of these useful aspects of the various $K$ functions, Preisendorfer (1961) has suggested that they be classified as *apparent optical properties* and we shall so treat them in this book. We shall not, however, adopt his suggestion that reflectance, $R$, and the distribution functions (the reciprocals of $\bar{\mu}_d$ and $\bar{\mu}_u$) also be regarded as apparent optical properties, partly because they are dimensionless, but also because they do not in fact seem to be generally treated as such by workers in this area. We shall continue to classify $R$ and the average cosines strictly as properties of the radiation field.

The two fundamental inherent optical properties – the coefficients for absorption and scattering – are, as we saw earlier, defined in terms of the behaviour of a parallel beam of light incident upon a thin layer of medium. Analogous coefficients can be defined for incident light streams having any specified angular distribution. In particular, such coefficients can be defined for incident light streams corresponding to the upwelling and downwelling streams that exist at particular depths in real water bodies.

We shall refer to these as the *diffuse* absorption and scattering coefficients for the upwelling or downwelling light streams at a given depth. Although related to the normal coefficients, the values of the diffuse coefficients are a function of the local radiance distribution, and therefore of depth.

The *diffuse absorption coefficient* for the downwelling light stream at depth $z$, $a_d(z)$, is the proportion of the incident radiant flux which would be absorbed from the downwelling stream by an infinitesimally thin horizontal plane parallel layer at that depth, divided by the thickness of the layer. The diffuse absorption coefficient for the upwelling stream, $a_u(z)$, is defined in a similar way. Absorption of a diffuse light stream within the thin layer will be greater than absorption of a normally incident parallel beam because the pathlengths of the photons will be greater. The average pathlength of the downwelling and upwelling photons will be in proportion to $1/\bar{\mu}_d$ and $1/\bar{\mu}_u$, respectively. The diffuse absorption coefficients are therefore related to the normal absorption coefficients by

$$a_d(z) = \frac{a}{\bar{\mu}_d(z)} \tag{1.43}$$

$$a_u(z) = \frac{a}{\bar{\mu}_u(z)} \tag{1.44}$$

where $\bar{\mu}_d(z)$ and $\bar{\mu}_u(z)$ are the values of $\bar{\mu}_d$ and $\bar{\mu}_u$ which exist at depth $z$.

So far as scattering of the upwelling and downwelling light streams is concerned, it is mainly the backward scattering component which is of importance. The diffuse backscattering coefficient for the downwelling stream at depth $z$, $b_{bd}(z)$, is the proportion of the incident radiant flux from the downwelling stream which would be scattered backwards (i.e. upwards) by an infinitesimally thin, horizontal plane parallel layer at that depth, divided by the thickness of the layer: $b_{bu}(z)$, the corresponding coefficient for the upwelling stream is defined in the same way in terms of the light scattered downwards again from that stream. Diffuse total $(b_d(z), b_u(z))$ and forward $(b_{fd}(z), b_{fu}(z))$ scattering coefficients for the downwelling and upwelling streams can be defined in a similar manner. The following relations hold

$$b_d(z) = b/\bar{\mu}_d(z), \qquad b_u(z) = b/\bar{\mu}_u(z)$$
$$b_d(z) = b_{fd}(z) + b_{bd}(z), \qquad b_u(z) = b_{fu}(z) + b_{bu}(z)$$

Attenuation coefficients for the downwelling and upwelling light streams, $c_d(z)$ and $c_u(z)$, can also be defined

$$c_d(z) = a_d(z) + b_d(z), \qquad c_u(z) = a_u(z) + b_u(z)$$

The relation between a diffuse backscattering coefficient and the normal backscattering coefficient, $b_b$, is not simple but may be calculated from the volume scattering function and the radiance distribution existing at depth $z$. The calculation procedure is discussed later (§4.2).

Preisendorfer (1961) has classified the diffuse absorption, scattering and attenuation coefficients as *hybrid optical properties* on the grounds that they are derived both from the inherent optical properties and certain properties of the radiation field. I prefer the term *quasi-inherent optical properties*, on the grounds that it more clearly indicates the close relation between these properties and the inherent optical properties. Both sets of properties have precisely the same kind of definition: they differ only in the characteristics of the light flux that is imagined to be incident upon the thin layer of medium. Baker & Smith (1979) have applied the term 'quasi-inherent' to $K_d$, but as discussed above we shall here classify $K_d$ as an 'apparent' optical property.

### 1.6    Optical depth

As we have already noted, but will discuss more fully later, the downward irradiance diminishes in an approximately exponential manner with depth. This may be expressed by the equation

$$E_d(z) = E_d(0)\, e^{-K_d z} \tag{1.45}$$

where $E_d(z)$ and $E_d(0)$ are the values of downward irradiance at $z$ m depth, and just below the surface, respectively, and $K_d$ is the average value of the vertical attenuation coefficient over the depth interval 0 to $z$ m. We shall now define the *optical depth*, $\zeta$, by the eqn

$$\zeta = K_d z \tag{1.46}$$

It can be seen that a specified optical depth will correspond to different physical depths but to the same overall diminution of irradiance, in waters of differing optical properties. Thus in a coloured turbid water with a high $K_d$, a given optical depth will correspond to a much smaller actual depth than in a clear colourless water with a low $K_d$. Optical depth, $\zeta$, as defined here is distinct from *attenuation length*, $\tau$ (sometimes also called optical depth or optical distance), which is the geometrical length of a path multiplied by the beam attenuation coefficient ($c$) associated with the path.

Optical depths of particular interest in the context of primary production are those corresponding to attenuation of downward irradiance to $10\%$ and $1\%$ of the subsurface values: these are $\zeta = 2.3$ and $\zeta = 4.6$, respectively. These optical depths correspond to the mid-point and the lower limit of the euphotic zone, within which significant photosynthesis occurs.

## 1.7    Radiative transfer theory

Having defined the properties of the light field and the optical properties of the medium we are now in a position to ask whether it is possible to arrive, on purely theoretical grounds, at any relations between them. Although, given a certain incident light field, the characteristics of the underwater light field are uniquely determined by the properties of the medium, it is nevertheless true that explicit, all-embracing analytical relations, expressing the characteristics of the field in terms of the inherent optical properties of the medium, have not yet been derived. Given the complexity of the shape of the volume scattering function in natural waters (see Chapter 4), it may be that this will never be achieved.

It is, however, possible to arrive at a useful expression relating the absorption coefficient to the average cosine and the vertical attenuation coefficient for net downward irradiance. In addition, relations have been derived between certain properties of the field and the diffuse optical properties. These various relations are all arrived at by making use of the equation of transfer for radiance. This describes the manner in which radiance varies with distance along any specified path at a specified point in the medium.

Assuming a horizontally stratified water body (i.e. with properties everywhere constant at a given depth), with a constant input of mono-

Fig. 1.6. The processes underlying the equation of transfer of radiance. A light beam passing through a distance, $dr$, of medium, in the direction $\theta, \phi$, loses some photons by scattering out of the path and some by absorption by the medium along the path, but also acquires new photons by scattering of light initially travelling in other directions $(\theta', \phi')$ into the direction $\theta, \phi$.

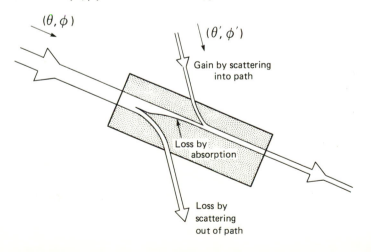

chromatic unpolarized radiation at the surface, and ignoring fluorescent emission within the water, the equation may be written

$$\frac{dL(z, \theta, \phi)}{dr} = -c(z)\,L(z, \theta, \phi) + L^*(z, \theta, \phi) \tag{1.47}$$

The term on the left is the rate of change of radiance with distance, $r$, along the path specified by zenith and azimuthal angles $\theta$ and $\phi$, at depth $z$. The net rate of change is the resultant of two opposing processes: loss by attenuation along the direction of travel ($c(z)$ being the value of the beam attenuation coefficient at depth $z$), and gain by scattering (along the path $dr$) of light initially travelling in other directions ($\theta', \phi'$) into the direction $\theta, \phi$ (Fig. 1.6). This latter term is determined by the volume scattering function of the medium at depth $z$ (which we write $\beta(z, \theta, \phi; \theta', \phi')$ to indicate that the scattering angle is the angle between the two directions $\theta, \phi$ and $\theta', \phi'$), and by the distribution of radiance, $L(z, \theta', \phi')$. Each element of irradiance, $L(z, \theta', \phi')\,d\omega(\theta', \phi')$, (where $d\omega(\theta', \phi')$ is an element of solid angle forming an infinitesimal cone containing the direction $\theta', \phi'$) incident on the volume element along $dr$ gives rise to some scattered radiance in the direction $\theta, \phi$. The total radiance derived in this way is given by

$$L^*(z, \theta, \phi) = \int_{4\pi} \beta(z, \theta, \phi; \theta', \phi')\,L(z, \theta', \phi')\,d\omega(\theta', \phi') \tag{1.48}$$

If we are interested in the variation of radiance in the direction $\theta, \phi$ as a function of depth, then since $dr = dz/\cos\theta$, we may rewrite eqn 1.47 as

$$\cos\theta \frac{dL(z, \theta, \phi)}{dz} = -c(z)L(z, \theta, \phi) + L^*(z, \theta, \phi) \tag{1.49}$$

By integrating each term of this equation over all angles

$$\int_{4\pi} \cos\theta \frac{dL(z, \theta, \phi)}{dz}\,d\omega = -\int_{4\pi} c(z)\,L(z, \theta, \phi)\,d\omega$$

$$+ \int_{4\pi} L^*(z, \theta, \phi)\,d\omega$$

we arrive at the relation

$$\frac{d\vec{E}}{dz} = -cE_0 + bE_0 = -aE_0 \tag{1.50}$$

originally derived by Gershun (1936).

It follows that

$$a = K_{\mathrm{E}} \frac{\vec{E}}{E_0} \tag{1.51}$$

and

$$a = K_{\mathrm{E}} \bar{\mu} \tag{1.52}$$

Thus we have arrived at a relation between an inherent optical property and two of the properties of the field. Eqn 1.52, as we shall see later (§3.2), can be used as the basis for determining the absorption coefficient of a natural water from *in situ* irradiance and scalar irradiance measurements.

Preisendorfer (1961) has used the equation of transfer to arrive at a set of relations between certain properties of the field and the *diffuse* absorption and scattering coefficients. One of these, an expression for the vertical attenuation coefficient for downward irradiance,

$$K_{d}(z) = a_{d}(z) + b_{bd}(z) - b_{bu}(z) R(z) \tag{1.53}$$

we will later (§6.7) find of assistance in understanding the relative importance of the different processes underlying the diminution of irradiance with depth.

# 2

.........................................................................................

# *Incident solar radiation*

## 2.1     Solar radiation outside the atmosphere

The intensity and the spectral distribution of the radiation received by the Earth are a function of the emission characteristics and the distance of the Sun. Energy is generated within the Sun by nuclear fusion. At the temperature of about $20 \times 10^6$K existing within the Sun, hydrogen nuclei (protons) fuse to give helium nuclei, positrons and energy. A number of steps is involved but the overall process may be represented by

$$4\,{}^1_1\mathrm{H} \rightarrow {}^4_2\mathrm{He} + 2\,{}^0_{+1}\mathrm{e} + 25.7\,\mathrm{Mev}$$

The energy liberated corresponds to the slight reduction in mass which takes place in the fusion reactions. Towards its periphery the temperature of the Sun greatly diminishes and at its surface it is only about 6000K.

For any physical system with the properties of a black body (or 'full radiator') the amount of radiant energy emitted per unit area of surface and the spectral distribution of that radiation are determined by the temperature of the system. The radiant flux*, $M$, emitted per unit area, is proportional to the fourth power of the absolute temperature, in accordance with the Stefan–Boltzmann Law

$$M = \sigma T^4$$

where $\sigma$ is $5.67 \times 10^{-8}$ W m$^{-2}$ K$^{-4}$. The Sun appears to behave approximately as a full radiator, or black body, and so the radiant flux emitted per m$^2$ of its surface is about $73.5 \times 10^6$ W, corresponding (assuming a diameter of $1.39 \times 10^6$ km) to a total solar flux of about $446 \times 10^{24}$ W. At the distance of the Earth's orbit ($149 \times 10^6$ km) the solar flux per unit area facing the Sun should therefore be about 1600 W m$^{-2}$. In fact, since the

---

\* Throughout this chapter radiant flux and irradiance will be in energy units rather than quanta.

Sun is not a perfect full radiator, the solar irradiance in our region of space is somewhat lower than this: the best current estimate, combining the measurements carried out in spacecraft, high-altitude aircraft and a balloon, is that the total solar irradiance outside the Earth's atmosphere (sometimes called the *solar constant*) is 1353 W m$^{-2}$.[499]

The Earth, having a diameter of 12756 km, presents a cross-sectional area of $1.278 \times 10^8$ km$^2$ to the Sun. The solar radiation flux onto the whole of the Earth is therefore about $1730 \times 10^{14}$ W, and the total radiant energy received by the Earth from the Sun each year is about $5.46 \times 10^{24}$ J.

For a full radiator, the spectral distribution of the emitted energy has a certain characteristic shape, rising steeply from the shorter wavelengths to a peak, and diminishing more slowly towards longer wavelengths. As the temperature, $T$, of the radiator increases, the position of the emission peak ($\lambda_{max}$) moves to shorter wavelengths in accordance with Wien's Law

$$\lambda_{max} = \frac{w}{T}$$

where $w$ (Wien's displacement constant) is equal to $2.8978 \times 10^{-3}$ m K. The Sun, with a surface temperature of about 6000K has, in agreement with Wien's Law, maximum energy per unit wavelength at about 480 nm, as may be seen in the curve of solar spectral irradiance above the Earth's atmosphere in Fig. 2.1. It can be seen that the spectral distribution of the emitted solar flux is, compared to the theoretical curve for a full radiator

Fig. 2.1. The spectral energy distribution of solar radiation outside the atmosphere compared with that of a black body at 6000K, and with that at sea level (zenith Sun). (By permission, from *Handbook of geophysics*, revised edition, U.S. Air Force, Macmillan, New York, 1960.)

at 6000K, somewhat irregular in shape at the short-wavelength end. The dips in the curve are due to the absorption bands of hydrogen in the Sun's outer atmosphere. Photosynthetically available radiation (commonly abbreviated to PAR), 400–700 nm, constitutes 38% of the extraterrestrial solar irradiance.

## 2.2     Transmission of solar radiation through the Earth's atmosphere

Even when the sky is clear, the intensity of the solar beam is significantly reduced during its passage through the atmosphere. This reduction in intensity is due partly to scattering by air molecules and dust particles and partly to absorption by water vapour, oxygen, ozone and carbon dioxide in the atmosphere. With the Sun vertically overhead, the total solar irradiance on a horizontal surface at sea level is reduced by about 14% with a dry, clean atmosphere and by about 40% with a moist, dusty atmosphere, compared to the value above the atmosphere.[323] The proportion of the incident solar flux removed by the atmosphere increases as the solar elevation (the angle of the Sun's disc to the horizontal) decreases, in accordance with the increase in pathlength of the solar beam through the atmosphere. The atmospheric pathlength is approximately proportional to the cosecant of the solar elevation: it is, for example, twice as long with a solar elevation of 30° as with the Sun at the zenith.

### *Effect of scattering*

Since the air molecules are much smaller than the wavelengths of solar radiation, the efficiency with which they scatter light is proportional to $1/\lambda^4$, in accordance with Rayleigh's Law. Scattering of solar radiation is therefore much more intense at the short-wavelength end of the spectrum, and most of the radiation scattered by the atmosphere is in the visible and ultraviolet ranges.

Some of the radiation scattered from the solar beam is lost to space, and some finds its way to the Earth's surface. In the case of 'pure' Rayleigh scattering (scattering entirely due to air molecules, with a negligible contribution from dust), there is as much scattering in a forward, as in a backward, direction (Fig. 2.2; Chapter 4). Thus, ignoring for simplicity the effects of multiple scattering, half the light scattered from a sunbeam in a dust-free atmosphere will be returned to space and half will continue (at various angles) towards the Earth's surface. The total solar energy reflected (by scattering) back to space by a cloudless dust-free atmosphere alone, after allowing for absorption by ozone, is about 7%.[402]

The atmosphere over any part of the Earth's surface always contains a certain amount of dust, the quantity varying from place to place and with time at any given place. Dust particles scatter light, but are generally not sufficiently small relative to the wavelength of most of the solar radiation for scattering to obey the Rayleigh Law. They exhibit instead a type of scattering known as Mie scattering (see Chapter 4) which is characterized by an angular distribution predominantly in the forward direction, and a much weaker dependence on wavelength, although scattering is still more intense at shorter wavelengths. Although dust particles scatter light mainly in a forward direction, they do also scatter significantly in a backward direction; furthermore, a proportion of the light scattered forward, but at large values of scattering angle, will be directed upwards if the solar beam is not vertical. Since the particle scattering is additive to the air molecule scattering, a hazy (dusty) atmosphere reflects more solar energy to space than does a clean atmosphere.

That fraction of the solar flux which is scattered by the atmosphere in the direction of the Earth's surface constitutes skylight. Skylight appears blue because it contains a high proportion of the more intensely scattered, blue, light from the short-wavelength end of the visible spectrum. In a hazy atmosphere the increased scattering increases skylight at the expense of the direct solar flux. The proportion of the total radiation received at the Earth's surface which is skylight varies also with the solar elevation. As the atmospheric pathlength of the solar beam increases with decreasing solar elevation, so more of the radiation is scattered: as a consequence the direct solar flux diminishes more rapidly than does skylight.

Fig. 2.2. Polar plot of intensity as a function of scattering angle for small particles ($r \simeq 0.025\ \mu$m) for green ($r \simeq 0.5\ \mu$m) and red ($r \simeq 0.7\ \mu$m) light. (By permission, from *Solar radiation*, N. Robinson, Elsevier, Amsterdam, 1966.)

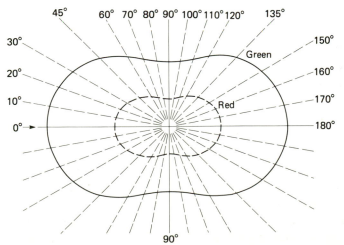

At very low solar elevations (0°–20°) in the absence of cloud, the direct solar beam and the diffuse flux (skylight) contribute approximately equally to irradiance at the Earth's surface. As solar elevation increases, the irradiance due to the direct beam rises steeply but the irradiance due to skylight levels off above about 30°. At high solar elevations under cloudless conditions, skylight commonly accounts for 15–25% of the irradiance and the direct solar beam for 75–85%:[321] in very clear dry air the contribution of skylight can be as low as 10%.

### Spectral distribution of irradiance at the Earth's surface

The scattering and absorption processes that take place within the atmosphere not only reduce the intensity, but also change the spectral distribution of the direct solar beam. The lowest curve in Fig. 2.1 shows the spectral distribution of solar irradiance at sea level for a zenith Sun and a clear sky. The shaded areas represent absorption, and so the curve forming the upper boundary of these shaded areas corresponds to the spectral distribution as it would be if there was scattering but no absorption. It is clear that the diminution of solar flux in the ultraviolet band (0.2–0.4 $\mu$m) is largely due to scattering, with a contribution from absorption by ozone. In the visible/photosynthetic band (0.4–0.7 $\mu$m), attenuation is mainly due to scattering, but with absorption contributions from ozone, oxygen and, at the red end of the spectrum, water vapour. In the long, infrared tail of the distribution, scattering becomes of minor importance and the various absorption bands of water vapour are mainly responsible for the diminution in radiant flux.

The proportion of infrared radiation removed from the solar beam by absorption during its passage through the atmosphere is variable since the amount of water vapour in the atmosphere is variable. Nevertheless it is generally true that a higher proportion of the infrared is removed than of the photosynthetic waveband. As a consequence, photosynthetically available radiation (0.4–0.7 $\mu$m) is a higher proportion of the solar radiation that reaches the Earth's surface than of the radiation above the atmosphere. PAR constitutes about 45% of the energy in the direct solar beam at the Earth's surface when the solar elevation is more than 30°.[322] Skylight, consisting as it does only of scattered and therefore mainly short-wave radiation, is predominantly in the visible/photosynthetic range. Combining the direct and the diffuse (sky) components, we may assume that PAR constitutes about 50% of the total irradiance (energy units) at the Earth's surface.[321]

Since the efficiency with which the photosynthetic apparatus captures light energy varies with wavelength, the usefulness of a given sample of solar radiation for primary production depends on the proportion of different wavelengths of light present, i.e. on its spectral distribution. Fig. 2.3 shows data for the spectral distribution of total solar irradiance at the Earth's surface under clear sunny skies, within a few hours of solar noon at three different locations. As the solar elevation diminishes, the ratio of short- (blue) to long- (red) wavelength light in the direct solar beam decreases because of the intensified removal of the more readily scattered, short-wavelength light in the longer atmospheric path. On the other hand, as solar elevation diminishes, the relative contribution of skylight to total irradiance increases, and skylight is particularly rich in the shorter

Fig. 2.3. Spectral distribution of solar quantum irradiance at the Earth's surface at three geographical locations (plotted from the data of Tyler & Smith, 1970). (*a*) Crater Lake, Oregon, USA (42°56′N, 122°07′W). Elevation 1882 m. 1100–1125 h, 5th August 1966. (*b*) Gulf Stream, Bahamas, Atlantic Ocean (25°45′N, 79°30′W). 1207–1223 h, 3rd July 1967. (*c*) San Vicente reservoir, San Diego, California, USA (32°58′N, 116°55′W). 0937–0958 h, 20th January 1967. All the measurements were made under clear skies.

wavelengths: there is therefore no simple relation between solar elevation and the spectral distribution of total irradiance.

### Effect of cloud

In addition to the effects of the gaseous and particulate components of the atmosphere, the extent and type of cloud cover are of great importance in determining the amount of solar flux which penetrates to the Earth's surface. We follow here the account given by Monteith (1973).

A few isolated clouds in an otherwise clear sky increase the amount of diffuse flux received at the Earth's surface but, provided they do not obscure the Sun, they have no effect on the direct solar beam. Thus, a small amount of cloud can increase total irradiance by 5–10%. A continuous sheet of cloud, however, will always reduce irradiance. Under a thin sheet of cirrus, total irradiance may be about 70% of that under a clear sky. A deep layer of stratus cloud, on the other hand, may transmit only 10% of the solar radiation, about 70% being reflected back to space by its upper surface and 20% being absorbed within it. On a day with broken cloud, the irradiance at a given point on the Earth's surface is intermittently varying from the full Sun's value to perhaps 20–50% of this as clouds pass over the Sun.

In desert regions there is rarely enough cloud to affect surface irradiance, but in the humid parts of the globe, cloud cover significantly reduces the average solar radiation received during the year. In much of Europe, for example, the average insolation (total radiant energy received $m^{-2}$ $day^{-1}$) in the summer is 50–80% of the insolation that would be obtained on cloudless days.[321]

### Angular distribution of solar radiation under different atmospheric conditions

That part of the solar radiation at the Earth's surface which is the direct solar beam, of necessity consists of a parallel flux of radiation at an angle to the vertical equal to the solar zenith angle. Radiance in the direction of the solar disc is very high and falls off sharply just beyond the edge of the disc.

The angular distribution of skylight from a clear sky is complex; it is shown diagrammatically in Fig. 2.4. Because scattering by dust particles, and to a lesser extent by air molecules, is strong in a forward direction, skylight is particularly intense at angles near to the angle of the Sun.[321] Radiance from the sky decreases markedly with increasing angular distance from the Sun; it reaches its minimum in that region of the celestial hemisphere approximately 90° from the Sun and then rises again towards the horizon.

Contrary to what casual observation might suggest, the radiance distribution beneath a heavily overcast sky is not uniform. In fact, the radiance at the zenith is about three times that at the horizon. To provide an approximation to the radiance distribution under these conditions, a Standard Overcast Sky has been defined[321] under which the radiance at any angle to the horizontal, $\beta$, is proportional to $(1+2\sin\beta)/3$. If the total irradiance, $E_d$ (W m$^{-2}$), on a horizontal surface is known, then the radiance, $L(\beta)$ (W m$^{-2}$ sr$^{-1}$), at any angle of elevation, $\beta$, is given by

$$L(\beta) = 3E_d(1+2\sin\beta)/7\pi \qquad (2.1)$$

The radiance under a sky with broken cloud must necessarily vary with angle in a highly discontinuous manner, and no useful general description can be given.

### Average transmission characteristics of the atmosphere

A valuable summary of the effects of the atmosphere as a whole on the transmission of solar radiation from space to the Earth's surface has been provided by Gates (1962) in terms of the average fate of the

Fig. 2.4. Angular distribution of luminance (approximately proportional to radiance) in a clear sky. (By permission, from *Solar radiation*, N. Robinson, Elsevier, Amsterdam, 1966.)

radiant flux incident on the northern hemisphere. Over a year, 34% of the incoming solar radiation is reflected to space by the atmosphere; this is made up of 25% reflected by clouds and 9% scattered out to space by other constituents of the atmosphere. Another 19% of the incoming radiation is absorbed by the atmosphere: 10% within clouds, 9% by other components. This leaves 47% of the solar flux which, on average, reaches the Earth's surface. Of this 47%, 24% consists of the direct solar beam and 23% of diffuse light scattered from clouds (17%) and from the air (6%).

In the cloudy, dirty, moist atmosphere of London, about a third of the incident solar radiation per year is scattered back to space, about a third is absorbed within the atmosphere, and the remainder penetrates to the Earth's surface.[321] Of the 30% or so which is absorbed, 13% is by water vapour, 9% within clouds, and 8% by dust and smoke.

## 2.3    Diurnal variation of solar irradiance

For a given set of atmospheric conditions, the irradiance at any point on the Earth's surface is determined by the solar elevation, $\beta$. This rises during the day from zero (or its minimum value in the Arctic or Antarctic during the summer) at dawn to its maximum value at noon, and then diminishes in a precisely symmetrical manner to zero (or the minimum value) at dusk. The exact manner of the variation of $\beta$ with time of day depends on the latitude, and on the solar declination, $\delta$, at the time. The declination is the angle through which a given hemisphere (north or south) is tilted towards the Sun (Fig. 2.5). In summer it has a positive value, in winter a negative one: at the spring and autumn equinoxes its value is zero. Its maximum value, positive or negative, is 23°27'. Published tables exist

Fig. 2.5. Variation of the solar declination throughout the year.

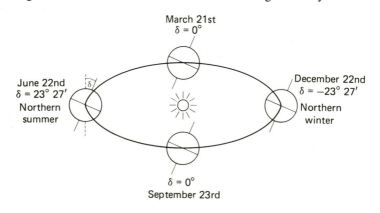

in which the solar declination for the northern hemisphere may be found for any day of the year. Alternatively, a relation derived by Spencer (1971) may be used

$$\delta = 0.39637 - 22.9133 \cos \psi + 4.02543 \sin \psi$$
$$- 0.3872 \cos 2\psi + 0.052 \sin 2\psi \quad (2.2)$$

where $\psi$ is the date expressed as an angle ($\psi = 360°.d/365$; $d =$ day number, ranging from 0 on January 1st to 364 on December 31st): both $\delta$ and $\psi$ are in degrees. The declination for the southern hemisphere on a given date has the same numerical value as that for the northern hemisphere, but the opposite sign.

The solar elevation, $\beta$, at any given latitude, $\gamma$, varies with the time of day, $\tau$, (expressed as an angle) in accordance with the relation

$$\sin \beta = \sin \gamma \sin \delta - \cos \gamma \cos \delta \cos \tau \quad (2.3)$$

where $\tau$ is $360°.t/24$ ($t$ being the hours elapsed since 00.00 h). If we write this in the simpler form

$$\sin \beta = c_1 - c_2 \cos \tau \quad (2.4)$$

where $c_1$ and $c_2$ are constants for a particular latitude and date ($c_1 = \sin \gamma \sin \delta$, $c_2 = \cos \gamma \cos \delta$), then it becomes clear that the variation of $\sin \beta$ with the time of day is sinusoidal. Fig. 2.6 shows the variations in both $\beta$ and $\sin \beta$ throughout a 24-h period corresponding to the longest summer day (December 21st) and shortest winter day (June 21st) at the latitude of Canberra, Australia (35° S). For completeness, the values of $\beta$ and $\sin \beta$ during the hours of darkness are shown: these are negative and correspond to the angle of the Sun below the horizontal plane. The variation of $\sin \beta$ is sinusoidal with respect to time measured within a 24-h cycle, but not with respect to hours of daylight.

The sine of the solar elevation $\beta$ is equal to the cosine of the solar zenith angle $\theta$. Irradiance due to the direct solar beam on a horizontal surface is, in accordance with the Cosine Law, proportional to $\cos \theta$. We might therefore expect that in the absence of cloud, solar irradiance at the Earth's surface will vary over a 24-h period in much the same way as $\sin \beta$ (Fig. 2.6), except that its value will be zero during the hours of darkness. Exact conformity between the behaviour of irradiance and $\sin \beta$ is not to be expected since, for example, the effect of the atmosphere varies with solar elevation. The greater attenuation of the direct beam at low solar elevation due to increased atmospheric pathlength, although in part balanced by a (proportionately) greater contribution from skylight, does reduce irradiance early in the morning and towards sunset below the values which might be anticipated on the basis of eqn 2.4. This has the useful effect of making the curve of daily variation of irradiance approximately sinusoidal with respect to daylight hours, even though, as we have seen, the variation of $\sin \beta$ is strictly speaking only sinusoidal with respect to hours since 00.00 h. Fig. 2.7 gives some examples of the diurnal variation of total irradiance at different times of year and under different atmospheric conditions.

A smooth curve is of course only obtained when cloud cover is absent, or is constant throughout the day. Broken cloud imposes short-term irregularities on the underlying sinusoidal variation in irradiance (Fig. 2.7). If $E(t)$ is the total irradiance at time $t$ h after sunrise, then by integrating $E(t)$ with respect to time we obtain the total solar radiant energy received

Fig. 2.6. Variation in solar elevation ($\beta$) and $\sin \beta$ during a 24-h period on the longest summer day and the shortest winter day at 35° latitude.

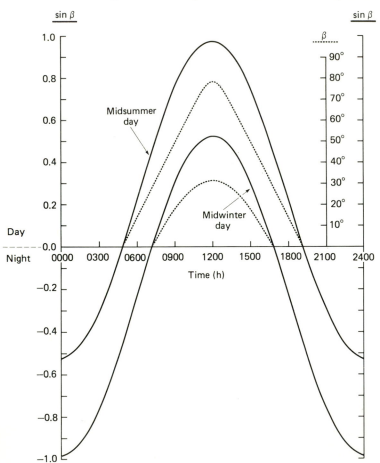

per unit horizontal area during the day. This is referred to as the *daily insolation* and we shall give it the symbol, $Q_s$

$$Q_s = \int_0^N E(t)\,\mathrm{d}t \qquad (2.5)$$

where $N$ is the daylength.

Fig. 2.7. Diurnal variation of total solar irradiance at different times of year and under different atmospheric conditions. The measurements were made at Krawaree, NSW, Australia (149°27′E, 35°49′S; 770 m above sea level) on dates close to (*a*) the shortest winter day, (*b*) the longest summer day, (*c*) the autumn equinox. (—— clear sky; ...... intermittent cloud; ------generally overcast.) Data provided by Mr F. X. Dunin, CSIRO, Canberra.

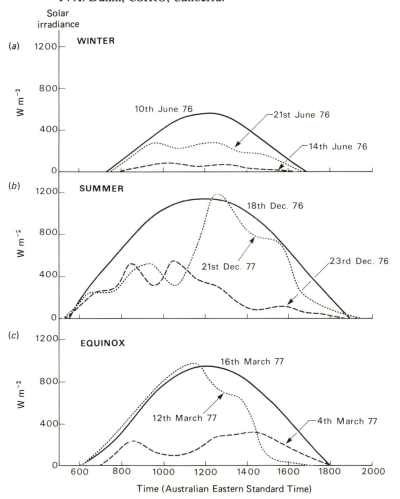

If $E(t)$ has the units W m$^{-2}$ and $N$ is in s, then $Q_s$ is in J m$^{-2}$. On a day of no, or constant, cloud cover the diurnal variation in $E(t)$ is expressed approximately by

$$E(t) = E_m \sin(\pi t/N) \qquad (2.6)$$

where $E_m$ is the irradiance at solar noon.[321] By integrating this expression in accordance with eqn 2.5, it can be shown that the daily insolation on such a day is related to the maximum (noon) irradiance by

$$Q_s = 2NE_m/\pi \qquad (2.7)$$

Thus if, as for example in Fig. 2.7c (16th March), the maximum irradiance is 940 W m$^{-2}$ in a 12-h day, then the total energy received is about 26 MJ.

On days with broken cloud, the degree of cloud cover is varying continuously throughout any given day. Nevertheless, in many parts of the world, the cloud cover averaged over a period as long as a month is approximately constant throughout the day.[520] Therefore, the diurnal variation in total irradiance averaged over a month will be approximately sinusoidal and will conform with eqn 2.6. From the average maximum irradiance, the average daily insolation can be calculated using eqn 2.7.

## 2.4     Variation of solar irradiance and insolation with latitude and time of year

At any given point on the Earth's surface the daylength and the solar elevation reach their maximum values in the summer and their minimum values in the winter. Substituting into eqn 2.3 we can obtain expressions for the noon solar elevation on the longest summer day ($\delta = 23°27'$)

$$\sin\beta = 0.39795 \sin\gamma + 0.91741 \cos\gamma \qquad (2.8)$$

and the shortest winter day ($\delta = -23°27'$)

$$\sin\beta = -0.39795 \sin\gamma + 0.91741 \cos\gamma \qquad (2.9)$$

as a function of latitude. At the latitude of Canberra (35° S) for example, the corresponding solar elevations are 78.5 and 31.5°: at the latitude of London (51.5° N) they are 62 and 15°.

Maximum and minimum daylengths can also be obtained using eqn 2.3. If the time (expressed as an angle) at sunrise is $\tau_s$, then (since $\sin\beta = 0$ at sunrise) at any time of year

$$\cos\tau_s = \tan\gamma \tan\delta \qquad (2.10)$$

Daylength expressed as an angle is $(360° - 2\tau_s)$, which is equal to $2\cos^{-1}(-\tan\gamma \tan\delta)$. Expressed in hours, daylength is given by

$$N = 0.133 \cos^{-1}(-\tan\gamma \tan\delta) \qquad (2.11)$$

The longest day is therefore $0.133 \cos^{-1}(-0.43377 \tan \gamma)$ h and the shortest day is $0.133 \cos^{-1}(0.43377 \tan \gamma)$ h.

With increasing latitude, the solar elevation at noon and therefore the maximum value of solar irradiance ($E_m$) decrease, which, in accordance with eqn 2.7, decreases the daily insolation. In the summer, however, this effect is counteracted by the increase in daylength ($N$ in eqn 2.7) with increasing latitude, and the net result is that high-latitude regions can have a slightly greater daily insolation in midsummer than tropical regions. In the winter, of course, the high-latitude regions have shorter days (zero daylength in the polar regions) as well as lower solar elevations, so their daily insolation is much less than at low latitudes. Throughout most of the year, in fact, the rule-of-thumb holds that the higher the latitude, the lower the daily insolation. Fig. 2.8 shows the change in daily insolation throughout the year, calculated ignoring atmospheric losses, for a range of latitudes. Fig. 2.9 shows the true (measured) daily insolation as a

Fig. 2.8. Change in calculated daily insolation (ignoring the influence of the atmosphere) throughout the year at different latitudes in the northern hemisphere. The latitude is indicated above each curve. Plotted from data of Kondratyev (1954).

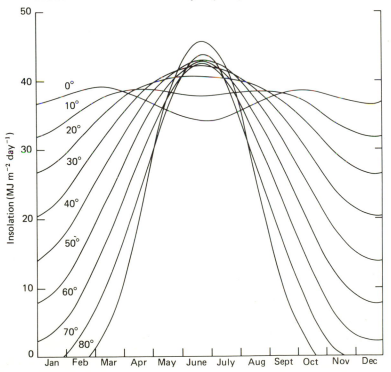

function of time of year at a southern hemisphere site, near Canberra, Australia, averaged over a 3-yr period. These data, unlike those in Fig. 2.8, include the effects of cloud and atmospheric haze.

## 2.5    Transmission across the air–water interface

To become available in the aquatic ecosystem, the solar radiation which has penetrated through the atmosphere must now find its way across the air–water interface. Some of it will be reflected back into the atmosphere. The proportion of the incident light which is reflected by a flat water surface increases from 2% for vertically incident light towards 100% as the beam approaches grazing incidence. The dependence of reflectance, $r$, of unpolarized light on the zenith angle of the incident light in air ($\theta_a$), and of the angle to the downward vertical ($\theta_w$) of the transmitted beam in water is given by Fresnel's Equation

$$r = \frac{1}{2}\frac{\sin^2(\theta_a - \theta_w)}{\sin^2(\theta_a + \theta_w)} + \frac{1}{2}\frac{\tan^2(\theta_a - \theta_w)}{\tan^2(\theta_a + \theta_w)} \tag{2.12}$$

The angle, $\theta_w$, in water is itself determined by $\theta_a$ and the refractive index, as we shall shortly see. The percentage reflectance from a flat water surface

Fig. 2.9. Change in measured daily insolation throughout the year at the CSIRO Ginninderra Experiment Station, (35°S, 149°E). The curve corresponds to the daily average for each calendar month, averaged over the period 1978–80.

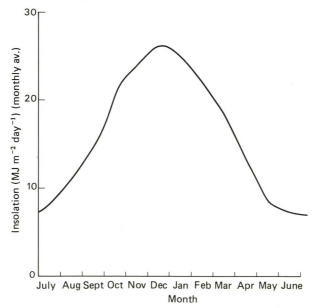

Table 2.1. *Reflectance of unpolarized light from a flat water surface. The values of reflectance have been calculated using eqns 2.12 and 2.13, assuming that the water has a refractive index of 1.33*

| Zenith angle of incidence, $\theta_a$ (degrees) | Reflectance (%) | Zenith angle of incidence, $\theta_a$ (degrees) | Reflectance (%) |
|---|---|---|---|
| 0.0 | 2.0 | 50.0 | 3.3 |
| 5.0 | 2.0 | 55.0 | 4.3 |
| 10.0 | 2.0 | 60.0 | 5.9 |
| 15.0 | 2.0 | 65.0 | 8.6 |
| 20.0 | 2.0 | 70.0 | 13.3 |
| 25.0 | 2.1 | 75.0 | 21.1 |
| 30.0 | 2.1 | 80.0 | 34.7 |
| 35.0 | 2.2 | 85.0 | 58.3 |
| 40.0 | 2.4 | 87.5 | 76.1 |
| 45.0 | 2.8 | 89.0 | 89.6 |

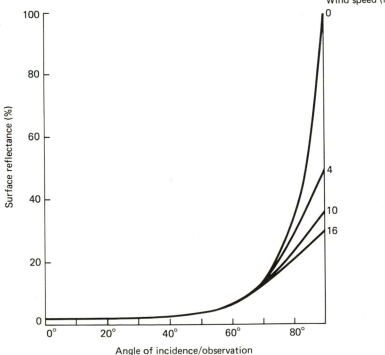

Fig. 2.10. Reflectance of water surface as a function of zenith angle of light (incident from above), at different wind speeds (data of Gordon, 1969; Austin, 1974*a*).

as a function of zenith angle is shown in Fig. 2.10, and in tabular form in Table 2.1. It will be noted that reflectance remains low, increasing only slowly, up to zenith angles of about 50°, but rises very rapidly thereafter.

Roughening of the water surface by wind has little effect on the reflectance of sunlight from high solar elevations. At low solar elevations, on the other hand, reflectance is significantly lowered by wind since the roughening of the surface on average increases the angle between the light direction and the surface at the point of entry. The three lower curves in Fig. 2.10 show the effect of wind at different velocities on the reflectance.[161, 16]

Because of the complex angular distribution of skylight, the extent to which it is reflected by a water surface is difficult to determine. If the very approximate assumption is made that the radiance is the same from all directions, then a reflectance of 6.6% for a flat water surface is obtained.[214] For an incident radiance distribution such as might be obtained under an overcast sky, the reflectance is calculated to be about 5.2%.[374] Roughening of the surface by wind will lower reflectance for the diffuse light from a clear or an overcast sky.

As the unreflected part of a light beam passes across the air–water interface, it changes its direction to the vertical (while remaining, if the surface is flat, in the same vertical plane), due to refraction. The phenomenon of refraction is the result of the differing velocities of light in the two media, air and water. The change in angle (Fig. 2.11) is governed by Snell's Law

$$\frac{\sin \theta_a}{\sin \theta_w} = \frac{n_w}{n_a} \tag{2.13}$$

Fig. 2.11. Refraction and reflection of light at the air–water boundary. (*a*) A light beam incident from above is refracted downwards within the water: a small part of the beam is reflected upwards at the surface. (*b*) A light beam incident from below at a nadir angle of 40° is refracted away from the vertical as it passes through into the air: a small part of the beam is reflected downwards again at the water–air boundary. (*c*) A light beam incident from below at a nadir angle greater than 49° undergoes complete internal reflection at the water–air boundary.

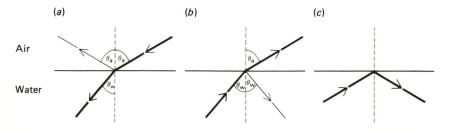

where $n_w$ and $n_a$ are the refractive indices of water and air, respectively. The ratio of the refractive index of water to that of air is a function of temperature, salt concentration and the wavelength of the light in question. For our purposes a value of 1.33 for $n_w/n_a$ is close enough for both sea and fresh water at normal ambient temperatures and for light of any wavelength within the photosynthetic range. As Fig. 2.11 shows, the effect of refraction is to move the light closer to the vertical in the aquatic medium than it was in air. It will be noted that even light reaching the surface at grazing incidence ($\theta_a$ approaching 90°) is refracted downwards so that $\theta_w$ should not be greater than about 49° for a calm surface. When the water is disturbed, some of the light will be found at angles greater than 49° after passing through the surface: nevertheless most of the downwelling light will be at values of $\theta_w$ between 0° and 49°. Statistical relations between the distribution of surface slopes and wind speed were derived by Cox & Munk (1954) from aerial photographs of the Sun's glitter pattern. Calculations using these relations show that, as might be expected, as wind speed increases, the underwater light becomes more diffuse.[165]

Snell's Law works in reverse also. A light beam passing upward within the water at angle $\theta_w$ to the downward vertical will emerge through the calm surface into the air at angle $\theta_a$ to the zenith, in accordance with eqn 2.13. There is, however, a very important difference between transmission from water to air, and from air to water, namely that in the former case complete internal reflection can occur. If, in the case of the upward-directed light beam within the water, $\theta_w$ is greater than 49°, then all the light is reflected down again by the water–air interface. Disturbance of the surface by wind decreases water-to-air transmission for upwelling light at angles within the range $\theta_w = 0°$ to $\theta_w = 49°$, but, by ensuring that not all the light is internally reflected, it increases transmission within the range $\theta_w = 49°$ to $\theta_w = 90°$.

# 3

..................................................................................................

# Absorption of light within the aquatic medium

Having discussed how the solar radiation gets through the water surface, we shall now consider what happens to it within the water. Sooner or later most of the photons are absorbed: how this happens, and which components of the medium are responsible, form the subjects of this chapter.

## 3.1    The absorption process

The energy of a molecule can be considered to be part rotational, part vibrational, part electronic. A molecule can only have one of a discrete series of energy values. Energy increments corresponding to changes in a molecule's electronic energy are large, those corresponding to changes in vibrational energy are intermediate in size, and those corresponding to changes in rotational energy are small. This is indicated diagrammatically in Fig. 3.1. When molecules collide with each other in the liquid or gaseous state, or are in contact with each other in the solid state, there can be transfer of rotational or vibrational energy between molecules and this is accompanied by transitions from one rotational or vibrational energy level to another within each molecule.

Molecules can obtain energy from radiation as well as from other molecules. When a photon passes within the vicinity of a molecule, there is a finite probability that it will be captured by that molecule, i.e. be absorbed. If the photon is captured, then the energy of the molecule must increase by an amount corresponding to the energy of the photon. If the photon is of long wavelength ($> 20 \, \mu$m), in the far infrared/microwave region of the spectrum, then its energy is low and its absorption can only cause a transition in the energy of the molecule from one rotational energy level to another. If the photon is in the infrared region ($< 20 \, \mu$m), then its absorption will cause a transition from one vibrational level to another.

Photons in the visible/photosynthetic part of the spectrum have sufficient energy to bring about, when they are absorbed, transitions from one electronic energy level (usually the ground state) to another. The first event in the absorption of light by the aquatic medium is, therefore, the capture of a photon by some molecule in the medium and the simultaneous transition of an electron in that molecule from the ground state to an excited state.

Within a complex molecule such as chlorophyll or any of the other photosynthetic pigments, there is usually more than one possible electronic energy transition which can occur. For example, in Fig. 3.1 excitation can occur up to electronic energy level I or to level II: to be more precise, excitation will occur to one of the many vibrational/rotational levels belonging to the given electronic energy level. Any given electronic transition is preferentially excited by light which has an amount of energy per photon corresponding to the energy required for the transition. Chlorophyll, as we shall see later, has two main absorption bands, in the red and in the blue regions of the spectrum. Absorption of a blue photon

Fig. 3.1. Absorption of a photon raises an electron from the ground state to one of two possible excited singlet states, depending on the wavelength (and therefore energy) of the photon. Wavelengths $\lambda_1$ and $\lambda_2$ are each within a separate absorption band in the absorption spectrum of the molecule ($\lambda_2 > \lambda_1$).

leads to transition to a substantially higher electronic energy level than absorption of a red photon. Because of the multiplicity of vibrational/rotational states, these two electronic energy states nevertheless overlap: i.e. the highest vibrational/rotational levels of the lower electronic state can have energies as high as the lowest level of the upper electronic state (Fig. 3.1). As a consequence, immediately after absorption of a blue photon there is a very rapid series of transitions downwards through the various rotational/vibrational levels (associated with transfer of small increments of rotational/vibrational energy to adjoining molecules) until the lower electronic energy state (referred to as the lowest excited singlet state) is reached.

It is the energy in this lowest excited singlet state which is used in photosynthesis and it is because an excited chlorophyll molecule usually ends up in this state anyway, that all *absorbed* visible quanta are equivalent, i.e. a red quantum, or even a green quantum if it succeeds in being absorbed, brings about as much photosynthesis as a blue quantum. This is also the reason why, in the context of primary production, it is more meaningful to express irradiance in quanta $m^{-2} s^{-1}$ than in $W m^{-2}$.

In the case of light-absorbing molecules in the aquatic medium which are not part of the photosynthetic system, and which cannot therefore transfer their energy to a reaction centre for use in photosynthesis, there is likely to be a transition from the lowest excited singlet state to the corresponding triplet state by interaction with a paramagnetic molecule such as oxygen. In a singlet state, the two members of every pair of electrons in the molecule have opposite spin $(+\frac{1}{2}$ and $-\frac{1}{2})$ so that the resultant spin is zero: in a triplet state, the members of a pair of electrons in the molecule have the same spin so that the resultant spin is one. The excited triplet state is at a lower energy level and is also much longer lived than the excited singlet state, the average lifetime of the excited molecules being $10^{-5}$ to $10^{-4}$ s in the former case, compared to $10^{-9}$ to $10^{-8}$ s in the latter. Eventually, by interaction with a paramagnetic molecule such as oxygen, the excited molecule is returned to a singlet state by spin reversal and, by vibrational/rotational interactions with the surrounding molecules, it undergoes a downwards transition to one of the upper vibrational levels of the ground state. The electronic energy is thus dissipated as heat energy.

Excitation energy, whether in the photosynthetic system or in a molecule outside it, can be lost by re-emission of radiation. An electronically excited molecule, after undergoing the radiationless transition to the lowest level of the lowest excited singlet state, can then undergo a transition to one of the vibrational/rotational levels of the ground state by re-emitting a photon of light. This phenomenon is referred to as *fluorescence*. In living, photosynthesizing algal cells, only a very small proportion, $\sim 1\%$, of the absorbed light is lost in this way. Most of the absorbed energy, whether captured initially by chlorophyll, carotenoid or biliprotein, is transferred by inductive resonance to the reaction centres where it is used to bring about biochemical changes (see Chapter 8). If photosynthesis is inhibited, say with DCMU (dichlorophenylmethylurea), then fluorescence increases to about 3% of the absorbed light. The non-photosynthetic, light-absorbing material of the aquatic medium can also re-emit some of its absorbed energy as fluorescence, but the fluorescence yield (quanta emitted/quanta absorbed) is again very low: most of the excited molecules are converted to the triplet state (see above), and from there to the ground state before they can emit a photon.

Thus, most of the light energy absorbed by the aquatic ecosystem, after existing for a very brief period as electronic excitation energy, ends up either as heat (vibrational/rotational energy distributed amongst all the molecules of the system) or as chemical energy in the form of photosynthetically produced biomass. Only a tiny proportion is turned back into light again by fluorescence, and even this is for the most part re-absorbed before it can escape from the system.

### 3.2    The measurement of light absorption

We saw in Chapter 1 how the light absorption properties of the aquatic medium are characterized in terms of the absorption coefficient, $a$. We saw also that although $a$ is defined in terms of the behaviour of light passing through an infinitesimally thin layer of medium, its value can nevertheless be obtained from the observed absorptance, $A$, of a layer of medium of finite thickness, provided that the effects of scattering can be overcome.

Commercially available equipment (spectrophotometers) used for the measurement of light absorption, do so in terms of an optical parameter known as the *absorbance* or *optical density*: although there is no standard symbol we shall indicate it by $D$. Although absorbance is one of the most frequently measured parameters in biochemistry and chemistry it does not, unfortunately, appear to have a rigorous definition. It is commonly defined as the logarithm to the base 10 of the ratio of the light intensity, $I_0$, incident on a physical system to the light intensity, $I$, transmitted by the system

$$D = \log_{10} \frac{I_0}{I}$$

The meaning of the word 'intensity' is not usually given: we shall equate it to radiant flux, $\Phi$. The incident beam is generally assumed to be parallel.

The exact meaning of 'transmitted' is also not stated. If we make the simplest assumption that any light which re-emerges from the system, in any direction, is 'transmitted', then $I$ includes all scattered light as well as that neither absorbed nor scattered. Thus, $I$ is equal to the incident flux minus the absorbed flux $(\Phi_0 - \Phi_a)$, from which, given that $\Phi_a/\Phi_0$ is absorptance, $A$, it follows that

$$D = \log_{10}\frac{1}{(1-A)} = -\log_{10}(1-A)$$

If we can arrange that all the scattered light is included in the measured value of $I$, then scattering may be ignored (except in so far as it increases the pathlength of the light within the system), and so eqn 1.32 may be considered to apply. Therefore,

$$D = 0.4343\, ar$$

i.e. absorbance is equal to the product of the absorption coefficient of the medium and the pathlength ($r$) through the system, multiplied by the factor for converting base e to base 10 logarithms. The absorption coefficient may be obtained from the absorbance value using

$$a = 2.303\, D/r$$

In a typical spectrophotometer, monochromatic light beams are passed through two transparent glass or quartz cells of known pathlength, one (the sample cell) containing a solution or suspension of the pigmented material, the other (the blank cell) containing pure solvent or suspending medium. The intensity of the light beam after passing through each cell is measured with an appropriate photoelectric device such as a photomultiplier. The intensities of the beams which have passed through the blank cell and the sample cell are taken to be proportional to $I_0$ and $I$, respectively, and so the logarithm of the ratio of these intensities is displayed as the absorbance of the sample.

When used in their normal mode (Fig. 3.2a), however, spectrophotometers have light sensors with a collection angle of only a few degrees. Thus any light scattered by the sample outside this collection angle will not be detected. Such spectrophotometers do not, therefore, measure true absorbance: indeed if their collection angle is very small, in effect they measure attenuance (absorption plus scattering, see §1.4). If, as is often the case, the sample measured consists of a clear coloured solution with negligible scattering relative to absorption, then scattering losses are trivial and the absorbance displayed by the instrument will not differ significantly from the true absorbance. If, however, the pigmented material is particulate, then a substantial proportion of the light transmitted by the sample may be scattered outside the collection angle of the photomultiplier, the value of $I$ registered by the instrument will be too low, and the absorbance displayed will be higher than the true absorbance.

This problem may be solved by so arranging the geometry of the instrument that nearly all the scattered light is collected and measured by the photomultiplier. The simplest arrangement is to have the cells placed close to a wide photomultiplier so that the photons scattered up to quite wide angles are detected. A better solution is to place each cell at an entrance window in an integrating sphere (Fig. 3.2*b*): this is a spherical cavity, coated with white, diffusely reflecting material inside. By multiple reflection, a diffuse light field is set up within the sphere from all the light which enters it, whether scattered or not. A photomultiplier at another window receives a portion of the flux from this field, which may be taken to be proportional to $I$ or $I_0$ in accordance with whether the beam is passing through the sample, or the blank cell. With this system virtually all the light scattered in a forward direction, and this will generally mean nearly all the scattered light, is collected.

Another procedure, which can be used with some normal spectrophotometers, is to place a layer of scattering material, such as opal glass, behind the blank and the sample cell.[425] This ensures that the blank beam and the sample beam are both highly scattered after passing through the cells. As a result, the photomultiplier, despite its small collection angle, harvests about the same proportion of the light transmitted through the sample cell, as of the light transmitted through the blank cell, and so the registered values of intensity are indeed proportional to the true values of $I$ and $I_0$, and so can give rise to an approximately correct value for absorbance.

Fig. 3.2. Principles of measurement of absorption coefficient of non-scattering and scattering samples. The instrument is assumed to be of the double-beam type, in which a rotating mirror is used to direct the monochromatic beam alternately through the sample and the blank cells. (*a*) Operation of spectrophotometer in normal mode for samples with negligible scattering. (*b*) Use of integrating sphere for scattering samples.

With all these procedures for measuring the absorbance of particulate materials, it is important that the suspension should not be so concentrated that, as a result of multiple scattering, there is a significant increase in the pathlength that the photons traverse in passing through the sample cell. This would have the effect of increasing the absorbance of the suspension and lead to an estimate of absorption coefficient which would be too high. Even with an integrating sphere, some of the light is scattered by the sample at angles too great for collection: a procedure for correction for this error has been described.[258]

As we shall see later, the absorption coefficients of the different constituents of the aquatic medium can be determined separately. The value of the absorption coefficient of the medium as a whole, at a given wavelength, is equal to the sum of the individual absorption coefficients of all the components present. Furthermore, providing that no changes in molecular state or physical aggregation take place with changes in concentration, then the absorption coefficient due to any one component is proportional to the concentration of that component (Beer's Law).

The total absorption coefficient of a natural water at a given wavelength can be determined from a measurement made within the water body itself: an instrument designed for this purpose is known as an absorption meter. The simplest arrangement is to have, immersed in the water, a light source giving a collimated beam and, at a suitable distance, a detector with a wide angle of acceptance. Most of the light scattered, but not absorbed, from the beam is detected: diminution of radiant flux between source and detector is thus almost entirely due to absorption and can be used to give a value for the absorption coefficient.

An alternative approach is to have, not a collimated beam, but a point source of light radiating equally in all directions.[30] The detector does not have to have a particularly wide angle of acceptance. For every photon which is initially travelling towards the detector and is scattered away from the detector, there will, on average, be another photon, initially not travelling towards the detector, which is scattered towards the detector. Thus diminution of radiant flux between source and detector is entirely due to absorption. A problem with this and the previous type of absorption meter is that although scattering may not directly prevent photons reaching the detector, it increases the pathlength of the photons and thus increases the probability of their being absorbed, i.e. spuriously high values of absorption coefficient may be obtained. This error can be significant in waters with high ratios of scattering to absorption.

To avoid the necessity of comparing readings in the water body with readings in pure water or other standard medium, a variant on the second type of absorption meter has been developed which still uses a point light source but has two detectors at different distances from the source: from the difference in radiant flux on the two detectors, the absorption coefficient may be obtained.[129] By means of interference filters, this particular instrument carries out measurements of absorption coefficient at 50 nm intervals from 400 to 800 nm.

With all these absorption meters, if the light source is modulated and the detector is designed to measure only the modulated signal, then the contribution of ambient daylight to radiant flux on the detector can be eliminated.

Absorption coefficients of natural waters can also be determined *in situ* from measurements of irradiance and scalar irradiance. The instrumentation for such measurements we shall discuss in a later chapter (see §5.1). Determination of absorption coefficient makes use of the relation (see §1.7)

$$a = K_E \vec{E}/E_0$$

where $E_0$ is the scalar irradiance, $\vec{E}$ is the *net* downward irradiance $(E_d - E_u)$, and $K_E$ is the vertical attenuation coefficient for $\vec{E}$. Thus $a$ can be obtained from measurements of net downward and scalar irradiance at two, or a series of, depths (measurements at more than one depth are required to give $K_E$). Absorption meters based on these principles have been built.[189, 461]

### 3.3 The major light-absorbing components of the aquatic system

Essentially all the light absorption which takes place in natural waters is attributable to four components of the aquatic ecosystem: the water itself, dissolved yellow pigments, the photosynthetic biota (phytoplankton, and macrophytes where present) and inanimate particulate matter (*tripton*). We shall now consider the spectral absorption properties of each of these, and their relative contributions to the absorption of photosynthetically available radiation.

#### Water

Pure water, although it appears colourless in the quantities we handle in everyday life, is in fact a blue liquid: the blue colour is clearly apparent under sunny conditions in oceanic waters or in coastal waters which are infertile and have little input from rivers. The colour of pure water arises from the fact that it absorbs only very weakly in the blue and the green regions of the spectrum, but its absorption begins to rise as wavelength increases above 550 nm and is quite significant in the red region: a 1-m thick layer of pure water will absorb about 35% of incident light of wavelength 680 nm.

Because of the very weak absorption it is very difficult to measure the absorption coefficient of pure water in the blue/green spectral region, and the values reported in the literature, determined by normal spectrophotometric procedures, vary widely. Smith & Baker (1981) have arrived at a set of values (Table 3.1) based partly on their own measurements of vertical attenuation coefficient for irradiance in the clearest ocean waters ($K_d$ in such waters is slightly greater than $a$) and partly on what they consider to be the best laboratory measurements. Fig. 3.3 shows the absorption spectrum of pure water from the near ultraviolet, through the photosynthetic range, to the near infrared, based on the compilation of Smith & Baker in the range 310 to 790 nm, and the measurements of Palmer & Williams (1974) at longer wavelengths.

It can be seen that the absorption by water at the red end of the visible

Table 3.1. *Absorption coefficients for pure water (Smith & Baker, 1981)*

| $\lambda$ (nm) | $a$ (m$^{-1}$) | $\lambda$ (nm) | $a$ (m$^{-1}$) |
|---|---|---|---|
| 310 | 0.105 | 560 | 0.0708 |
| 320 | 0.0844 | 570 | 0.0799 |
| 330 | 0.0678 | 580 | 0.108 |
| 340 | 0.0561 | 590 | 0.157 |
| 350 | 0.0463 | 600 | 0.244 |
| 360 | 0.0379 | 610 | 0.289 |
| 370 | 0.0300 | 620 | 0.309 |
| 380 | 0.0220 | 630 | 0.319 |
| 390 | 0.0191 | 640 | 0.329 |
| 400 | 0.0171 | 650 | 0.349 |
| 410 | 0.0162 | 660 | 0.400 |
| 420 | 0.0153 | 670 | 0.430 |
| 430 | 0.0144 | 680 | 0.450 |
| 440 | 0.0145 | 690 | 0.500 |
| 450 | 0.0145 | 700 | 0.650 |
| 460 | 0.0156 | 710 | 0.839 |
| 470 | 0.0156 | 720 | 1.169 |
| 480 | 0.0176 | 730 | 1.799 |
| 490 | 0.0196 | 740 | 2.38 |
| 500 | 0.0257 | 750 | 2.47 |
| 510 | 0.0357 | 760 | 2.55 |
| 520 | 0.0477 | 770 | 2.51 |
| 530 | 0.0507 | 780 | 2.36 |
| 540 | 0.0558 | 790 | 2.16 |
| 550 | 0.0638 | 800 | 2.07 |

spectrum is really the tail end of a series of much stronger absorption bands in the infrared: there are even more intense absorption bands beyond 1000 nm. The two shoulders in the visible spectrum – the distinct one at ~ 604 nm and the weak one at ~ 514 nm – have been identified as corresponding to the fifth and sixth harmonics, respectively, of the O—H stretch vibration of liquid water;[491] the peaks at ~ 960 and ~ 745 nm in the infrared correspond to the third and fourth harmonics (the fundamental is at about 3 $\mu$m).

The contribution of water itself to the attenuation of PAR by absorption of quanta is of importance only above about 550 nm. The absorption properties of water at the short-wavelength/ultraviolet end of the spectrum are nevertheless of considerable importance since, in oceanic and other waters low in dissolved colour, they have a marked influence on the penetration of ultraviolet radiation which, as we shall discuss later (§ 10.1) has strongly inhibitory effects on phytoplankton photosynthesis in the surface layer.

### Gilvin (yellow substance)

When plant tissue decomposes in the soil or in a water body, most of the organic matter is broken down by microbial action within days or weeks to, ultimately, carbon dioxide and inorganic forms of nitrogen,

Fig. 3.3. Absorption spectrum of pure water. Absorption coefficient values have been taken from Table 3.1 (data of Smith & Baker, 1981) for the range 310 to 790 nm, and from the data of Palmer & Williams (1974) for the range 790 to 1000 nm.

sulphur and phosphorus. In the course of the decomposition process there is formed, however, a complex group of compounds loosely referred to as 'humic substances'. The chemistry of humic substances has been reviewed by Schnitzer (1978).

Degradation and other studies indicate that these substances are polymers of which the building blocks are phenolic and benzenecarboxylic acids, together with aliphatic di- and monocarboxylic acids. Fig. 3.4 shows some of the many different compounds produced on chemical degradation of humic substances. The bonding which holds the subunits together is still not well understood but ether, ester, alkyl and hydrogen bonds may all be involved. A humic substance sample from the water of the Okefenokee Swamp, Georgia, USA, was found to have[502] an average atomic composition of $C_{74}H_{72}O_{46}N_{0.7}$.

Humic substances vary in size from freely soluble compounds with molecular weights (relative molecular masses) of a few hundred, to insoluble macromolecular aggregates with molecular weights in the hundreds of thousands and perhaps ranging up to the millions. Soil chemists classify humic substances into three main fractions on the basis of solubility behaviour. The soil is first extracted with dilute alkali. The humic material that does not dissolve is referred to as *humin*. Of the alkali-soluble fraction, some is precipitated on acidification: this is called *humic acid*. The humic material remaining in solution is called *fulvic acid*. In fact all three fractions are chemically very similar and differ mainly in molecular weight. All are dark brown (the soluble ones giving rise to yellow-brown solutions), hydrophilic and acidic (due to the presence of carboxyl and phenolic groups). Fulvic acid has a higher content of oxygen-containing groups such as carboxyl and hydroxyl. The insolubility of the humin fraction may be due to its being firmly adsorbed or bonded to mineral particles and/or to its having a very high molecular weight.

While some humic material may be formed by oxidation and polymerization directly from the existing phenolic compounds (particularly lignin) in the decomposing plant tissue, it is also true that some saprophytic fungi excrete large amounts of phenolic substances when grown on

Fig. 3.4. Structures of some of the products liberated on chemical degradation of humic substances.

(a) 3,5-Dihydroxybenzoic acid  (b) 1, 2, 4-Benzene tricarboxylic acid  (c) Vanillin  (d) Catechol  (e) Aliphatic dicarboxylic acid, $n = 0$–$8$

carbohydrate, and these phenolic substances can undergo oxidation and polymerization to give humic-like material.[88] Thus it seems likely that some of the aromatic subunits of the humic materials originate in the plant and some are generated *de novo* during microbial breakdown.

From the point of view of aquatic ecology, the significance of the soil humic material is that as water, originating as rainfall, drains through soil and into rivers and lakes, and ultimately into estuaries and the sea, it extracts from the soil some of the water-soluble humic substances and these impart a yellow colour to the water, with major consequences for the absorption of light, particularly at the blue end of the spectrum. James & Birge (1938) in the USA and Sauberer (1945) in Austria carried out extensive quantitative studies on the absorption spectra of lake waters with varying degrees of yellowness. The light absorption properties of these dissolved humic materials in natural inland waters can be determined relatively easily by measuring the absorption spectrum of a water sample which has been filtered (0.2–0.4 $\mu$m pore size) in 5 cm or 10 cm pathlength cells.[254] Fig. 3.5 shows the absorption spectrum of this material from some Australian inland waters. These are typical humic substance absorption spectra, with absorption being very low or absent at the red end of the visible spectrum, and rising steadily with decreasing wavelength towards the blue: absorption in the ultraviolet region is higher still.

The presence of dissolved yellow material in inland waters is often easily apparent to the eye. Its presence in marine waters (where its concentration is much lower) is not so readily apparent, and the fact that it is important in these ecosystems too was pointed out by Kalle (1937, 1966). Fig. 3.5 also shows the absorption spectra of soluble yellow material in estuarine and coastal waters.

It seems likely that most of the dissolved yellow colour in inland waters is due to soluble humic substances leached from the soils in the catchment areas. Yellow material of the humic type can also be generated by decomposition of plant matter within the water: this could be of significance in productive water bodies. Much of the soluble humic material in river water is precipitated when it comes into contact with sea water.[214, 430] Nevertheless, a fraction of the material remains in solution and most of the dissolved yellow colour in coastal sea water is due to humic substances derived from the land in river discharge. For example, Monahan & Pybus (1978) found that in regions with major river discharge off the west coast of Ireland, the concentration of soluble humic material diminished linearly with increasing salinity: the results indicated that essentially all the humic material in these coastal North Atlantic waters originated in the rivers.

As in fresh waters, some formation of dissolved yellow materials may take place within marine waters. Brown seaweeds, for example, actively excrete phenolic compounds which, probably as a consequence of oxidation and polymerization, give rise to yellow-brown materials of the humic type within the water.[431] It seems possible that this phenomenon might make a significant contribution to the amount of dissolved yellow material in the water near luxuriant brown algal beds.

It is not certain to what extent the yellow substance (present at low

Fig. 3.5. Absorption spectra of soluble yellow material (gilvin) in various Australian natural waters (from Kirk, 1976*b*). The lowest curve (Batemans Bay, NSW) is for coastal sea water near the mouth of a river; the next curve (Clyde River, NSW) is for an estuary; the remainder are for inland water bodies in the southern tablelands of New South Wales/Australian Capital Territory. The ordinate scale corresponds to the true *in situ* absorption coefficient due to gilvin.

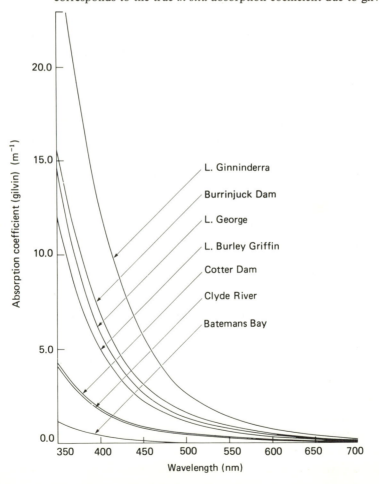

concentrations but still optically significant) in oceanic waters away from the coastal zone is land-derived humic material or is generated within the sea by decomposition of the phytoplankton. Højerslev (1979), on the basis of his measurements in the Baltic and North Atlantic, considers it unlikely that significant amounts of yellow substance are formed in the sea and attributes it, even in oceanic areas, to river discharge. Jerlov (1976), however, argues that the presence of yellow substance in the upwelling region west of South America proves its immediate marine origin, as this area is practically devoid of fresh-water supply from land drainage. Bricaud, Morel & Prieur (1981), on the basis of their measurements in a variety of waters, suggest that in the ocean away from regions of river discharge, the concentration of yellow substance is determined by biological activity averaged over a long period. Kopelevich & Burenkov (1977) observed a strong correlation between the concentration of yellow substance and the level of phytoplankton chlorophyll in productive oceanic waters. They proposed that oceanic yellow substance is of two kinds: a component resulting from the recent decomposition of phytoplankton, and a more stable component of much greater age. This latter 'conservative' component would predominate in oligotrophic waters and might, in agreement with Bricaud *et al.*, reflect average biological activity over a long time.

The humic material that ends up in the dissolved state in natural waters occurs in a wide range of molecular weights, as indicated by gel filtration. The dissolved humic substances in water from a Norwegian moorland area were found by Gjessing (1965) to occur mainly in two size classes: 100000–200000 and somewhat below 10000. In water from a Wisconsin lake, fractions ranging from < 700 to > 50000 were found.[166] In waters from streams and lakes in northwestern North America, the dissolved yellow substances were mainly in the molecular weight range of 700–10000.[144] Stuermer & Harvey (1974) found that soluble humic material from the Sargasso Sea and Massachusetts coastal waters consisted (by mass) mainly (78–97%) of fulvic acid, the rest being classified as humic acid: about 55% of the coastal and 73% of the Sargasso Sea fulvic acids had apparent molecular weights less than 700, the remainder in each case being about equally distributed between the 700–1500 and 1500–5000 molecular weight categories. Wheeler (1976) used ultrafiltration through calibrated membranes to study the molecular weight distribution of dissolved organic matter in Georgia coastal waters. He found that most of the dissolved colour of salt-marsh, estuarine, nearshore (3 km) and offshore (15 km) waters was due to substances with molecular weights between 1000 and 30000.

The dissolved yellow material in natural waters has been variously referred to as 'yellow substance', 'gelbstoff' (the same name, in German), 'yellow organic acids', 'humolimnic acid', 'fulvic acid', 'humic acid', etc. Dissolved yellow materials from different waters vary not only in molecular size but also in chemical composition.[241, 476, 477] In the context of light attenuation it would be useful to have a general name, applicable to all or any of these compounds regardless of their chemical nature, which simply indicated that they preferentially absorbed light at the blue end of the spectrum. 'Yellow substance' or 'gelbstoff' are too non-specific and could apply to anything from butter to ferric chloride. I have suggested the word 'gilvin', a noun derived from the Latin adjective *gilvus* meaning pale yellow.[254] 'Gilvin' would thus be defined as a general term to be applied to those soluble yellow substances, whatever their chemical structure, which occur in natural waters, fresh or marine, at concentrations sufficient to contribute significantly to the attenuation of photosynthetically available radiation. I shall use the word 'gilvin' when referring to the dissolved yellow substances in natural waters, in the remainder of this book.

As the absorption spectra in Fig. 3.5 show, the concentration of gilvin varies markedly, not only between marine and fresh waters, but also amongst different inland waters. A convenient parameter by means of which the concentration of gilvin may be indicated is the absorption coefficient at 440 nm due to this material within the water: this we shall indicate by $g_{440}$. This wavelength is chosen because it corresponds approximately to the mid-point of the blue waveband peak that most classes of algae have in their photosynthetic action spectrum.

In many waters, certainly most inland waters, the gilvin absorption is sufficiently high for $g_{440}$ to be measurable with reasonable accuracy, using 50 or 100 mm pathlength cells. When gilvin absorption is low, as in most marine waters, the absorption coefficient can be measured in the near-ultraviolet (350–400 nm) where absorption is higher, and $g_{440}$ determined by proportion from a typical gilvin absorption spectrum, or using the approximate relationship[59]

$$a(\lambda) = a(\lambda_0)e^{-0.014(\lambda - \lambda_0)}$$

where $a(\lambda)$ and $a(\lambda_0)$ are the absorption coefficients at wavelengths $\lambda$ and $\lambda_0$ nm, respectively.

Surprisingly few data on gilvin absorption in inland waters are available. Table 3.2 contains most of the published data in the form of values of $g_{440}$ for marine as well as fresh waters in various geographical regions. This compilation gives some idea of typical gilvin concentrations that may be expected in natural waters, but the lack of measurements in what are

Table 3.2. *Values of absorption coefficient at 440 nm due to dissolved colour* ($g_{440}$) *and particulate colour* ($p_{440}$) *in various natural waters. Where several measurements have been taken, the mean value, the standard deviation, the range and the time period covered are in some cases indicated. An asterisk* (*) *indicates values measured at a wavelength less than 440 nm and converted to* $g_{440}$ *on the basis of an appropriate gilvin absorption spectrum. For some waters, only the absorption coefficient for gilvin plus particles* (g+p) *was available.* L. = lake, R. = river

| Water body | $g_{440}$ (m$^{-1}$) | | $p_{440}$ (m$^{-1}$) | Reference |
|---|---|---|---|---|
| I. *Oceanic waters* | | | | |
| *Atlantic Ocean* | | | | |
| Sargasso Sea | ~0 | | 0.01 | 276 |
| Off Bermuda | 0.01* | | – | 199 |
| Caribbean Sea | | 0.03(g+p) | | 214 |
| Romanche Deep | | 0.05(g+p) | | 214 |
| Mauritanian upwelling | 0.034–0.075* | | – | 59 |
| Gulf of Guinea | 0.024–0.113* | | – | 59 |
| *Pacific Ocean* | | | | |
| Galapagos Islands | 0.02* | | – | 70 |
| Galapagos Islands | | 0.16(g+p) | | 214 |
| Central Pacific | | 0.04(g+p) | | 214 |
| Off Peru | 0.05* | | – | 70 |
| *Indian Ocean* | | | | |
| Oligotrophic water | 0.02 | | – | 272 |
| Mesotrophic water | 0.03 | | – | 272 |
| Eutrophic water | 0.09 | | – | 272 |
| II. *Coastal and estuarine waters* | | | | |
| *Mediterranean* | | | | |
| Villefranche Bay | 0.060–0.161* | | – | 59 |
| Marseilles drainage outfall | 0.073–0.646* | | – | 59 |
| R. Var mouth | 0.136* | | – | 59 |
| R. Rhone mouth | 0.086–0.572* | | – | 59 |
| *North Atlantic/North Sea/Baltic system* | | | | |
| North Atlantic | 0.02* | | – | 240 |
| North Sea | 0.07* | | – | 240 |
| Baltic Sea | 0.24* | | – | 240 |
| Kattegat | 0.04* | | – | 214 |
| South Baltic Sea | 0.26* | | – | 214 |
| Bothnian Gulf | 0.41* | | – | 214 |
| *Southeast Australia* | | | | |
| (a) Jervis Bay | | | | |
| 3 stations | 0.09–0.14* | | 0.03–0.04 | a |
| (b) Tasman Sea/Clyde R. system | | | | |
| Tasman Sea | 0.02* | | – | 255 |
| Batemans Bay | 0.18 | | – | 254 |
| Clyde R. estuary | 0.64 | | – | 255 |

Table 3.2. (*cont.*)

| Water body | $g_{440}$ $(m^{-1})$ | $p_{440}$ $(m^{-1})$ | Reference |
|---|---|---|---|
| (c) Gippsland (estuarine) lakes system | | | |
| L. King | 0.58 | 0.25 | a |
| L. Victoria | 0.65 | 0.22 | a |
| L. Wellington | 1.14 | 2.27 | a |
| Latrobe R. | 1.89 | 2.78 | a |
| *North America* | | | |
| Georgia salt marsh | 1.52 | – | 542 |
| III. *Inland waters* | | | |
| *Europe* | | | |
| Carmean Quarry, Ireland | 0.23 | – | 219 |
| Killea Reservoir, Ireland | 0.5 | – | 219 |
| Lough Neagh, N. Ireland | 1.9 | – | 219 |
| Lough Fea, Ireland | 4.7 | – | 219 |
| Lough Erne, Ireland | 5.3 | – | 219 |
| Loughnagay, Ireland | 6.4 | – | 219 |
| Lough Bradan, Ireland | 17.4 | – | 219 |
| Lough Napeast, Ireland | 19.1 | – | 219 |
| Loch Leven, Scotland | 1.2 | – | 485 |
| Neusiedlersee, Austria | $\sim 2.0 \pm 0.4$* | – | 93 |
| (8 months) | 1.4–2.8 | | |
| Blaxter L. (bog lake), | 9.65* | – | 310 |
| England | | | |
| *Africa* | | | |
| L. George, Uganda | 3.7 | – | 485 |
| *North America* | | | |
| Crystal L., Wisc., USA | 0.16 | – | 203 |
| Adelaide L., Wisc., USA | 1.85 | – | 203 |
| Bluff L., N.S., Canada | 0.94 | – | 162 |
| Punch Bowl, N.S., Canada | 6.22 | – | 162 |
| *South America* | | | |
| Guri reservoir, Venezuela | 4.84 | – | 289 |
| Carrao R., Venezuela | 12.44 | – | 289 |
| *Australia* | | | |
| (a) Southern tablelands | | | |
| Cotter Dam | 1.28–1.46 | 0.77 | 258, a |
| Corin Dam | 1.19–1.61 | 0.11 | 258, a |
| L. Ginninderra | $1.54 \pm 0.78$ | 0.16–0.58 | 254, 255, 258, a |
| (3-year range) | 0.67–2.81 | | |
| L. George | $1.80 \pm 1.06$ | 3.73–4.21 | 254, 255, 258, a |
| (5-year range) | 0.69–3.04 | | |
| Burrinjuck Dam | $2.21 \pm 1.13$ | 0.63–1.44 | 254, 255, 258, a |
| (5-year range) | 0.81–3.87 | | |
| L. Burley Griffin | $2.95 \pm 1.70$ | 2.91–2.96 | 254, 255, 258, a |
| (5-year range) | 0.99–7.00 | | |
| Googong Dam | 3.42 | 0.83 | 258 |
| Queanbeyan R. | 2.42 | – | a |
| Molonglo R. | 0.44 | – | a |

Table 3.2. (*cont.*)

| Water body | $g_{440}$ (m$^{-1}$) | $p_{440}$ (m$^{-1}$) | Reference |
|---|---|---|---|
| Molonglo R. below confluence with Queanbeyan R. | 1.84 | – | a |
| Creek draining boggy ground | 11.61 | – | a |
| (b) Murrumbidgee Irrigation Area | | | |
| L. Wyangan | 1.13 | 0.38 | a |
| Griffith Reservoir | 1.34 | 3.73 | a |
| Barren Box swamp | 1.59 | 2.55 | a |
| Main canal | 1.11 | 5.35 | a |
| Main drain | 2.12 | 10.34 | a |
| (c) Northern Territory (Magela Creek billabongs) | | | |
| Mudginberri | 1.11 | 1.13 | b |
| Leichardt | 1.77 | 2.43 | b |
| Georgetown | 1.99 | 18.00 | b |
| (d) Tasmania (lakes) | | | |
| Perry | 0.06 | – | 463 |
| Ladies Tarn | 0.40 | – | 463 |
| Risdon Brook | 0.98 | – | 463 |
| Barrington | 3.05 | – | 463 |
| Gordon | 8.29 | – | 463 |

a Kirk, J. T. O., unpublished data
b Tyler, P. A. & Kirk, J. T. O., unpublished data

otherwise limnologically well-characterized parts of the world is apparent. In view of the great importance of gilvin in determining the penetration of light in natural waters, and in view of the ease (in most inland waters) with which its optical concentration can be measured,[254] it is to be hoped that the spectrophotometric assessment of dissolved colour will become a routine part of limnological methodology.

The data in Table 3.2 show that marine waters generally have much less dissolved colour than inland waters, and the greater the distance from land, the lower the concentration. The high concentration (for a marine water) within the Baltic Sea is noteworthy: it decreases from the Bothnian Gulf southwards, as the proportion of river water in the sea diminishes. The increase in concentration with distance from the sea upstream in estuarine systems can be seen in the data for Clyde River–Batemans Bay and Latrobe River–Gippsland Lakes (Australia).

Although gilvin is chemically rather stable – its concentration in a stored sample usually shows only small changes over a few weeks – its concentration within any inland water body changes, in either direction, with time, in accordance with rainfall events in different parts of the catchment and consequent changes in the concentration of gilvin in the inflowing waters. Some of the data in Table 3.2 give an indication of the extent of this variability. For example, in Lake Burley Griffin (Canberra, Australia) the value of $g_{440}$ varied seven-fold over a 5-year period. Nevertheless, for any given water body, variation does tend to be around a certain mean value and the water body can usefully be regarded as typically high, low or intermediate in gilvin concentration. The category in which a particular lake, impoundment or river falls is determined by the drainage pattern, vegetation, soils and climate in the catchment. The factors governing the concentration of gilvin in surface waters are not well understood and in view of the great influence of this material on aquatic primary production they deserve intensive investigation. One generalization that can be made is that gilvin concentration is high in drainage water from bogs or swamps: this can readily be seen, for example, in the peat bogs of northern Europe, and is shown quantitatively in the $g_{440}$ value for an English bog lake and a creek draining boggy ground in the Australian southern tablelands (Table 3.2). Gilvin concentration is also high in the waters draining from humid tropical forests,[289] as the $g_{440}$ values for two Venezuelan waters in Table 3.2 show. The lack of oxygen in the permanently, or frequently, waterlogged soil of such areas leads to a build-up of partially decomposed organic matter, and gilvin is derived from the soluble component of this. The other side of the coin is that water draining limestone-rich catchments tends to be low in gilvin. The effects of vegetation type, soil mineralogy, agricultural practices, climate and other environmental parameters on gilvin concentration are little, if at all, understood.

### Tripton

The inanimate particulate matter, or tripton, of natural waters is the fraction whose light absorption properties have received the least attention because they are so difficult to measure. At typical concentrations the material does not absorb light strongly but scatters quite intensely and so its absorption properties cannot be characterized by normal spectrophotometry with long-pathlength cells. To overcome these problems a procedure has been developed[258] in which the particulate matter from a natural water sample is collected on a filter and resuspended in a much smaller volume: the absorption spectrum of the concentrated material is then measured in a short-pathlength cell, using an integrating sphere, and from the absorbance values the absorption coefficients due to the particulate matter, in the original water body, may be calculated.

The particulate fraction will of course include phytoplankton as well as tripton. If, however, the phytoplankton concentration is low then the particulate fraction spectrum can be attributed to tripton. Fig. 3.6 shows the absorption spectra of the particulate and soluble fractions of several Australian inland waters. In all cases except (*c*) and (*f*) the absorption by the particulate fraction is almost entirely due to tripton, although in two of these ((*d*) and (*g*)) a small 'shoulder' at about 670 nm due to phytoplankton chlorophyll may be seen.

The tripton absorption spectra all have much the same shape: absorption is low or absent at the red end of the spectrum and rises steadily as wavelength decreases into the blue and ultraviolet. These are typical humic

Fig. 3.6. Comparison of the spectral absorption properties of the particulate and soluble fractions of various Australian inland waters in the southern tablelands of New South Wales/Australian Capital Territory (from Kirk, 1980*b*). The absorption spectrum of pure water is included in (*a*) for comparative purposes. The phytoplankton chlorophyll *a* contents ($C_a$) and nephelometric turbidities ($T_n$) of the different water bodies at the time of sampling were as follows: (*a*) Corin Dam, 8th June 1979, $C_a = 2.0$ mg m$^{-3}$, $T_n = 0.51$ NTU; (*b*) Lake Ginninderra, 6th June 1979, $C_a = 1.5$ mg m$^{-3}$, $T_n = 1.1$ NTU; (*c*) Cotter Dam, 8th June 1979, $C_a = 9.0$ mg m$^{-3}$, $T_n = 1.6$ NTU; (*d*) Lake Burley Griffin, 6th June 1979, $C_a = 6.3$ mg m$^{-3}$, $T_n = 17.0$ NTU; (*e*) Googong Dam, 21st June 1979, $C_a = 1.7$ mg m$^{-3}$, $T_n = 5.8$ NTU; (*f*) Burrinjuck Dam, 7th June 1979, $C_a = 16.1$ mg m$^{-3}$, $T_n = 1.8$ NTU; (*g*) Lake George, 28th November 1979, $C_a = 10.9$ mg m$^{-3}$, $T_n = 49$ NTU.

The ordinate scale corresponds to the true *in situ* absorption coefficient due to soluble or particulate colour: note that the scale in (*d*), (*e*) and (*g*) is different to that in (*a*), (*b*), (*c*) and (*f*).

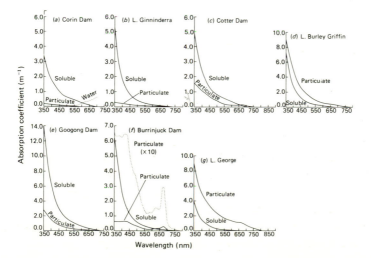

substance absorption spectra and indeed have much the same shape as those of the dissolved yellow materials. Furthermore, typical tripton samples collected on a filter are brown in colour. The most plausible supposition is that the yellow-brown tripton colour is largely due to particulate humic material existing either bound to mineral particles or as free particles of humus. It seems likely that in inland waters it arises, together with the soluble humic material (gilvin), from the soils in the catchment. In productive waters or in oceanic waters well away from land drainage, some of the light-absorbing inanimate matter could arise by decomposition of the phytoplankton: it is interesting that Yentsch (1962) found that the particulate fraction from the deeper layers of the sea (200–1000 m) had absorption spectra (measured on filters) similar in shape to the fresh-water tripton spectra in Fig. 3.6.

The *in situ* absorption coefficient due to particulate matter at 440 nm is a convenient general measure of particulate colour in any water. A suitable symbol would be $p_{440}$, which is analogous to $g_{440}$ previously defined as a measure of soluble colour. Table 3.2 lists some observed values of $p_{440}$ for various natural waters. In turbid waters containing large amounts of suspended material derived from soil erosion in the catchment or (in shallow waters) wind resuspension of sediments, non-living particulate absorption can exceed absorption due to dissolved colour. The waters of the Murrumbidgee Irrigation Area (Table 3.2) are examples of the first situation (see also Fig. 3.6*d*) and Lake George, NSW, is an example of the second (Fig. 3.6*g*, Table 3.2).

It should be realized that $p_{440}$ is not as good a general guide to particulate colour as $g_{440}$ is to soluble colour. Whereas all gilvin spectra have approximately the same shape, the shape of the particulate fraction spectrum can change significantly in accordance with the proportion of humic material or phytoplankton (see below) in it. Nevertheless, in those waters (of common occurrence) with a substantial particulate humic component, it is a useful parameter.

### Phytoplankton

The absorption of light by the photosynthetic pigments – chlorophylls, carotenoids, biliproteins – of the phytoplankton contributes to the attenuation of PAR with depth. Indeed in productive waters, the algae may be present in concentrations such that by self-shading they limit their own growth.

Light absorption by algal cells grown in laboratory culture has received a great deal of attention because of the use of such cultures as experimental material in fundamental photosynthesis research. The light absorption properties of phytoplankton populations as they occur in nature have been hardly studied at all, which is unfortunate since we cannot assume that the natural populations have absorption spectra identical to those of the same species grown in culture. The total concentration of pigments and the relative amounts of the different pigments can all be affected by environmental variables such as nitrogen concentration in the medium, light intensity and spectral distribution, etc. Much research therefore needs to be done on the absorption spectra of natural phytoplankton populations: measurement techniques which overcome the problem of light scattering by the suspensions will of course have to be used. Ultimately what is required is not only the shape of the absorption spectrum, but the actual values of the absorption coefficients due to phytoplankton in the original water body. Data on phytoplankton populations consisting mainly of single known species would be particularly valuable.

Yentsch (1960) measured the absorption spectrum of a natural population

Fig. 3.7. Comparison of the spectral absorption properties of the different fractions in an estuarine water from southeast Australia – Lake King, Victoria (Kirk, unpublished data). Phytoplankton were present at a level corresponding to 3.6 mg chlorophyll *a* m$^{-3}$ and the turbidity of the water was $\sim$ 1.0 NTU.

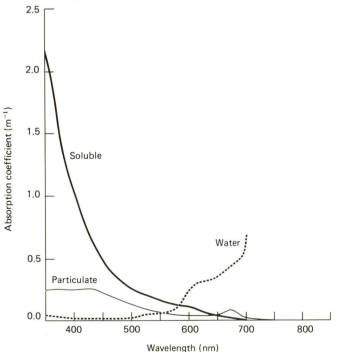

of marine phytoplankton from waters off Woods Hole, Massachusetts, USA: the cells were in a layer on a cleared membrane filter. This method provides valuable information on the shape of the absorption spectrum but, because of the intensification of absorption resulting from multiple scattering within the layer, cannot be used to calculate *in situ* absorption coefficients. Absorption spectra measured on suspensions, either with an integrating sphere or by the opal glass method, can be used to obtain *in situ* absorption coefficients. Fig. 3.6*f* shows the spectrum (integrating sphere method) of the particulate fraction from Burrinjuck Dam, a eutrophic impoundment in New South Wales, Australia, at a time when it had a mixed bloom of *Melosira* sp. (a diatom) and *Anacystis cyanea* (a blue-green alga). The fraction consisted mainly of algal biomass and so the spectrum provides approximate values (somewhat too high in the blue due to the presence of some tripton) for the *in situ* absorption coefficients due to phytoplankton. Fig. 3.7 shows the particulate fraction spectrum – again largely due to phytoplankton – of the estuarine water of Lake King (Gippsland Lakes, Australia).

It is possible to estimate the absorption coefficient of the medium at a given wavelength from the vertical attenuation coefficient for irradiance at that wavelength. By carrying out such calculations for pairs of stations (in the Atlantic off Northwest Africa) where the scattering coefficients were about the same but the phytoplankton population varied, Morel & Prieur (1977) were able to arrive at an absorption spectrum for the natural phytoplankton population present. The absorption coefficients corresponding to 1 mg chlorophyll *a* m$^{-3}$ are plotted against wavelength in Fig. 3.8.

Fig. 3.8. Specific absorption coefficient (*in situ*), corresponding to 1 mg chlorophyll *a* m$^{-3}$, for oceanic phytoplankton (after Morel & Prieur, 1977).

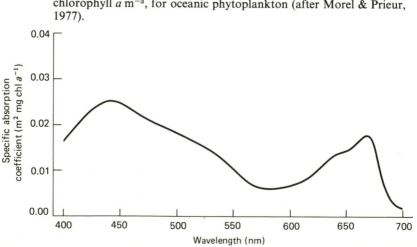

The amount of light harvested by the phytoplankton component of the aquatic medium depends not only on the total amounts of the photosynthetic pigments present, but also on the size and shape of the algal cells or colonies within which the pigments are located. This subject is dealt with later (§9.3).

### Total absorption spectra

At any wavelength, the aquatic medium has a total absorption coefficient, which is the sum of the absorption coefficients of all the light-absorbing components, at that wavelength. The variation of this total absorption coefficient with wavelength is the absorption spectrum of the

Fig. 3.9. Total absorption spectra of various natural waters in southeastern Australia (Kirk, 1981a and unpublished data). (a) Lake George, NSW. (b) Lake Burley Griffin, ACT. (c) Burrinjuck Dam, NSW. (d) Lake King, Victoria. (e) Jervis Bay, NSW. Waters a, b and c are inland, d is estuarine, e is marine (Tasman Sea). Chlorophyll and turbidity data for waters a–d are given in the legends to Figs 3.6 and 3.7. The Jervis Bay water (e) contained 0.2 mg phytoplankton chlorophyll a m⁻³ and was optically intermediate between Jerlov oceanic water types II and III.

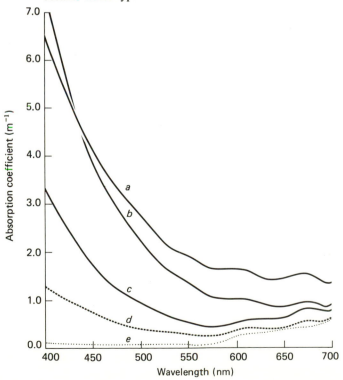

medium as a whole. For any given water body, the total absorption coefficient at each wavelength is obtained by adding together the known absorption coefficient of pure water at that wavelength and the absorption coefficients due to dissolved and particulate colour, determined as described above. Fig. 3.9 shows the total absorption spectra of five Australian waters, three inland, one estuarine and one marine.

### 3.4     Optical classification of natural waters

Natural waters vary greatly in the extent to which they transmit solar radiation and it is useful to have some broad indication of the optical character of a water without having to fully specify all the inherent optical properties. Jerlov (1951, 1976) has classified marine waters into a number of different categories on the basis of the curve of percent transmittance of downward irradiance against wavelength. He recognized three basic types of oceanic water – I, II and III – and nine types of coastal water (1 to 9), in order of decreasing transmittance: the spectral variation of percentage transmittance for some of these water types is shown in Fig. 3.10.

Jerlov's pioneering measurements were, however, made with broad-band colour filters, and the curves obtained with modern submersible spectro-radiometers are in some cases in poor agreement with his.[359] Pelevin & Rutkovskaya (1977) have proposed that, instead, ocean waters be classified in terms of the vertical attenuation coefficient (base 10 logarithm) for

Fig. 3.10. Transmittance per metre of various optical types of marine water for downward irradiance. The waters are Jerlov's oceanic types I and III, and coastal types 1, 5 and 9 (data from Table XXVI of Jerlov, 1976).

irradiance at 500 nm, multiplied by 100. Since the $K_d$ versus $\lambda$ curves vary in a fairly systematic manner from one oceanic water to another, the value of $K_d$ (500 nm) would indeed convey quite a lot of information about any such water. However, given that all definitions of optical properties for natural waters are now standardized to use logarithms to the base e rather than base 10,[198] it would be preferable to modify Pelevin & Rutkovskaya's proposal so that the water type was specified by 100 $K_d$ (500 nm), where $K_d$ is based on $\log_e$, in accordance with the definition in eqn 1.17. This has the advantage that the number obtained is approximately equal (if attenuation is not too intense) to the percentage of downwelling light which is removed. For example, Jerlov's oceanic waters I and III, with irradiance transmittance values at 500 nm of 97.3% and 89%, corresponding to irradiance diminution of 2.7% and 11%, have 100 $K_d$ (500 nm) values of 2.7 and 11.6.

Smith & Baker (1978*b*), on the basis of their measurements of the spectral variation of the vertical attenuation coefficient ($K_d$) for irradiance in various ocean waters, have concluded that in regions away from terrigenous influences, the attenuation (apart from that due to water) is mainly due to the phytoplankton and the various pigmented detrital products that covary with it. They suggest that for such ocean waters, the total content of chlorophyll-like pigments provides a sufficient basis for optical classification, since on the basis of the pigment content, the curve of $K_d$ against wavelength can be calculated.

A crude optical classification applicable mainly to inland waters has been proposed by Kirk (1980*b*) on the basis of measurements of the absorption spectra of the soluble and particulate fractions from Australian water bodies. In type G waters, the dissolved colour (gilvin), at all wavelengths in the photosynthetic range, absorbs light more strongly than the particulate fraction: examples are shown in Figs 3.6*a*, *b* and *e*. In type GA waters, gilvin absorbs more strongly than the particulate fraction throughout the shorter wavelength part of the spectrum, but the absorption coefficients of the particulate fraction exceed those of the soluble fraction at the red end of the spectrum due to the presence of substantial levels of algal chlorophyll: Figs 3.6*c* and *f* are examples. In type T waters, the particulate fraction, consisting mainly of inanimate material (tripton) absorbs light more strongly than the soluble fraction at all wavelengths: turbid waters with large amounts of suspended silt particles (Figs 3.6*d* and *g*) are in this category. In type GT waters, absorption by the soluble and particulate fractions is roughly comparable throughout the photosynthetic range: the waters in certain billabongs in tropical Australia have been observed to be in this category. In highly productive waters, the absorption coefficients due to algal biomass may exceed those due to dissolved colour (and water itself). Measurements by Talling (1970) of the spectral variation of the vertical attenuation coefficient of water from Loch Leven, Scotland, suggest that this eutrophic lake, dominated by *Synechococcus* sp., was in this category (type A) at the time: the waters of the upwelling area off West Africa, studied by Morel & Prieur (1975), also appear to be of type A. Most oceanic, and some coastal, waters (Fig. 3.9e) are so non-productive, and so free of silt and dissolved colour, that water itself is the dominant light absorber: they may be categorized as type W. Such waters are rare inland: one example is Crater Lake, Oregon, USA.[447] In certain estuarine and the more coloured coastal waters, the gilvin absorption at the blue end of the spectrum can be roughly similar in magnitude to the water absorption at the red end. Such waters, for example that in Fig. 3.7, may be classified as WG.

A water body can change from one optical type to another. Heavy rain in the catchment with consequent soil erosion could, for example, quickly change a type G water (gilvin-dominated) to a type T water (tripton-dominated); development of an algal bloom could change it to a type GA water. Nevertheless, some water bodies have water of a particular type most of the time. Bog lakes are typically type G all the time. Shallow, wind-exposed lakes with unconsolidated sediments are likely to be of type T all the time. Marine waters are, apart from the effects of the yearly phytoplankton cycle (in non-tropical areas), generally constant in their optical properties.

Prieur & Sathyendranath (1981) have proposed an optical classification scheme for sea water similar in some respects to that described above for fresh water. It is based on the relative importance of absorption by the three major components other than water: algal pigments, dissolved yellow materials and particulate matter other than phytoplankton. As a numerical indication of importance the absorption coefficient due to each component – $C'$, $Y'$ and $P'$, respectively – is used. Different waters are classified as $C'$ type, $Y'$ type, $P'$ type, or as hybrid types such as $C'Y'$, $C'P'$ or $P'Y'$. Most of the marine waters studied by these authors were of the $C'$ type.

## 3.5     Contribution of the different components of the aquatic medium to absorption of PAR

Apart from the small amount of light scattered back out of the water, attenuation of PAR in water bodies is due to absorption, although the extent of this absorption within a given depth may be greatly amplified by scattering which increases the average pathlength of the photons within that depth. The relative contribution of different components of the system to this absorption at a given wavelength is in proportion to their absorption coefficients at that wavelength. Absorption coefficients for any particular component and the intensity of the incident solar radiation vary independently with wavelength throughout the photosynthetic range, and so the contribution of the different components to absorption of total PAR can only be assessed by carrying out the appropriate calculations at a series of narrow wavebands and summing the results. In this way it is possible to calculate what proportion of those photosynthetic quanta which are absorbed are captured by each of the different absorbing components of the system.[254, 258] Table 3.3 presents the results of such calculations for 12 Australian water bodies.

It is clear that the three major components into which the light-absorbing material of the system has been divided – the water itself, the dissolved

Table 3.3. *Calculated distribution of absorbed photosynthetic quanta between the particulate fraction, soluble fraction and water in Australian water bodies. The calculations have in each case been carried out for the euphotic zone of the water body in question. Data from Kirk (1980b) and unpublished*

| Water body | Optical type | Quanta absorbed (% of total) | | |
| | | By water | By soluble fraction[a] | By particulate fraction[b] |
|---|---|---|---|---|
| *Coastal – Oceanic* | | | | |
| Jervis Bay[c] | W | 68.1 | 23.9 | 8.0 |
| *Estuarine* | | | | |
| Lake King[d] | WG | 41.9 | 40.4 | 17.7 |
| *Inland impoundments* | | | | |
| Corin Dam[e] | G | 34.8 | 60.0 | 5.2 |
| Lake Ginninderra[e] | G | 39.1 | 50.4 | 10.5 |
| Googong Dam[c] | G | 22.0 | 60.4 | 17.6 |
| Cotter Dam[e] | GA | 26.2 | 49.8 | 24.0 |
| Burrinjuck Dam[c] | GA | 28.2 | 45.5 | 26.3 |
| Lake Burley Griffin[e] | T | 19.4 | 22.2 | 58.4 |
| *Natural inland waters* | | | | |
| Latrobe River[d] | T | 17.5 | 28.1 | 54.5 |
| Lake George[c] | T | 12.4 | 8.3 | 79.3 |
| Long Harry billabong[f] | GT | 20.0 | 37.0 | 43.0 |
| Georgetown billabong[f] | T | 5.9 | 7.5 | 86.6 |

[a]Gilvin. [b]Tripton/phytoplankton. [c]New South Wales. [d]Victoria. [e]Australian Capital Territory. [f]Northern Territory.

yellow substances (gilvin), and the particulate fraction (tripton/phyto-plankton) – can all be substantial light absorbers. In type W waters, most of the photosynthetic quanta are captured by water itself. In oceanic waters not affected by terrigenous material (generally type W), those quanta not absorbed by water are captured mainly by the pigments of the living and dead phytoplankton.[441] In coastal waters, with low but significant amounts of gilvin, the contributions of the phytoplankton and the dissolved yellow material to absorption are likely to be small and comparable. In estuarine waters, such as Lake King (type WG), with more substantial levels of soluble yellow substances, the water and the gilvin capture most of the photosynthetic quanta. In the type G-inland waters (plenty of dissolved colour, but low turbidity) described in Table 3.3, gilvin captures the most quanta, followed by water. In the type GA waters (dissolved colour, rather low turbidity, plentiful phytoplankton), gilvin is still the most important component but the particulate fraction now captures about as many quanta as does the water. In the type T waters (high turbidity due to tripton), most of the quanta are absorbed by the particulate fraction. The particulate fraction and gilvin between them take most of the quanta in type GT waters (intense soluble colour, high tripton turbidity). In highly productive, type A, waters, dominated by phytoplankton, we may assume that the algal biomass captures more photosynthetic quanta than the other components of the medium.

# 4

..................................................................................................

# *Scattering of light within the aquatic medium*

We have seen that most of the solar photons which enter the water are absorbed. Many of these photons – most, in some waters – undergo scattering one or more times before they are absorbed. Scattering does not by itself remove light – a scattered photon is still available for photosynthesis. The effect of scattering is to impede the vertical penetration of light. It makes the photons follow a zig-zag path as they ricochet from one scattering particle to the next. This increases the total pathlength which the photons must follow in traversing a certain depth, and so increases the probability of their being captured by one of the absorbing components of the medium. In addition, some of the photons are actually scattered back in an upwards direction. Thus the effect of scattering is to intensify the vertical attenuation of the light.

In this chapter we shall consider the nature of the scattering process and the scattering properties of natural waters.

## 4.1 The scattering process

What do we mean by scattering? We say that a photon is scattered when it interacts with some component of the medium in such a way that it is caused to diverge from its original path. There are two kinds of scattering to be considered – density fluctuation scattering and particle scattering.

### *Density fluctuation scattering*

In understanding the basis of density fluctuation scattering in liquids, it is helpful to begin with a consideration of molecular, or Rayleigh, scattering by gases such as air. According to the Rayleigh theory, within any particle, such as an air molecule, in a light field, a dipole is induced by the electrical vector of the field. As the dipole oscillates at the

71

frequency of the exciting radiation, it emits radiation of the same frequency in all directions. It is this radiation which is the scattered light.

The Rayleigh molecular theory of scattering does not apply to liquids: the strong interactions between the molecules make it impermissible to consider the interaction of the radiation with molecules on an individual basis. In any liquid, however, the continual random motion of the molecules leads to localized microscopic fluctuations of density and therefore of dielectric constant and, in the Einstein–Smoluchowski theory, the interaction of the radiation field with these inhomogeneities – each of which can be regarded as a dipole – rather than with the individual molecules, is considered. The predicted angular distribution of scattering is similar to that given by the Rayleigh theory for gases, i.e. it is identical in the forward and backward directions (Figs 2.2 and 4.8). Again, as in Rayleigh scattering by gases, scattering by a pure liquid is predicted to vary inversely with the fourth power of the wavelength.

### Particle scattering

The Rayleigh and Einstein–Smoluchowski theories of scattering apply only when the scattering centres are small relative to the wavelength of light: this is true in the case of gas molecules and of the tiny density fluctuations in pure liquids. Even the most pristine natural waters, however, are not, optically speaking, pure and they invariably contain high concentrations of particles – mineral particles derived from the land or from bottom sediments, phytoplankton, bacteria, dead cells and fragments of cells, etc. – all of which scatter light. The particles which occur in natural waters have a continuous size distribution which is roughly hyperbolic,[22] i.e. the number of particles with diameter greater than $D$ is proportional to $1/D^{\gamma}$, where $\gamma$ is a constant for a particular water body, but varies widely from 0.7 to 6 in different water bodies.[214] Although a hyperbolic distribution implies that smaller particles are more numerous than big ones, nevertheless most of the particle cross-sectional area that would be encountered by light in natural waters is due to particles of diameter greater than 2 $\mu$m[214] which is not small relative to the wavelengths of visible light, and so scattering behaviour different to the density fluctuation type must be anticipated. The smaller particles, although numerous, have a lower scattering efficiency.

A theoretical basis for predicting the light-scattering behaviour of spherical particles of any size was developed by Mie (1908). The physical basis of the theory is similar to that of Rayleigh in that it considers the oscillations set up within a polarizable body by the incident light field and the light re-radiated (i.e. scattered) from the body as a result of these oscillations. Instead of (as in Rayleigh theory) equating the particle to a single dipole, the Mie theory considers the additive contributions of a series of electrical and magnetic multipoles located within the particle. The advantage of the Mie theory is that it is all-embracing – for very small particles, for example, it leads to the same predictions as the Rayleigh theory; the disadvantage is that the analytical expressions are complex and do not lend themselves to easy numerical calculations. For particles larger than the wavelength of the light, Mie theory predicts that most of the scattering is in the forward direction within small angles of the beam axis (Fig. 4.1). A series of maxima and minima is predicted at increasing scattering angle, but these are smoothed out when a mixture of particle sizes is present.

In the case of particles larger than a few wavelengths of light, a reasonable understanding of the mechanism of scattering can be obtained

Fig. 4.1. Angular distribution of scattered intensity from transparent spheres calculated from Mie theory (Ashley & Cobb, 1958) or on the basis of transmission and reflection, or diffraction, transmission and reflection (Hodkinson & Greenleaves, 1963). The particles have a refractive index (relative to the surrounding medium) of 1.20, and have diameters 5–12 times the wavelength of the light. After Hodkinson & Greenleaves (1963).

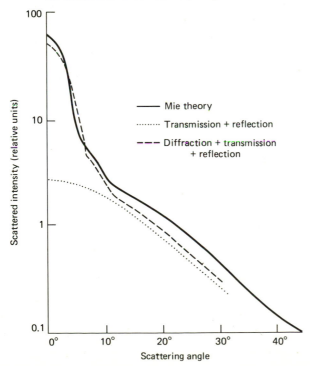

on the basis of diffraction and geometrical optics, without recourse to electromagnetic theory. When an object is illuminated by a plane light wave, the shadow of an object on a screen placed behind it is not quite precisely defined: just outside it a series of concentric faint dark bands will be present; and lighter concentric bands, where clearly some light is falling, will be present within the area of the geometric shadow. This phenomenon – *diffraction* – is due to interference (destructive in the dark rings, constructive in the light rings) between parts of the wave coming from different points around the edge of the illuminated object, and arriving simultaneously but out of phase (because of the different distances traversed) at particular points on the screen. In the case of a round object, superimposed on the circular shadow, there is a bright spot in the centre, surrounded by alternate dark and light rings. In fact most of the light diffracted by a particle is propagated in the forward direction within a small angle of the initial direction of the beam (giving rise to the bright spot). With increasing angular distance away from the axis, the diffracted intensity goes through a series of minima and maxima (dark and light rings) which diminish progressively in height.

Applying geometrical optics to these larger particles, it can readily be appreciated that some of the light will be reflected at the external surface and some will pass through the particle and undergo refraction, or internal reflection as well as refraction (Fig. 4.2). In all cases the photons involved will be made to deviate from their initial direction, i.e. will be scattered. Light scattered by these mechanisms is still predominantly in the forward direction: scattered intensity diminishes continuously with increasing angle but does not show the same degree of concentration at small angles as does scattering due to diffraction. Calculations by Hodkinson & Greenleaves (1963) for suspensions of spherical particles of mixed sizes show that most of the scattering at small angles (up to about 10–15°) can be attributed

Fig. 4.2. Scattering of light by a particle: reflection and refraction processes.

to diffraction, whereas most of the scattering at larger angles is due to external reflection and transmission with refraction (Fig. 4.1). The angular variation of scattering calculated on the basis of diffraction and geometrical optics is reasonably close to that derived by the Mie electromagnetic theory. Such discrepancies as exist may be largely due to the fact that as a result of the phase change induced in the transmitted ray by passage through a medium of higher refractive index, there are additional interference effects between the diffracted and the transmitted light:[519] this phenomenon is known as anomalous diffraction.

Any particle in a beam of light will scatter a certain fraction of the beam and the radiant flux scattered will be equivalent to that in a certain cross-sectional area of the incident beam. This area is the *scattering cross-section* of the particle. The efficiency factor for scattering, $Q_{scat}$, is the scattering cross-section divided by the geometrical cross-sectional area of the particle ($\pi r^2$ for a spherical particle of radius $r$). Similarly, in the case of an absorbing particle, the radiant flux absorbed is equivalent to that in a certain cross-sectional area of the incident beam: this area is the *absorption cross-section* of the particle. The efficiency factor for absorption, $Q_{abs}$, is the absorption cross-section divided by the geometrical cross-sectional area of the particle. The efficiency factor, $Q_{att}$, for attenuation (absorption and scattering combined) is thus given by

$$Q_{att} = Q_{scat} + Q_{abs}$$

The attenuation efficiency of a particle can be greater than unity: that is, a particle can affect the behaviour of more light in the incident beam than will be intercepted by its geometrical cross-section. This can be true of absorption and scattering separately: i.e. it is possible for a particle to absorb, or to scatter, more light than its geometrical cross-section would intercept. In terms of electromagnetic theory we may say that the particle can perturb the electromagnetic field well beyond its own physical boundary. Mie theory, based as it is upon electromagnetism, can be used to calculate the absorption and scattering efficiencies of particles. The simpler anomalous diffraction theory of van de Hulst (1957) can also be used to calculate the scattering efficiency of particles with refractive index up to about twice that of the surrounding medium. The relation is

$$Q_{att} = 2 - \frac{4}{\rho}\sin\rho + \frac{4}{\rho^2}(1 - \cos\rho)$$

where $\rho = (4\pi a/\lambda)(m-1)$, $m$ being the refractive index of the particle relative to that of the surrounding medium, and $a$ being the radius of the particle. For a non-absorbing particle, $Q_{scat} = Q_{att}$. Fig. 4.3 shows the way in which the scattering efficiency for green light of a spherical non-absorbing particle of refractive index relative to water of 1.17 (a typical value for inorganic particles in natural waters), varies with particle size. It can be seen that scattering efficiency rises steeply from very low values for very small particles to about 3.2 at a diameter of 2 $\mu$m. With increasing diameter, it first decreases and then increases again and undergoes a series of oscillations of diminishing amplitude to level off at a $Q_{scat}$ value of 2.0 for very large particles. A similar general pattern of variation of $Q_{scat}$ with size would be exhibited by any scattering particle of the types found in natural waters at any wavelength in the photosynthetic range.

## 4.2    Measurement of scattering

### Beam transmissometers

In the absence of absorption, the scattering coefficient could in principle be determined by measuring the loss of intensity of a narrow parallel beam passing through a known pathlength of medium. If absorption as well as scattering occurs, then the parameter measured by the instrument would in fact be the beam attenuation coefficient, $c$, rather than the scattering coefficient. If it is possible also to measure the absorption coefficient, $a$, of the water at the appropriate wavelength, then the scattering coefficient, $b$, may be calculated ($b = c - a$).

Beam attenuation meters – or beam transmissometers as they are more commonly called – for the *in situ* measurement of $c$ have long been important tools in hydrological optics. In principle all they have to do is

Fig. 4.3. Scattering efficiency of a non-absorbing spherical particle as a function of size. Calculated using the equation of van de Hulst (1957) – see text. The particle has a refractive index, relative to water, of 1.17. $\lambda = 550$ nm.

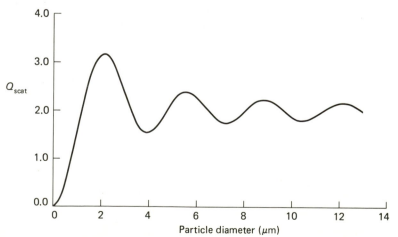

measure the proportion, $C$, of the incident beam which is lost by absorption and scattering in a pathlength, $r$, the beam attenuation coefficient being equal to $-[\ln(1-C)]/r$ (§1.4). In practice, the construction of instruments which accurately measure $c$ is difficult. The problem is that most of the scattering by natural waters is at small angles. Therefore, unless the acceptance angle of the detector of the transmissometer is very small ($< 1°$), significant amounts of scattered light remain in the beam and so attenuation is underestimated. The principles of design of beam transmissometers and the variation of the size of the error with the design parameters have been discussed by Austin & Petzold (1977). One possible optical system is shown schematically in Fig. 4.4. Commercial instruments are available. Although *in situ* measurements are to be preferred, it is possible to measure the beam attenuation coefficient in the laboratory with a spectrophotometer, provided that a very small acceptance angle and long-pathlength cells are used. Bochkov, Kopelevich & Kriman (1980) have developed such an instrument for the shipboard determination of $c$ of sea water samples over the range 250–600 nm.

On the basis of laboratory measurements of media varying widely in their scattering and absorbing properties, Timofeeva & Gorobets (1967) were able to devise an empirical relation between $K/c$ and $b/c$, where $K$ is the vertical attenuation coefficient of radiance or irradiance in the deep, 'asymptotic' (see §6.6) region. Thus if, in any water body, both $K$ and $c$ are measured, the Timofeeva–Gorobets curve can be used to determine $b/c$. Since $c$ is known, $b$ can be found (and then also $a$, using $c = a+b$).

### Variable-angle scattering meters

The scattering properties of natural waters are best determined by directly measuring the scattered light. The general principle is that a parallel beam of light is passed through the water, and the light scattered

Fig. 4.4. Beam transmissometer optical system (adapted from Austin & Petzold, 1977). To confine measurements to a narrow spectral waveband, a filter can be included in the light path within the receiver.

from a known volume at various angles is measured. In the ideal case, the volume scattering function, $\beta(\theta)$, is measured from 0 to 180°: this provides not only the angular distribution of scattering for that water but also, by integration, the total, forward and backward scattering coefficients (§1.4). Such measurements are in reality difficult to carry out and relatively few natural waters have been completely characterized in this way. The problem is that most of the scattering occurs at small angles (typically 50% between 0 and 2–6°) and it is hard to measure the relatively faint scattering signal so close to the intense illuminating beam. We shall consider *in situ* scattering meters first.

An instrument developed by Petzold (1972) for very low angles (Fig. 4.5) uses a highly collimated beam of light traversing a 0.5 m pathlength in water and then being brought to a focus by a long-focal-length lens. Light which has been neither scattered nor absorbed comes to a point in this plane. Light which has been caused to deviate by scattering arrives at this plane displaced a certain distance (proportional to the scattering angle) to one side. A field stop is placed in the focal plane, opaque except for a clear annular ring which allows only light corresponding to a certain narrow (scattering) angular range to pass through and be detected by a photomultiplier behind the stop. Three such field stops are used, each with the annular ring a different radial distance from the centre: these correspond to scattering angles of 0.085, 0.17 and 0.34°. To measure the intensity of the incident beam a fourth stop is used which has a calibrated neutral-density filter at the centre.

Kullenberg (1968) used a He–Ne laser to provide a collimated light beam traversing a pathlength of 1.3 m in water. At the receiving end of the instrument, the central part of the beam was occluded with a light trap and a system of conical mirrors and annular diaphragms was used to isolate the light which had been scattered at 1, 2.5 or 3.5°.

Fig. 4.5. Optical system of low-angle scattering meter (after Petzold, 1972). The vertical dimensions of the diagram are exaggerated.

Bauer & Morel (1967) used a central stop to screen off the collimated incident beam and all light scattered at angles up to 1.5°. Light scattered between 1.5 and 14° was collected by a lens and brought to a focus on a photographic plate: $\beta(\theta)$ over this angular range was determined by densitometry.

Instruments for measuring scattering at larger angles are constructed so that either the detector or the projector can rotate relative to the other. The general principle is illustrated in Fig. 4.6. It will be noted that the detector 'sees' only a short segment of the collimated beam of light in the water, and the length of this segment varies with the angle of view. Instruments of this type have been developed by Tyler & Richardson (1958) (20–180°), Jerlov (1961) (10–165°) and Petzold (1972) (10–170°).

The scattering properties of natural waters can be measured in the laboratory but, at least in cases in which scattering values are low, as in many marine waters, there is a real danger that the scattering properties may change in the time between sample collection and measurement. In the case of the more turbid waters commonly found in inland, estuarine and some coastal systems, such changes are less of a problem but it is still essential to keep the time between sampling and measurement to a minimum and to take steps to keep the particles in suspension. Commercial light scattering photometers were developed primarily for studying macromolecules and polymers in the laboratory, but have been adapted by a number of workers for the measurement of $\beta(\theta)$ in natural waters.[34, 458] The water sample is placed in a glass cell illuminated with a collimated light beam. The photomultiplier is on a calibrated turntable and can be positioned to measure the light scattered at any angle within the range of the instrument (typically 20–135°). Laboratory scattering photometers for very small angles have been developed[15, 459] and commercial instruments are available.

Fig. 4.6. Schematic diagram of optical system of general-angle scattering meter (after Petzold, 1972).

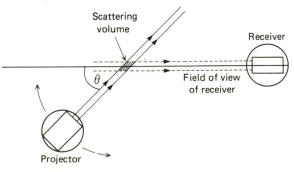

From the definition of volume scattering function (§1.4) it follows that to calculate the value of $\beta(\theta)$ at each angle it is necessary to know not only the radiant intensity at the measuring angle, but also the value of the scattering volume 'seen' by the detector (this varies with angle as we noted above), and the irradiance incident upon this scattering volume. In the case of laboratory scattering meters, scattering by a water sample can be related to that from a standard scattering medium such as pure benzene.[325]

Once the volume scattering function has been measured over all angles, the value of the scattering coefficient can be obtained by summation (integration) of $2\pi\beta(\theta) \sin \theta$ in accordance with eqn 1.36. The forward and backward scattering coefficients, $b_f$ and $b_b$, are obtained by integration from 0 to 90°, and from 90 to 180°, respectively.

### Fixed-angle scattering meters

As an alternative to measuring the whole volume scattering function in order to determine $b$, $\beta(\theta)$ can be measured at one convenient fixed angle, and by making reasonable assumptions about the likely shape of the volume scattering function in the type of water under study (see below), an approximate value of $b$ can be estimated by proportion. In marine waters, for example, the ratio of the volume scattering function at 45° to the total scattering coefficient ($\beta$ (45°)$/b$) is commonly in the range 0.021–0.035 sr$^{-1}$.[214] From an analysis of their own measurements of $\beta(\theta)$ and $b$ in the Pacific and Indian Oceans and in the Black Sea, Kopelevich & Burenkov (1971) concluded that the error in estimating $b$ from single-angle measurements of $\beta(\theta)$ is lower for angles less than 15°: 4° was considered suitable. A linear regression of the type

$$\log b = c_1 \log \beta(\theta) + c_2$$

where $c_1$ and $c_2$ are constants, was found to give more accurate values for $b$ than a simple proportionality relation such as $b = constant \times \beta(\theta)$.

### Turbidimeters

A specialized and simplified laboratory form of the fixed-angle scattering meter is the nephelometric turbidimeter. The word 'turbidity' has been used in a general sense to indicate the extent to which a liquid lacks clarity, i.e. scatters light as perceived by the human eye. In the most common type of instrument, a beam of light is directed along the axis of a cylindrical glass cell containing the liquid sample under study. Light scattered from the beam within a rather broad angle centred on 90° is measured by a photomultiplier located at one side of the cell (Fig. 4.7). The 'turbidity' ($T_n$) of the sample in nephelometric turbidity units (NTU) is measured relative to that of an artificial standard with reproducible light-scattering properties. The standard can be a suspension of latex particles, or of the polymer formazin, made up in a prescribed manner. Turbidimeters, as at present constituted, do not attempt to provide a direct estimate of any fundamental scattering property of the water and the nephelometric turbidity units are essentially arbitrary in nature. Nevertheless, the turbidity measured in this way should be directly related to the average volume scattering function over an angular range centred on 90°, and so for waters of a given optical type (e.g. waters with moderate to high turbidity due to inorganic particles) should bear an approximately linear relation to the scattering coefficient. Since turbidimetric measurements are so easily made, comparative studies on a variety of natural waters to determine the empirical relation between nephelometric turbidity and scattering coefficient would be valuable. Some existing indirect measurements (see below) suggest that for turbid waters $b/T_n$ is about $1 \text{ m}^{-1} \text{NTU}^{-1}$.

Fig. 4.7. Schematic diagram of optical system of nephelometric turbidimeter.

Light trap

Sample cell

Cell holder

Photomultiplier

Condensing lens

Lamp

Turbidimeters are however, not well suited for characterizing waters with very low scattering values, such as the clear oceanic types.

### Indirect estimation of scattering properties

While few aquatic laboratories carry out measurements of the fundamental scattering properties of natural waters, many routinely measure underwater irradiance (§5.1). Since irradiance at any depth is in part determined by the scattering properties of the water, there is the possibility that information on the scattering properties might be derived from the measured irradiance values. For water with a specified volume scattering function and with incident light at a given angle, then at any given optical depth (see §1.6 for definition), the irradiance reflectance, $R$, and the average cosine, $\bar{\mu}$, of the light are functions only of the ratio, $b/a$, of scattering coefficient to absorption coefficient. Conversely, if the value of reflectance at a certain optical depth is given, then the values of $\bar{\mu}$ and $b/a$ are fixed, and in principle determinable. Kirk (1981a, b) used Monte Carlo simulation (§§5.5 and 6.7) to determine the relations between $b/a$ for the medium and $R$ and $\bar{\mu}$ at a fixed optical depth, for water with a normalized volume scattering function identical to that determined by Petzold (1972) for the turbid water of San Diego harbour. It was considered that these relations would be approximately valid for most natural waters of moderate to high turbidity. Using the computer-derived curves it is possible, given a measured value of irradiance reflectance at a specified optical depth, such as $\zeta = 2.3$ (irradiance reduced to 10% of the subsurface value) to read off the corresponding values of $\bar{\mu}$ and $b/a$. The irradiance values are used to estimate $K_E$, the vertical attenuation coefficient for *net* downward irradiance (§1.3), and the value of $a$ is calculated from the relation $a = \bar{\mu}K_E$ (§1.7). Knowing $a$ and $b/a$, the value of $b$ may then be obtained.

Values of scattering coefficient obtained in this way by Kirk (1981b) for various water bodies in Southeastern Australia were found to correlate very closely with the values of nephelometric turbidity ($T_n$), a parameter which (see above) we might reasonably expect to be linearly related to the scattering coefficient: the average ratio of $b$ to $T_n$ was 0.92 m$^{-1}$ NTU$^{-1}$. Using literature data for Lake Pend Oreille in Idaho, USA,[376] the method was found to give a value of $b$ differing by only 5% from that derived from the beam attenuation and absorption coefficients.

Di Toro (1978) using radiation transfer theory with a number of simplifying assumptions has arrived at an approximate relation between the scattering coefficient and a certain function of the reflectance and vertical attenuation coefficient for irradiance. For the turbid waters of San

Francisco Bay, he observed a good linear relation between the values of $b$ calculated in this way and the nephelometric turbidity: the actual ratio of scattering coefficient to turbidity was $1.1 \, \text{m}^{-1} \, \text{FTU}^{-1}$, which is in reasonable agreement with the value of $0.92 \, \text{m}^{-1} \, \text{NTU}^{-1}$ (NTU and FTU being equivalent) obtained by Kirk (1981*b*).

### *Diffuse scattering coefficients*

As we saw in our discussion of radiation transfer theory (§1.7), certain quasi-inherent optical properties of the medium – the diffuse scattering coefficients – play an essential role in our understanding of the underwater light field. These scattering coefficients have been defined previously (§1.5) in terms of the light scattered backwards or forwards by a thin layer of medium from an incident light field which is not a parallel beam at right angles (as in the case of the normal scattering coefficients) but has a radiance distribution identical to the downwelling or upwelling parts of a given underwater light field existing at a certain depth in a certain water body. There is no way at present of directly measuring the diffuse scattering coefficients. If, however, the radiance distribution at a given depth in the water body and the volume scattering function for the water are known, then it is possible to calculate the diffuse scattering coefficients.[257]

To calculate, for example, the backscattering coefficient for downwelling flux, the downwelling radiance distribution is distributed into a manageable number, e.g. 15, of angular intervals. Using the volume scattering function the proportion of the incident flux in each angular interval which is scattered in an upwards or a downwards direction by a thin layer of medium is calculated. The summed flux (for all incident angles) travelling upwards after scattering, expressed as a proportion of the total downwelling flux, divided by the thickness of the layer of medium, is $b_{bd}(z)$, the diffuse backscattering coefficient for downwelling flux at depth $z$ m in that particular water body: $b_{bd}(z)$ is commonly 2- to 5-fold greater than $b_b$ in natural water bodies.[257]

The radiance distribution data necessary for such calculations may be obtained by direct measurement (§5.1), or derived from the inherent optical properties by computer modelling (§5.2).

## 4.3    The scattering properties of natural waters

The scattering properties of pure water provide us with a suitable baseline from which to go on to the properties of natural waters. We shall make use here of a valuable review by Morel (1974) of the optical properties of pure water and pure sea water. For measurements of the scattering properties, water purified by distillation *in vacuo* or by repeated filtration through small-pore-size filters must be used: ordinary distilled water contains too many particles. Scattering by pure water is of the density fluctuation type, and so varies markedly with wavelength. Experimentally, scattering is found to vary in accordance with $\lambda^{-4.32}$ rather than $\lambda^{-4}$ as predicted by density fluctuation theory alone: this is a result of the variation of the refractive index of water with wavelength. Pure sea water (35–38‰ salinity) scatters about 30% more intensely than pure water. Table 4.1 lists values of the scattering coefficient for pure water and pure sea water at a number of wavelengths. The volume scattering function of pure water or pure sea water has, as predicted by density fluctuation theory, its minimum at 90° and rises symmetrically towards greater or lesser angles (Fig. 4.8).

The scattering coefficients of natural waters are invariably much higher than those of pure water. Table 4.1 lists a selection of values from the literature. Even the lowest value – 0.016 m$^{-1}$ at 546 nm for water from 1000 m depth in the Tyrrhenian Sea[326] – is 10 times as high as the value for pure water at that wavelength. Unproductive oceanic waters away from land have low values. Coastal and semi-enclosed marine waters have higher values due to the presence of terrigenous particulate material plus phytoplankton. High levels of phytoplankton can give rise to quite high

Fig. 4.8. Volume scattering function of pure water for light of wavelength 550 nm. The values are calculated on the basis of density fluctuation scattering, assuming that $\beta(90°) = 0.93 \times 10^{-4}$ m$^{-1}$ sr$^{-1}$ and that $\beta(\theta) = \beta(90°)(1 + 0.835\cos^2\theta)$ (following Morel, 1974).

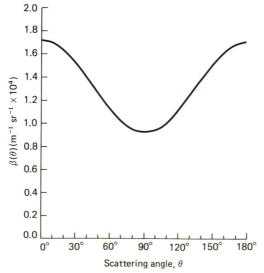

Table 4.1. *Scattering coefficient values for various waters*

| Water | Wavelength (nm) | Scattering coefficient, $b$ ($m^{-1}$) | Reference |
|---|---|---|---|
| Pure water | 400 | 0.0058 | 327 |
| | 450 | 0.0035 | 327 |
| | 500 | 0.0022 | 327 |
| | 550 | 0.0015 | 327 |
| | 600 | 0.0011 | 327 |
| Pure sea water | 450 | 0.0045 | 327 |
| | 500 | 0.0019 | 327 |
| *Marine Waters* | | | |
| *Atlantic Ocean* | | | |
| Sargasso Sea | 633 | 0.023 | 275 |
| | 440 | 0.04 | 214 |
| Caribbean Sea | 655 | 0.06 | 214 |
| 3 oceanic stations | 544 | 0.06–0.30 | 271 |
| Bahama Islands | 530 | 0.117 | 363 |
| Mauritanian upwelling[a] | 550 | 0.4–1.7 | 330 |
| Mauritanian coastal[b] | 550 | 0.9–3.7 | 330 |
| *Pacific Ocean* | | | |
| Central (Equator) | 440 | 0.05 | 214 |
| Galapagos Islands | 655 | 0.07 | 214 |
| | 440 | 0.08 | 214 |
| 126 oceanic stations (av.) | 544 | 0.18 | 271 |
| Offshore, South California | 530 | 0.275 | 363 |
| Kieta Bay (Solomon Is.) | 544 | 0.54 | 271 |
| Lagoon, Tarawa atoll (Gilbert Is.) | 544 | 1.04 | 271 |
| San Diego harbour (California) | 530 | 1.21–1.82 | 363 |
| *Indian Ocean* | | | |
| 164 oceanic stations (av.) | 544 | 0.18 | 271 |
| *Mediterranean Sea* | | | |
| Tyrrhenian Sea (1000 m depth) | 546 | 0.016 | 326 |
| Western Mediterranean | 655 | 0.04 | 188 |
| Bay of Villefranche | 546 | 0.1 | 326 |
| *Baltic Sea* | | | |
| Kattegat | 655 | 0.15 | 214 |
| South Baltic | 655 | 0.20 | 214 |
| Bothnian Gulf | 655 | 0.28 | 214 |
| *Black Sea* | | | |
| 33 stations (av.) | 544 | 0.41 | 271 |
| *North Sea* | | | |
| English Channel | 546 | 0.65 | 326 |
| *Inland Waters* | | | |
| *USA* | | | |
| Lake Pend Oreille[c] | 480 | 0.29 | 376 |

Table 4.1. (*cont.*)

| Water | Wavelength (nm) | Scattering coefficient, $b$ ($m^{-1}$) | Reference |
|---|---|---|---|
| *Australia* | | | |
| Corin Dam[d] | 400–700 | 1.5 | 260 |
| Burrinjuck Dam[d] | 400–700 | 2.0–5.5 | 260 |
| Lake Ginninderra[d] | 400–700 | 4.4–21.6 | 260 |
| Lake Burley Griffin[d] | 400–700 | 2.8–52.6 | 260 |
| Lake George[d] | 400–700 | 55.3, 59.8 | 260 |

[a]Waters rich in phytoplankton
[b]High concentration of resuspended sediments
[c]Scattering coefficient calculated from values of $c$ and $a$
[d]Scattering coefficients calculated by indirect procedure (see text) using underwater irradiance values obtained with PAR quanta meter

values of the scattering coefficient in oceanic upwelling areas, such as the Mauritanian upwelling off the west coast of Africa.[330] There are very few data available for inland waters. However, on the basis of nephelometric turbidity data (of which there are quite a lot) and visual appearance, together with the few existing (indirect) measurements, it would be safe to say that scattering coefficient values tend on average to be higher in inland than in marine waters. Indeed the very high values that can occur in some turbid inland waters (see, for example, the Australian data in Table 4.1) are unlikely to be equalled in the sea because high ionic strength promotes the aggregation and precipitation of colloidal clay minerals.

Volume scattering functions for natural waters differ markedly in shape from that of pure water. They are invariably characterized, even in the clearest waters, by an intense concentration of scattering at small forward angles. This, as we saw earlier, is typical of scattering by particles of diameter greater than the wavelength of light, and scattering in natural waters is primarily due to such particles. Fig. 4.9 shows the volume scattering functions for a clear oceanic water and a moderately turbid, harbour water. They are quite similar in shape but there is a noticeable difference at angles greater than 90°: scattering by the oceanic water shows a greater tendency to increase between 100° and 180°. This is because at these larger angles density fluctuation scattering (which exhibits this kind of variation in this angular range – Fig. 4.8) becomes a significant proportion of the total in the case of the clear oceanic water (see below), but remains insignificant compared to particle scattering in the case of the harbour water.

A knowledge of the shape of the volume scattering function is essential for calculations on the nature of the underwater light field. Table 4.2 lists values for the normalized volume scattering function ($\tilde{\beta}(\theta) = \beta(\theta)/b$), and the cumulative scattering (between 0° and $\theta$, as a proportion of the total) at a series of values of $\theta$, for Tongue of the Ocean (Bahamas) and San Diego harbour water.[363] It will be noted that backscattering ($\theta > 90°$) constitutes 4.4% of total scattering in the case of the clear oceanic water but only 1.9% in the turbid harbour water. Waters of a given broad optical type appear, on the basis of published measurements to date, to have $\tilde{\beta}(\theta)$ curves of rather similar shape. We may therefore take these data sets as being reasonably typical for clear oceanic, and moderately turbid/turbid waters, respectively. Above a certain minimum level of turbidity, such that particle scattering is dominant at all angles, we would not expect the normalized volume scattering function to alter its shape with increasing turbidity, since the shape of the $\tilde{\beta}(\theta)$ curve is determined by the intrinsic scattering properties of the particles, not by their concentration. This is why the San Diego harbour data may reasonably be applied to much murkier waters. Indeed the $\tilde{\beta}(\theta)$ data in Table 4.2(*b*) are applicable to the majority of natural waters other than the very clear oceanic ones: Timofeeva (1971) has

Fig. 4.9. Volume scattering functions for a clear, and a moderately turbid, natural water. The curves for the clear oceanic water ($b = 0.037$ m$^{-1}$) of Tongue-of-the-Ocean, Bahama Islands, and the somewhat turbid water ($b = 1.583$ m$^{-1}$) of San Diego harbour, have been plotted from the data of Petzold (1972).

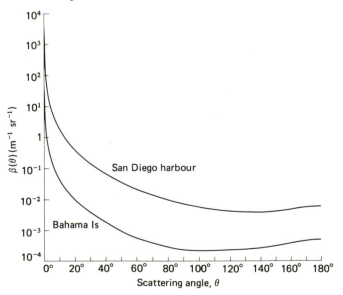

presented evidence that $\tilde{\beta}(\theta)$ is virtually the same for most natural waters. However, in the specialized situation of scattering being dominated by organic particles with a low refractive index, as might be the case in oceanic upwelling areas or eutrophic inland waters, a somewhat different set of $\tilde{\beta}(\theta)$ data might be appropriate.

The contribution of density fluctuation scattering to total scattering varies not only with the type of water body but also with the angular range of scattering and wavelength of light under consideration. In the moderately or very turbid waters such as are typically found in inland, estuarine and

Table 4.2. *Volume scattering data for green (530 nm) light for a clear and a turbid natural water.* $\tilde{\beta}(\theta)$ = *normalized volume scattering function* ($\beta(\theta)/b$). *Cumulative*

$$scattering = 2\pi \int_0^\theta \tilde{\beta}(\theta)\sin\theta\,d\theta. \text{ Data from Petzold (1972)}$$

(a) *Atlantic Ocean, Bahama Islands.* b = 0.037 m$^{-1}$
(b) *San Diego harbour, California, USA.* b = 1.583 m$^{-1}$

| | (a) | | (b) | |
|---|---|---|---|---|
| Angle $\theta$ (°) | $\tilde{\beta}(\theta)$ (sr$^{-1}$) | Cumulative scattering $0° \to \theta$ | $\tilde{\beta}(\theta)$ (sr$^{-1}$) | Cumulative scattering $0° \to \theta$ |
| 1.0 | 67.5 | 0.200 | 76.4 | 0.231 |
| 2.0 | 21.0 | 0.300 | 23.6 | 0.345 |
| 3.16 | 9.0 | 0.376 | 10.0 | 0.431 |
| 5.01 | 3.91 | 0.458 | 4.19 | 0.522 |
| 6.31 | 2.57 | 0.502 | 2.69 | 0.568 |
| 7.94 | 1.70 | 0.547 | 1.72 | 0.616 |
| 10.0 | 1.12 | 0.595 | 1.09 | 0.664 |
| 15.0 | 0.55 | 0.687 | 0.491 | 0.750 |
| 20.0 | 0.297 | 0.753 | 0.249 | 0.806 |
| 25.0 | 0.167 | 0.799 | 0.156 | 0.848 |
| 30.0 | 0.105 | 0.832 | 0.087 | 0.877 |
| 35.0 | 0.072 | 0.857 | 0.061 | 0.898 |
| 40.0 | 0.051 | 0.878 | 0.0434 | 0.916 |
| 50.0 | 0.0276 | 0.906 | 0.0234 | 0.940 |
| 60.0 | 0.0163 | 0.925 | 0.0136 | 0.956 |
| 75.0 | 0.0093 | 0.943 | 0.0073 | 0.971 |
| 90.0 | 0.0066 | 0.956 | 0.00457 | 0.981 |
| 105.0 | 0.0060 | 0.966 | 0.00323 | 0.987 |
| 120.0 | 0.0063 | 0.975 | 0.00276 | 0.992 |
| 135.0 | 0.0072 | 0.984 | 0.00250 | 0.995 |
| 150.0 | 0.0083 | 0.992 | 0.00259 | 0.997 |
| 165.0 | 0.0110 | 0.997 | 0.00323 | 0.999 |
| 180.0 | 0.0136 | 1.000 | 0.00392 | 1.000 |

some coastal water bodies, scattering at all angles and all visible wavelengths is predominantly due to the particles present. In contrast, in the very clear waters of the least fertile parts of the ocean, density fluctuation scattering can be a significant component of total scattering at short wavelengths and the major component at large scattering angles. In the Sargasso Sea, for example, Kullenberg (1968) found that density fluctuation scattering accounted for 3% of the total scattering coefficient at 655 nm and 11% at 460 nm. For scattering angles from 60–75 to 180°, however, the greater part of the scattering at both wavelengths was due to density fluctuation: in the case of blue (460 nm) light it accounted for only 0.3% of the scattering at 1°, but contributed 89% at 135°. Because of its importance as a component of backscattering in oceanic waters, density fluctuation scattering makes a disproportionately large contribution to the upwelling light stream. Morel & Prieur (1977) pointed out that even for a water with a total scattering coefficient of 0.29 m$^{-1}$ (rather turbid by oceanic standards), while density fluctuation scattering accounts for only 1% of the total scattering coefficient, it contributes about 33% of the backscattering. Thus, even in the not-so-clear oceanic waters, density fluctuation scattering initiates a substantial proportion of the upwelling light flux.

Compared to density fluctuation scattering, particle scattering is rather insensitive to wavelength. Since, even in very clear oceanic water, density fluctuation scattering makes up only a small proportion of the total scattering coefficient, the dependence of this scattering on $\lambda^{-4.3}$ does not bring about a marked variation of the value of $b$ for natural waters with wavelength. There is, however, some evidence that in particle-dominated natural waters, the value of $b$ varies approximately in accordance with $\lambda^{-1}$;[326] i.e. shorter wavelengths are scattered more intensely. When backscattering alone is considered, then in oceanic waters, in which as we saw above density fluctuation scattering is a major contributor for $\theta > 90°$, a more marked inverse dependence on wavelength is to be expected.

Since total light scattering in natural waters is dominated by the particulate contribution, it increases broadly in proportion to the concentration of suspended particulate matter. For example, Jones & Wills (1956), using suspensions of kaolin (which would scatter but not significantly absorb light), found a linear relation between the beam attenuation coefficient (approximately equal in this system to the scattering coefficient) for green (550 nm) light and the concentration of suspended matter. Approximately linear relations between scattering coefficient and concentration of suspended matter have been observed for natural aquatic particulates by other workers.[42, 471] The constant of proportionality can, however, vary from one kind of suspended matter to another:[102] the refractive index and the size distribution of the particles both influence the relation between $b$ and sediment concentration.

The concentration of particulate matter, and therefore the intensity of scattering, in inland and estuarine waters, and to some extent in coastal waters, is strongly influenced, not only by the nature of the physical environment (climate, topography, soil type, etc.) but also by the uses to which the land is put. The more thickly the ground is covered with vegetation, the less erosion and so the less transference of soil particles to surface waters occurs. Erosion is very low under undisturbed forest and is generally not high from permanent grassland. Logging activities and overgrazing of pastures can cause serious erosion even in these environments. Ground which is broken up for crop planting is susceptible to erosion until such time as the crop canopy is established. The extent to which erosion occurs in these different terrestrial systems depends not only on the extent to which the ground is protected by vegetation, but also on the nature of the soil (some types being more easily eroded than others), the average slope (the greater the slope, the greater the erosion), and the intensity of the rainfall (a short torrential downpour causes more erosion than the same amount of rain falling over a longer period). The average lifetime of the soil particles in suspension once they have been transferred into the aquatic sphere is quite variable and depends on their size and mineral chemistry, and on the ionic composition of the water: some fine clay particles for example, particularly in waters of low electrolyte content, can remain in suspension for long periods with drastic consequences for light attenuation in the water bodies which contain them.

It is clear that to understand the underwater light field, we must know not only about the water itself but also about the surrounding land forms which to a large degree confer upon the aquatic medium the particular optical properties it possesses.

# 5

............................................................................................

# Characterizing the underwater light field

Having described the nature of the incident radiation flux presented to the aquatic biosphere and the influences to which the light is subject within the water, we shall now consider the kind of light field that results. We shall begin, in this chapter, by examining the ways in which the properties of this light field are studied. The physical definitions of these properties are given in Chapter 1.

## 5.1    Irradiance

The most frequently and easily measured property of the underwater light field is irradiance. Knowledge of this parameter is valuable, not only because it provides information about how much light is available for photosynthesis, but also because irradiance plays a central role in the theory of radiation transfer in water. An irradiance meter, since it is meant to measure the radiant flux per unit area, must respond equally to all photons that impinge upon its collector, regardless of the angle at which they do so. With any given meter this can be tested by observing the way in which its response to a parallel light stream varies with its angle to that light stream.

As the angle of the radiant flux to the collector changes, the area of the collector projected in the direction of that radiant flux changes, and the proportion of the flux intercepted by the collector changes correspondingly (Fig. 5.1). Thus, the response of an irradiance meter to parallel radiant flux (wider than the collector) should be proportional to the cosine of the angle ($\theta$) between the normal to the collector surface and the direction of the flux. An irradiance collector which meets this criterion is known as a cosine collector.

The collector of an irradiance meter usually consists of a flat disc of translucent diffusing plastic, although opal glass may be used when measurements below 400 nm are required.[216] It has been found that a more faithful cosine response to light incident at large values of $\theta$ is obtained if the disc projects slightly up above the surrounding housing, allowing some collection of light through the edge.[438] Some of the light that penetrates into the diffusing plastic is scattered backwards and a part of this succeeds in passing up through the diffuser surface again. Because the difference between the refractive indices of plastic and water is less than that between those of plastic and air, there is less internal reflection of this light within the collector and therefore a greater loss of light when the collector is immersed in water than when it is in air: this is known as the immersion effect. To take account of this, if an irradiance meter has been calibrated in air, the underwater readings should be multiplied by the appropriate correction factor for that instrument. This factor is usually in the region of 1.3–1.4 in the case of a plastic collector.

Within the irradiance sensor, beneath the collecting disc, there is a photoelectric detector. In the older instruments this was usually a selenium photovoltaic cell: nowadays a silicon photodiode, which has a broader spectral response, is more commonly used. If a very narrow waveband of light is measured, as in certain spectroradiometers (see below), then a photomultiplier may be used as the detector to provide the necessary sensitivity. The electrical signal from the photodetector is transmitted by cable to the boat where, after amplification if required, it is displayed on a suitable meter or digital readout.

The holder in which the sensor is lowered into the water (Fig. 5.2) should be made so that the sensor can be facing upwards to measure downwards irradiance ($E_d$), or downwards to measure upward irradiance ($E_u$). To minimize shading effects, the sensor is always lowered on the sunny side of the boat. For measurements in rivers, the sensor must be attached to a rigid support.

Fig. 5.1. The area which the collector presents to the beam is linearly related to $XY$, which is proportional to $\cos\theta$.

Fig. 5.2. Submersible irradiance sensor (Li–Cor 192S Underwater Quantum Sensor: Li–Cor Inc., Lincoln, Nebraska, USA) in a lowering frame. The sensor can be placed in the frame facing upwards or downwards.

Measurements made with an uncorrected wide-band detector are unreliable because of the spectral variation of sensitivity: part of the diminution of irradiance with depth could be due to a shift of the wavelength of the predominating light to a part of the spectrum where the detector is less sensitive. It is therefore desirable to confine the measured light to a specific waveband with an appropriate filter in the sensor: narrow-band interference filters are best for this purpose. The alternative, and in the context of photosynthesis the best, solution (short of determining the complete spectral distribution of irradiance), is to use a meter designed to respond equally to all quanta within the photosynthetic range regardless of wavelength, but to be insensitive to quanta outside this range. In this way a measure of the total photosynthetically available radiation (PAR) is obtained.

It is better that the meter should be designed to respond equally to quanta regardless of wavelength rather than to respond equally to a given amount of radiant energy regardless of wavelength, because the usefulness of the prevailing light for photosynthesis is more closely related to the flux of quanta than to the flux of energy. This is because once a quantum has been absorbed by a cell, it makes the same contribution to photosynthesis regardless of its wavelength (see §3.1), although the probability of its being absorbed in the first place of course varies with wavelength. An absorbed red quantum for example is just as useful as an absorbed blue quantum even though it may contain only two-thirds of the energy.

An irradiance meter designed in this way is commonly referred to as a quanta meter. Since no existing photodetector in its normal state responds equally to all quanta throughout the photosynthetic range, ways of adjusting the relative response in different parts of the spectrum must be found. Jerlov & Nygard (1969) used a combination of three colour filters, each covering a different part of the photodetector (Fig. 5.3). The filters were covered with an opaque disc with a number of small holes drilled in it: the number, size and position of these holes determined the amount of light reaching each filter. By a suitable combination of this variable attenuation with the differing absorption properties of the filters it was possible to obtain an approximately constant quantum sensitivity throughout the photosynthetic range. Commercial instruments using filter combinations with a silicon photodiode, and measuring the total quantum flux in the range 400–700 nm, are now available.

The photosynthetically available radiation at any depth can now readily be measured with a submersible quanta meter. Smith (1968) has suggested that the vertical attenuation coefficient for downwards irradiance of PAR is the best single parameter by means of which different water bodies may be characterized with respect to the availability of photosynthetically useful radiant energy within them. As we shall see in the next chapter, $K_d$ for PAR is not exactly constant with depth even in a homogeneous water body. Nevertheless the variation is not great and knowledge of, say, the average value of $K_d(PAR)$ over a certain depth range such as the euphotic depth, is very useful. It is determined by measuring downwelling PAR at a series of depths within the depth interval of interest and making use of the approximate relation (see §§6.1, 6.3)

$$\ln E_d(z) = -K_d z + \ln E_d(0)$$

where $E_d(z)$ and $E_d(0)$ are the values of downward irradiance at $z$ m and just below the surface, respectively. The linear regression coefficient of $\ln E_d(z)$ with respect to depth gives the value of $K_d$. Or, less accurately, $K_d$ may be determined from measurements at just two depths, $z_1$ and $z_2$

$$K_d = \frac{1}{z_2 - z_1} \ln \frac{E_d(z_1)}{E_d(z_2)}$$

A crude visual method of estimating $K_d$, commonly used before the ready availability of photoelectric instruments and still sometimes used today, is based on the device known as the Secchi disc. A white disc, of diameter perhaps 20 cm, is lowered into the water and the depth at which it just disappears from view is noted. This is referred to as the *Secchi disc transparency*, or as the *Secchi depth*, $Z_{SD}$. On the basis of their measurements in marine waters, Poole & Atkins (1929) made the empirical observation that the Secchi depth is approximately inversely proportional to the

Fig. 5.3. Exploded view of a quanta meter together with its quantum sensitivity curve (after Jerlov & Nygard, 1969).

vertical attenuation coefficient for downwelling irradiance and pointed out
that $Z_{SD}$ could therefore be used to estimate $K_d$: the value obtained will
be that applicable to a rather broad waveband corresponding roughly to
the spectral sensitivity curve of the human eye. To calculate $K_d$, the relation
$K_d = 1.44/Z_{SD}$ may be used.[191]

Using the contrast transmittance theory of Duntley & Preisendorfer,
Tyler (1968) concluded that the Secchi depth is proportional to $(c + K_d)$,
the sum of the beam attenuation and vertical attenuation coefficients,
rather than to $K_d$ alone. On theoretical grounds, Tyler arrived at the
relation $(c + K_d) = 8.69/Z_{SD}$; from measurements in turbid coastal sea
water, Holmes (1970) obtained $(c + K_d) = 9.42/Z_{SD}$. Since, in natural
waters, $c$ is usually substantially greater than $K_d$, the Secchi depth is
determined more by $c$ than by $K_d$. The fact that $K_d$ and $c$ tend very roughly
to covary would account for the near constancy of the inverse relation
between $Z_{SD}$ and $K_d$. *A priori* considerations suggest, however, that use of
the Secchi disc could sometimes give highly inaccurate values for $K_d$. If,
for example, as a consequence of increased levels of particles, light
scattering in a water body increased much more than absorption, then $c$
would increase much more than $K_d$ and use of the relation $K_d = 1.44/Z_{SD}$
would overestimate $K_d$.

In all natural waters most of the light at any depth is travelling in a
downwards direction. Measurement of the irradiance of the upwelling flux
is nevertheless worthwhile. This is partly because, in some waters, upwelling
flux is a significant component (up to $\sim 20\%$) of the total available radiant
energy, but even more so because the ratio of upwelling to downwelling
flux (irradiance reflectance, $R = E_u/E_d$) can provide information about the
inherent optical properties of the water (see §§4.2, 6.4) and because the
properties of the upwelling flux are of central importance to the remote
sensing of the composition of natural waters. Thus, the more we know
about this flux and its relation to other characteristics of the medium,
the better.

## 5.2    Scalar irradiance

Although some multicellular aquatic plants have their photosynthetic tissue so arranged as to achieve maximum interception of light
incident from above, to the randomly oriented cells of the phytoplankton
all light, from whatever direction it comes, is equally useful. Scalar
irradiance $(E_0)$, as we saw earlier (§1.3) is defined as the integral of radiance
distribution at a point over all directions about the point: it is equivalent
to the total radiant flux per $m^2$ from all directions at a given point in the
medium. In the context of photosynthesis, information on the scalar
irradiance is therefore somewhat to be preferred to data on irradiance.

The technology of measurement of scalar irradiance is essentially the same as that of irradiance with the exception of the collector. Since this must respond equally to light from all directions, it has to be spherical. Fig. 5.4 illustrates the principles involved in collectors used for scalar irradiance and for its downwelling and upwelling components. Like the collector of a normal irradiance meter, the spherical collector is made of

Fig. 5.4. Principles involved in collection of light for determination of (*a*) scalar irradiance, (*b*) downward scalar irradiance and (*c*) upward scalar irradiance. In the cases of downward and upward scalar irradiance, the collector must have a shield wide enough relative to the collector to ensure that very little upward and downward light, respectively, reaches the collector.

Fig. 5.5. Submersible scalar irradiance sensors. (*a*) *Li–Cor* Spherical Quantum Sensor LI–193SB, in lowering frame (Li–Cor Inc., Lincoln, Nebraska, USA). (*b*) *Biospherical* Quantum Scalar Irradiance Sensor QSP-200, in lowering frame (courtesy of Biospherical Instruments Inc., San Diego, California, USA).

diffusing plastic or opal glass so that it collects all incident photons with equal efficiency, regardless of the angle at which they encounter its surface. The light which penetrates the spherical collector can be transmitted to the photodetector via another, flat collector in the bottom of the sphere[189] or, in the case of a solid diffusing sphere, can be transmitted from the centre of the sphere by a fibre-optic bundle or quartz light-conducting rod.[53] Scalar irradiance meters can use a filter to confine the measured light to a narrow waveband,[189] or can be designed as quanta meters covering the whole (400–700 nm) photosynthetic range.[53, 460] Fig. 5.5 shows two commercially available quantum (PAR) scalar sensors.

Højerslev (1975) describes a meter with two spherical collectors but with only half of each sphere exposed. One exposed hemisphere faces upwards, the other faces downwards. It is possible to show that the upper hemisphere collects radiant flux in proportion to $\frac{1}{2}(E_0 + \vec{E})$ and the lower hemisphere collects flux in proportion to $\frac{1}{2}(E_0 - \vec{E})$, where $\vec{E}$ is the net downward, or vector, irradiance $(E_d - E_u)$. The signals from the two collectors are recorded separately, and by adding and subtracting these measurements, the scalar irradiance and net downward irradiance, respectively, are determined. The information provided by this meter can also be used to determine the absorption coefficient of the aquatic medium at the wavelength determined by the interference filters incorporated in the instrument. From measurements of $\vec{E}$ above and below a given depth, $K_E$, the vertical attenuation coefficient for net downward irradiance is obtained. The absorption coefficient is then calculated from $a = K_E \vec{E}/E_0$ (§1.7) using the values of $\vec{E}$ and $E_0$ at the depth in question.

Karelin & Pelevin (1970) developed a meter with three collectors: upward and downward irradiance collectors and a spherical collector for scalar irradiance. Any one of five colour filters could, by remote control, be placed in front of the photodetector. They used the data to determine the absorption coefficients of the medium, as well as characterizing the underwater light field.

## 5.3    Spectral distribution of irradiance

Any photon with wavelength from 400 to 700 nm is in principle *available* for photosynthesis: however, the probability of a given photon being captured by the photosynthetic biomass within a water body will vary markedly with wavelength in accordance with the absorption spectrum of the array of photosynthetic pigments present. It is quite possible for two different light fields to have the same irradiance for PAR but, because of their differing spectral distributions, to differ significantly in their usefulness to a particular type of alga. Thus, a better assessment of the value of a particular light field for photosynthesis can be made if the way in which irradiance or scalar irradiance varies across the spectrum is known. An instrument which measures this is known as a spectroradiometer.

A spectroradiometer is an irradiance or scalar irradiance meter in which a variable monochromator is interposed between the collector and the photodetector. In its simplest form the monochromator can consist of an array of colour filters which can be sequentially moved in front of the photodetector. Burr & Duncan (1972) described a submersible spectroradiometer incorporating a set of 16 interference filters, 401–698 nm, mounted on a wheel: as the wheel rotates, a different filter is brought into position over the photodiode. In one commercially available spectroradiometer,[497] the instrument scans between 400 and 740 nm by having a wedge interference filter (in which the transmitted wavelength varies continuously with distance along the filter) move lengthways above the photodiode. In another commercial instrument, moving parts have been eliminated by having 12 separate photodiodes, each with its own interference filter: the array is scanned electronically.[41]

To provide data of greater spectral resolution than is possible with interference filters, which might typically have half-power bandwidths in the region of 10–18 nm, a grating monochromator is used. The Scripps Institution spectroradiometer described by Tyler & Smith (1970) and the instrument developed by Bauer & Ivanoff (1970) both achieve a spectral resolution of about 5 nm by this means. Narrowing the bandwidth of the monochromator, however, means reducing the energy available for measurement and so these instruments use a photomultiplier as the detector.

The Scripps spectroradiometer covers the range 360–740 nm. An improved version of this has been developed at the Scripps Institution to extend measurements into the middle ultraviolet;[446] this instrument can cover the range 250–750 nm, although in practice it is very difficult to make accurate underwater measurements at wavelengths below 310 nm.

## 5.4 Radiance distribution

To fully understand the underwater light field we need a detailed knowledge of the angular distribution of radiant flux at all depths. Some information on the angular structure of the field can be derived from the measurements of irradiance discussed above. Irradiance reflectance ($R = E_u/E_d$) is a crude measure of angular structure. More information is contained in the three average cosines: $\bar{\mu}$ for the total light field, $\bar{\mu}_d$ for the downwelling stream, and $\bar{\mu}_u$ for the upwelling stream, all of which can be derived from irradiance and scalar irradiance values

$$(\bar{\mu} = \vec{E}/E_0, \bar{\mu}_d = E_d/E_{0d}, \bar{\mu}_u = E_u/E_{0u}).$$

For a complete description, however, we need the value of radiance at all vertical and azimuthal angles, i.e. the radiance distribution. Radiance is a measure of the radiant intensity (per unit projected area) in a specified direction at a particular point in the field, and so should be measured with a meter which can be oriented in that direction and which ideally should have an infinitesimally small acceptance angle. In reality, to ensure that the meter collects enough light to measure with reasonable accuracy, particularly in the dim light fields existing at greater depths, acceptance angles of 4–7° may be used. The simplest type of radiance meter is known as a Gershun tube photometer, after the Russian physicist Gershun, who made notable contributions to our understanding of the structure of light fields in the 1930s. It uses a cylindrical tube to limit the acceptance angle; the photodetector, with a filter in front of it, is at the bottom (Fig. 5.6). To achieve the necessary sensitivity, a photomultiplier would normally be used as the detector.

Fig. 5.6. Schematic diagram of a radiance meter.

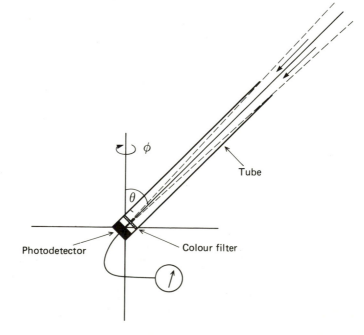

The most difficult problem encountered in measurement of underwater radiance is that of accurately controlling the zenith and azimuth angle of the photometer tube: quite complex systems involving built-in compasses, servo mechanisms and propellers to rotate the instrument are required.[214, 413, 511] Another problem is that to determine a complete radiance distribution with a radiance meter is likely to take quite a long time, during which the position of the Sun will change. For example, if radiance is measured at intervals of 10° for the zenith angle (0–180°) and 20° for the azimuth angle (0–180° is sufficient, the distribution being symmetrical about the plane of the Sun) then 190 separate values must be recorded. Smith, Austin & Tyler (1970) have developed a photographic technique by means of which radiance distribution data can be recorded quickly. Their instrument consists of two cameras placed back to back, each equipped with a 180°-field-of-view lens. One camera photographs the upper and the other photographs the lower hemisphere. By densitometry of the film, the relative radiance values over all zenith and azimuth angles can be determined, at whatever degree of resolution is required, at a later time.

## 5.5 Modelling the underwater light field

The nature of the underwater light field resulting from a given incident light field is determined by the inherent optical properties of the aquatic medium. In principle, therefore, it should be possible, if we know the inherent optical properties, to calculate the properties of the underwater light field. Such a calculation procedure could be used, for example, to explore in more detail than would be practicable by measurement the exact way in which the nature of the light field depends on the inherent optical properties. It could be used as a substitute for *in situ* measurement in environments in which such measurement might be difficult. It could also be used for predicting the effects on the underwater light field of anticipated changes in the optical properties of the water, resulting perhaps from the activities of man.

In reality, the complex behaviour of the photon population in the water caused by the combined effects of scattering and absorption precludes the establishment of an explicit analytical relation between the properties of the field and those of the medium. Nevertheless, a number of rather complicated mathematical treatments has been developed by means of which the characteristics of the radiation field within scattering/absorbing media may, with some difficulty, be calculated. This subject has been reviewed by Prieur & Morel (1973) and Prieur (1976). One such method which has been successfully applied to marine waters is to resolve the medium into a series of thin layers, find the diffuse transmission and reflection operators for a single layer, and then use these operators to determine the characteristics of the light field within the stack of layers. This procedure was used by Beardsley & Zaneveld (1969), and a particularly accurate variant of it, known as the matrix operator method, has been reviewed by Plass, Kattawar & Catchings (1973) and was used to study light in the ocean–atmosphere system by Plass, Humphreys & Kattawar (1981). Calculations on the atmosphere–ocean radiation field have also been carried out by Raschke (1972) using an iterative procedure based on an integrated form of the equation of radiation transfer. Another procedure – the successive orders of scattering method – has been used by Prieur (1976) to study the underwater light field.

The simplest calculation procedure, and the one which has been used most, is the Monte Carlo method. Its application to the behaviour of solar photons in the ocean was pioneered by Plass & Kattawar (1972), and it has also been put to valuable use by Gordon and coworkers.[155, 156] The principles are described in some detail in Kirk (1981a, c). A brief account will be given here.

The behaviour of individual photons within a scattering/absorbing medium is stochastic in nature. The lifetime of, and the geometrical path followed by, any given photon are governed by its random encounters with absorbing molecules or algal cells and with scattering particles. The inherent optical properties of the medium – the absorption and scattering coefficients, the volume scattering function – are not only (in accordance with their definitions) measures of the proportion of an incident flux absorbed, scattered and so on per unit depth by a thin layer of medium, but are also related in simple ways to the probability that any given photon will, within a certain distance, be absorbed, scattered, or scattered at a certain angle.

The Monte Carlo method takes advantage of the statistical nature of photon behaviour and uses a computer to follow the fate of a large number of photons, one at a time, passing into an imaginary body of water (corresponding, if desired, to some real body of water) with specified optical properties. Random numbers are used in conjunction with appropriate cumulative frequency distributions (based on the optical properties) to choose pathlengths between each interaction with the medium, to decide whether the interaction is one of scattering or absorption, to select the scattering angle, and so on. Photons are introduced from above the surface, at an angle or a selection of angles appropriate for the lighting conditions (direct sun, overcast sky, etc.) under consideration. Refraction at the surface is allowed for. If it is wished to take account of waves, then the surface slope/wind frequency distribution data of Cox & Munk (1954) can be used. Each photon is followed, its trajectory being recalculated after each scattering event until it is absorbed or scattered up through the surface again. Within the imaginary water body there is a series of depth markers. For every single trajectory of each photon the computer records which depth markers the photon passes, in what direction (upwards or downwards) and at what angle. When a large number (say $10^6$) of photons has been processed, the computer can calculate from the accumulated data the values at each depth of irradiance and scalar irradiance (upwards and downwards), reflectance, average cosines (downward, upward, total) and radiance distribution (at a set of azimuth angles or averaged over all azimuth angles). In this way a complete description of the underwater light field is obtained.

# 6

...................................................................................................

## The nature of the underwater
## light field




Having in earlier chapters considered the nature of the solar
radiation flux incident on the surface of the aquatic ecosystem, and the
influences to which the light is subject once it enters the water, we shall
now discuss the kind of underwater light field that results.

### 6.1    Downward irradiance – monochromatic

As a result of absorption and scattering of the solar flux, the
downward irradiance, $E_d$, of the light field diminishes with depth. In Fig.
6.1, $E_d$ for greenish-yellow light, expressed as a percentage of the value
just below the surface, is plotted against depth in a fresh water impoundment. Irradiance diminishes in an approximately exponential manner in
accordance with

$$E_d(z) = E_d(0) e^{-K_d z} \tag{6.1}$$

or

$$\ln E_d(z) = -K_d z + \ln E_d(0) \tag{6.2}$$

where $E_d(z)$ and $E_d(0)$ are the values of downward irradiance at $z$ m and
just below the surface, respectively, and $K_d$ is the vertical attenuation
coefficient for downward irradiance.

In Fig. 6.2$a$–$d$, the logarithm of $E_d$ is plotted against depth for red, green
and blue light in one oceanic and three inland waters. The graphs are
approximately linear, in accordance with eqn 6.2, but significant divergence
from linearity is in some cases apparent. In the oceanic water for example,
it can be seen that the green and the blue light are attenuated more rapidly
below about 10 m than above. This may be attributed to the downward
flux becoming more diffuse, less vertical in its angular distribution with
depth, with a consequent increase in attenuation (see §§6.6, 6.7).

The relative rates of attenuation in the different wavebands are determined largely by the absorption spectrum of the aquatic medium. In the tropical Pacific Ocean (Fig. 6.2*a*), and in non-productive oceanic waters generally, where water itself is the main absorber, blue and green light both penetrate deeply and to about the same extent, while red light, which water absorbs quite strongly, is attenuated much more rapidly. In the productive waters of oceanic upwelling areas, blue light is attenuated more strongly than green light, due to absorption by phytoplankton pigments,[330] but still not as strongly as red light. In coastal waters, which contain more yellow substances and phytoplankton than normal oceanic waters, green is again the most penetrating waveband. Only in the most coloured coastal waters, influenced by major river discharge (Jerlov's types 7–9), however, is blue light attenuated as strongly as red light.

In contrast to the sea, in fresh water the blue waveband is usually the most strongly attenuated (Fig. 6.2*b–d*), because of the higher levels of yellow substances that typically occur in inland waters. Green is usually the most penetrating waveband in inland waters, followed by red (Fig. 6.2*b*). When the concentration of yellow materials is high, however, red light may penetrate as far as green (Fig. 6.2*c*), and in very yellow waters red light penetrates best of all (Fig. 6.2*d*).

Fig. 6.1. Exponential diminution of downward irradiance of greenish-yellow (580 nm) light with depth in a fresh-water body (Burrinjuck Dam, NSW, Australia, 8th December 1977). (Kirk, unpublished data.)

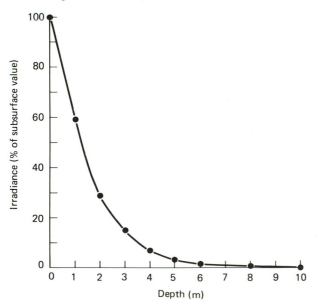

Fig. 6.2. Depth profiles for downward irradiance of blue (○), green
(●) and red (△) light in a tropical oceanic and three inland
fresh-water systems. (*a*) Pacific Ocean, 100 km off Mexican coast
(plotted from data of Tyler & Smith, 1970). (*b*) San Vicente Reservoir,
California, USA (plotted from data of Tyler & Smith, 1970). (*c*) Corin
Dam, ACT, Australia (Kirk, unpublished data). (*d*) Georgetown
billabong, Northern Territory, Australia (plotted from unpublished
data of P. A. Tyler).

All measurements were taken under sunny conditions. Note the
expanded depth scale in (*d*). The wavelength (nm) at which
measurements were made is indicated alongside each graph.

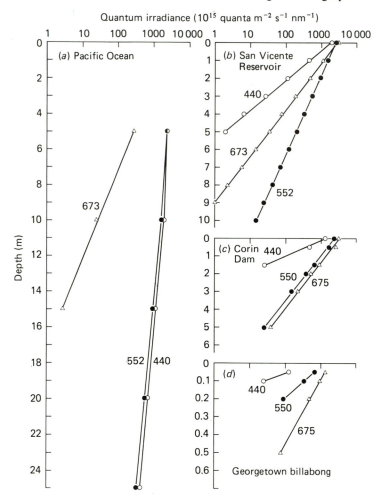

Fig. 6.3 compares the spectral variations of the vertical attenuation coefficient for irradiance across the photosynthetic range in an unproductive oceanic water, a productive (upwelling) oceanic water and an inland impoundment.

As the solar altitude decreases, so the pathlength of the unscattered solar beam within the water, per metre depth, increases in proportion to cosec $\alpha$, where $\alpha$ is the angle of the solar beam within the water. We may therefore expect $K_d$ to increase with diminishing solar altitude. Such an effect is observed but is small:[214] it is only of significance in the upper layer of very

Fig. 6.3. Spectral variation of vertical attenuation coefficient, $K_d(\lambda)$ for downward irradiance in (a) the Gulf Stream (Atlantic Ocean) off the Bahama Islands (plotted from data of Tyler & Smith, 1970). (b) The Mauritanian upwelling area (8.0–9.9 mg chl $a$ m$^{-3}$) off the West African coast (plotted from data of Morel, 1982). (c) Burrinjuck Dam, southern tablelands of NSW, Australia (plotted from data of Kirk, 1979).

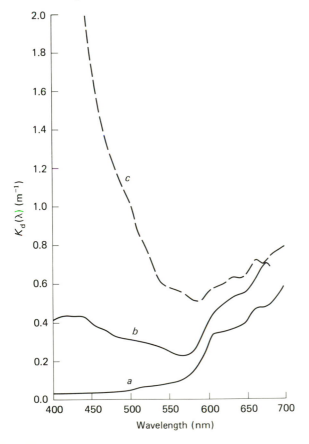

clear waters. Baker & Smith (1979) found that in an inland impoundment, $K_d$ at several wavelengths varied by not more than 5% between solar altitudes of 80 and 50°, and by not more than 18% between 80 and 10°.

## 6.2    Spectral distribution of downward irradiance

The data in Fig. 6.2 show that attenuation of solar radiation in the whole photosynthetic waveband takes place at widely different rates in different parts of the spectrum. As a consequence, the spectral composition of the downwelling flux changes progressively with increasing depth. Fig. 6.4a shows the changes down to 25 m depth in the clear oceanic water of the Gulf Stream. Most of the light below about 15 m is confined to the blue–green, 400–550 nm, waveband with the peak occurring in the blue region at about 440–490 nm. In the coastal/estuarine water of Batemans Bay, Australia, attenuation due to yellow substances in the blue region is comparable to that due to water at the red end of the spectrum, so that at 4 m depth, while there is still substantial radiant flux throughout the photosynthetic range, the distribution peaks markedly at about 570 nm (Fig. 6.4b). In European coastal waters, Halldal (1974) found spectral distributions rather similar to that in Fig. 6.4b, with a marked peak at about 570 nm.

In inland waters with their, usually, higher concentration of yellow substances, the rapid attenuation at the short-wavelength end of the spectrum means a virtually complete removal of blue light within quite shallow depths. In Lake Burley Griffin, Australia, for example, at a time of low water turbidity (Fig. 6.4c) there was essentially no blue light still present below about 2 m depth. In the lower half of the euphotic zone (defined in the following section) in such waters, the photosynthetically available flux typically consists of a broad band extending from the green to the red, often with a peak in the yellow at about 580 nm. Near the surface of the water there are significant levels of blue light available for photosynthesis, but taking the euphotic zone as a whole, the total amount of blue light available for photosynthesis is greatly reduced in such waters (see §6.5).

In those coloured, turbid waters in which there are high levels of humic material in the particulate fraction as well as in the soluble state, green light

Fig. 6.4. Spectral distribution of downward irradiance in marine and inland water. (a) The Gulf Stream (Atlantic Ocean) off the Bahama Islands (plotted from data of Tyler & Smith, 1970). (b) Batemans Bay, NSW, Australia (after Kirk, 1979). (c) Lake Burley Griffin, ACT, Australia, 29th September, 1977; water comparatively clear (turbidity = 3.7 NTU) (after Kirk, 1979). (d) Lake Burley Griffin, 6th April 1978; water turbid (turbidity = 69 NTU) (after Kirk, 1979).

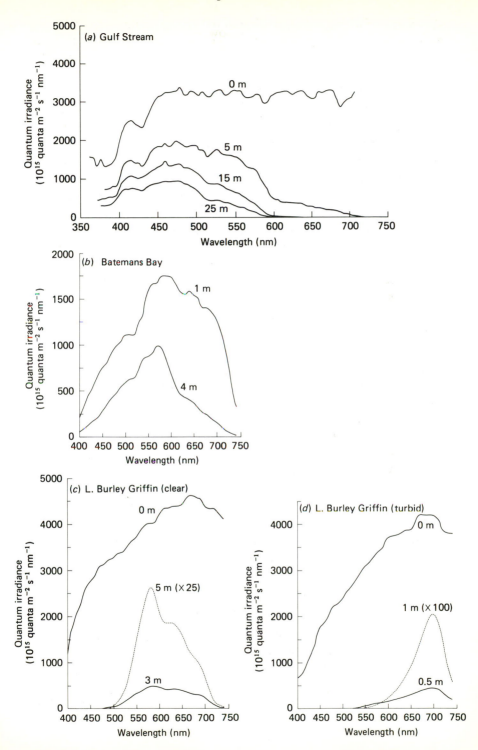

is also attenuated rapidly (Fig. 6.2*d*), with the result that the downward flux in the lower part of the, very shallow, euphotic zone consists largely of orange–red (600–700 nm) light. In the spectral distribution for Lake Burley Griffin under turbid conditions, shown in Fig. 6.4*d*, the peak is right over at the far-red end of the spectrum.

### 6.3    Downward irradiance – PAR

As a broad indication of the availability of light for photosynthesis in an aquatic ecosystem, information on the penetration of the whole photosynthetic waveband is of great value. As solar radiation penetrates a water body, it becomes progressively impoverished in those wavelengths which the aquatic medium absorbs strongly and relatively enriched in those wavelengths which are absorbed weakly. We would therefore expect the attenuation coefficient for total photosynthetically available radiation to be higher in the upper few metres and to fall to a lower value with increasing depth. This change in the rate of attenuation of PAR with depth can readily be observed in most marine waters and the cleaner inland systems: two of the curves in Fig. 6.5 – for the Tasman Sea, and for a relatively clear lake – show the increase in slope of the log $E_d$ curve with increasing depth. The curve eventually becomes approximately linear, indicating that the downward flux is now confined to wavebands all with about the same, relatively low, attenuation coefficient. In oceanic waters the light in this region is predominantly blue–green (Fig. 6.2*a*), whereas in inland waters the penetrating waveband is likely either to be green (Fig. 6.2*b*), to extend from the green to the red (Fig. 6.2*c*) or to be predominantly red (Fig. 6.2*d*).

A countervailing tendency, which exists at all wavelengths, is for attenuation to increase with depth as a result of the downward flux becoming more diffuse, due to scattering. By counteracting the effect of changes in spectral composition, it may partly explain why graphs of log $E_d$ against depth for turbid waters are so surprisingly linear (Fig. 6.5, L. Burley Griffin), and lack the biphasic character seen in the clearer waters. However, since high turbidity is commonly associated with increased absorption at the blue end of the spectrum (see §3.3) it is also true that in such waters the blue waveband is removed at even shallower depths than usual and so the change in slope of the curve occurs quite near the surface and is not readily detectable.

Even when, as in the clearer waters, the graph of log $E_d$ against depth is noticeably biphasic, the change of slope is usually not very great. Thus, the attenuation of total PAR with depth is nearly always approximately, and often accurately, exponential in agreement with eqns 6.1 and 6.2.

Attenuation of PAR in a given water body can therefore generally be characterized by a single value of $K_d$, or, at worst, by two values, one above and one below the change in slope.

Since the general availability of submersible quanta meters covering the photosynthetic waveband is a comparatively recent development in the aquatic sciences, the number of water bodies for which values of $K_d$ for PAR have been published is not large. Table 6.1 presents a selection of values, including some obtained by summation of spectral distribution data across the photosynthetic range. Oceanic waters have the lowest

Fig. 6.5. Attenuation of downward quantum irradiance of PAR with depth in a coastal water (Tasman Sea, off Batemans Bay, NSW) and two inland waters (Lake Burley Griffin, ACT; Burrinjuck Dam, NSW) in Australia (Kirk, 1977*a*, and unpublished). The marked decrease in rate of attenuation in Burrinjuck Dam below about 7 m is particularly noteworthy: spectroradiometric measurements showed that most of the light below this depth was confined to the 540–620 nm (green–yellow) waveband.

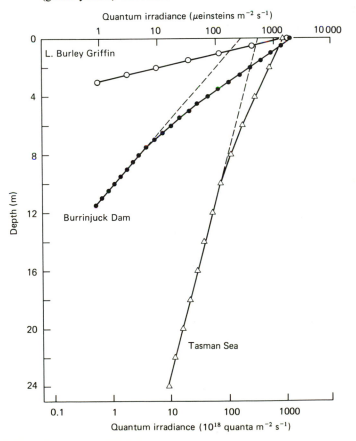

Table 6.1. *Vertical attenuation coefficients for downward quantum irradiance of PAR in some marine and fresh waters. Where several measurements have been taken, the mean value, the standard deviation, the range and the time period covered are in some cases indicated*

| Water body | $K_d(PAR)$ (m$^{-1}$) | Reference |
|---|---|---|
| I *Oceanic waters* | | |
| Sargasso Sea | 0.03 | 513 |
| Gulf Stream, off Bahamas | 0.08 | 517 |
| Pacific Ocean, 100 km off Mexico | 0.11 | 517 |
| II *Coastal and estuarine waters* | | |
| Bjornafjord, Norway | 0.15 | 216 |
| Gulf of California | 0.17 | 517 |
| Tasman Sea, Australian coast | 0.18 | 255 |
| Port Hacking estuary, Australia | 0.37 | 418 |
| Clyde R. estuary, Australia | 0.71 | 255 |
| III *Inland waters* | | |
| *N. America* | | |
| Crater Lake, Oregon | 0.06 | 517 |
| L. Ontario, Canada | 0.15–0.58 | 462 |
| San Vicente reservoir, Calif. | 0.64 | 517 |
| Lake Minnetonka, Minn. | 0.7–2.8 | 313 |
| McConaughy reservoir, Nebr. | 1.6 (av.) | 404 |
| Yankee Hill reservoir, Nebr. | 2.5 (av.) | 404 |
| Pawnee reservoir, Nebr. | 2.9 (av.) | 404 |
| *Europe* | | |
| Esthwaite Water, England | 0.8–1.6 | 177 |
| Loch Croispol, Scotland | 0.59 | 454 |
| Loch Uanagan, Scotland | 2.35 | 454 |
| Sea of Galilee | 0.5 | 100 |
| Sea of Galilee (*Peridinium* bloom) | 3.3 | 100 |
| *Africa* | | |
| L. Simbi, Kenya | 3.0–12.3 | 314 |
| L. Tanganyika | $0.16 \pm 0.02$ | 184 |
| *Australia* | | |
| (a) Southern Tablelands | | |
| Corin Dam | 0.87 | *a* |
| L. Ginninderra | $1.46 \pm 0.68$ | 255, *a* |
| (3 year range) | 0.84–2.74 | |
| Burrinjuck Dam | $1.65 \pm 0.81$ | 255, *a* |
| (6 year range) | 0.71–3.71 | |
| L. Burley Griffin | $2.81 \pm 1.45$ | 255, *a* |
| (6 year range) | 0.86–6.93 | |
| L. George | $15.1 \pm 9.3$ | |
| (5 year range) | 5.7–24.9 | |

Table 6.1. (*cont.*)

| Water body | $K_d(PAR)$ (m$^{-1}$) | Reference |
|---|---|---|
| (b) Northern Territory (Magela Creek billabongs) | | |
| Mudginberri | 1.24 | *b* |
| Leichardt | 1.68 | *b* |
| Georgetown | 8.50 | *b* |
| (c) Tasmania (lakes) | | |
| Perry | 0.21 | 463 |
| Ladies Tarn | 0.41 | 463 |
| Risdon Brook | 0.58 | 463 |
| Barrington | 1.13 | 463 |
| Gordon | 2.16 | 463 |

[a]Kirk, J. T. O., unpublished data
[b]Tyler, P. A. & Kirk, J. T. O., unpublished data

values of $K_d$ (PAR) as might be expected from their low absorption and scattering. Inland waters, with rare exceptions such as Crater Lake in Oregon, USA, have much higher values, with coastal and estuarine waters coming in between. The highest values are found in very turbid waters (e.g. L. George and Georgetown billabong in Australia) in which the suspended tripton strongly absorbs, as well as scatters the light. High values ($> 2.0$ m$^{-1}$) may also be associated with dense algal blooms (Sea of Galilee – *Peridinium*; L. Simbi, Kenya – *Spirulina*), with intense soluble yellow colour but low scattering (L. Gordon, Tasmania), or with a combination of high soluble colour and scattering (L. Burley Griffin, Australia).

A useful, if approximate, rule-of-thumb in aquatic biology is that significant phytoplankton photosynthesis takes place only down to that depth, $z_{eu}$, at which the downwelling irradiance of PAR falls to 1% of that just below the surface. That layer within which $E_d$ (PAR) falls to 1% of the subsurface value is known as the *euphotic zone*. Making the assumption that $K_d$ (PAR) is approximately constant with depth, the value of $z_{eu}$ is given by $4.6/K_d$. This, as we have seen, is a reasonable assumption for the more turbid waters and so will give useful estimates of the depth of the euphotic zone in many inland, and some coastal, systems. In the case of those clear marine waters in which there is a significant increase in slope of the log $E_d$ (PAR) *versus* depth curve, a value of $K_d$ (PAR) determined in the upper layer could give rise to a substantial underestimate of the euphotic depth.

Another useful reference depth is $z_m$, the mid-point of the euphotic zone.

This, by definition, is equal to $\frac{1}{2}z_{eu}$: given the approximately exponential nature of the attenuation of PAR with depth, it follows that $z_m \simeq 2.3/K_d$, and corresponds to that depth at which downward irradiance of PAR is reduced to 10% of the value just below the surface.

### 6.4    Upward irradiance

As a result of scattering within the water, at any depth where there is a downward flux there is also an upward flux. This is always smaller, usually much smaller, than the downward flux but at high ratios of scattering to absorption can contribute significantly to the total light available for photosynthesis. Furthermore, in any water the upwelling light is of crucial importance for the remote sensing of the aquatic environment (Chapter 7), since it is that fraction of the upward flux which penetrates the surface that is detected by the remote sensors.

At any depth, the upward flux can be regarded as that fraction of the downward flux at the same depth which at any point below is scattered upwards and succeeds in penetrating up to that depth again before being absorbed or scattered downwards. Thus we might expect the irradiance of the upward flux to be linked closely to the irradiance of the downward flux; this is found to be the case. Fig. 6.6 shows the upward and downward irradiance of PAR diminishing with depth in parallel together in an

Fig. 6.6. Parallel diminution of upward (●) and downward (○) irradiance of PAR with depth in an Australian lake (Burley Griffin, ACT) (after Kirk, 1977).

Australian lake. Changes in downward irradiance associated with variation in solar altitude, or cloud cover would also be accompanied by corresponding changes in upward irradiance. Given the close dependence of upward irradiance on downward irradiance, it is convenient to consider any effects that the optical properties of the water may have on the upwelling flux in terms of their influence on the ratio of upwelling to downwelling irradiance, $E_u/E_d$, i.e. irradiance reflectance, $R$.

A small proportion of the upward flux originates in forward scattering, at large angles, of downwelling light that is already travelling at some angular distance from the vertical. As solar altitude decreases and the solar beam within the water becomes less vertical, there is an increase in that part of the upward flux which originates in forward scattering. The shape of the volume scattering function is such that this more than counterbalances the fact that an increased proportion of backscattered flux is now directed downwards. The net result is that irradiance reflectance increases as solar altitude decreases, but the effect is not very large. In the Indian Ocean, $R$ for 450 nm light at 10 m depth increased from 5.2% to 7.0% as the solar altitude decreased from 80 to 31°.[214] For Lake Burley Griffin in Australia, irradiance reflectance of the whole photosynthetic waveband (400–700 nm) just below the surface increased from 4.6% at a solar altitude of 75° to 7.9% at a solar altitude of 27°. However, at 1 m, by which depth in this turbid water the light field is well on the way to reaching the asymptotic state (see §6.6), $R$ increased only from 7.4 to 8.6%.[255]

Although as we have seen, there is a contribution from forward scattering, most of the upwelling flux originates in backscattering. Thus we might expect $R$ to be approximately proportional to the backscattering coefficient, $b_b$, for the water in question. As the upwards-scattered photons travel up from the point of scattering to the point of measurement, their numbers are progressively diminished by absorption and – less frequently – by further backscatterings which redirect them downwards again. We may therefore expect that reflectance will vary inversely with the absorption coefficient, $a$, of the water. We might also expect that the dominant tendency of reflectance to increase with backscattering will be somewhat lessened by the contribution of backscattering to the diminution of the upward flux.

The actual manner in which irradiance reflectance varies with the inherent optical properties of the medium has been explored by mathematical modelling of the underwater light field. A simplified version of radiative transfer theory[104, 333] leads to the conclusion that $R$ is proportional to $b_b/(a+b_b)$, for media (such as most natural waters) in which $b_b \ll a$. This is the kind of relation that might be anticipated on the qualitative grounds outlined above. Gordon, Brown & Jacobs (1975) found that the relation between irradiance reflectance just below the surface, $R(0)$, and absorption and backscattering coefficients, which emerged from their Monte Carlo simulations, could be represented by a power series which can be simplified to

$$R = C b_b / (a + b_b) \qquad (6.3a)$$

where $C$ has the value 0.324 for a zenith Sun and 0.369 for a uniform sky. The successive order of scattering calculation method of Prieur (1976) yields an expression for $R(0)$ which can be simplified[333] to

$$R(0) = 0.33 \, b_b / a \qquad (6.3b)$$

In another Monte Carlo study covering waters with values of $b/a$ ranging from 0.2 to 30, Kirk (1981a) found a close linear relation between reflectance and $b_b/a$ which could be represented by

$$R(0) = C b_b / a \qquad (6.3c)$$

where $C$ is equal to 0.328, 0.391 and 0.449 at solar altitudes of 90, 45 and 30°, respectively. Since for most natural waters $b_b \ll a$ and therefore $(a + b_b) \simeq a$, eqn 6.3a is in fact approximately equivalent to eqns 6.3b and 6.3c.

Morel & Prieur (1977) compared their measurements of $R$ across the photosynthetic spectrum with the values calculated using eqn 6.3b, for a variety of oceanic and coastal waters. For clear blue oceanic waters, agreement between the observed and the calculated curves of spectral distribution of $R$ was good. For upwelling oceanic waters with high phytoplankton levels and for turbid coastal waters, agreement was satisfactory from 400 to 600 nm, but in some cases not good at longer wavelengths. Part of the problem in productive waters was a chlorophyll fluorescence emission peak at 685 nm (see § 7.5) in the upwelling flux which increased the observed reflectance over the calculated values in this region.

That part of the upwelling flux just below the surface which is directed approximately vertically upwards is of particular significance for remote sensing. The subsurface radiance in a vertically upward direction we shall refer to as $L_u$. The angular distribution of the upwelling flux is such that upward radiance does not change much with nadir angle in the range 0–20°. Thus, a measured value of $L_u(\theta)$ within this range, or an average value over this range, can be taken as a reasonable estimate of $L_u$. $L_u$, like $E_u$, varies for a given water, in parallel with $E_d$, and we shall refer to the ratio $L_u / E_d$ as the *radiance reflectance*.

The value of radiance reflectance, like that of irradiance reflectance, is a function of the inherent optical properties of the water. Given a relation between $E_u/E_d$ and inherent optical properties, such as one of those embodied in eqn 6.3, if we know the ratio of $E_u$ to $L_u$, we can relate $L_u/E_d$ to $b_b$ and $a$. The simplifying assumption is often made that the radiance distribution of the upwelling flux is identical to that above a Lambertian reflector (same radiance values at all angles). If this were so, then $E_u$ would be equal to $\pi L_u$. In fact, the radiance distribution is not Lambertian (see §6.6 and Fig. 6.12) and measurements in Lake Pend Oreille at a solar altitude of 57° [511] showed that $E_u$ was equal to 5.08 $L_u$ near the surface in this water body.[17] Monte Carlo modelling calculations (Kirk, unpublished) have yielded values of $E_u/L_u$, just below the surface, of about 4.9 for waters with $b/a$ values in the range 1.0 to 5.0, at a solar altitude of 45°. Thus, for intermediate solar altitudes we may reasonably assume that $E_u/L_u \simeq 5$, and the ratio is not likely to be very different at other solar altitudes. Thus, equations relating $L_u/E_d$ to $b_b$ and $a$ may be derived from each of eqns 6.3a, b and c by dividing the constant of proportionality (C) by 5. Monte Carlo calculations for waters with $b/a$ values in the range 1.0 to 5.0 give rise to the relation

$$L_u/E_d \simeq 0.083\, b_b/a \qquad (6.4)$$

for a solar altitude of 45°.

The spectral distribution of the upwelling flux must depend in part on that of the downwelling flux, but, as eqns 6.3 and 6.4 show, it is also

Fig. 6.7. Spectral distribution of upward irradiance and irradiance reflectance in an oceanic and an inland water (plotted from data of Tyler & Smith, 1970). (a) Gulf Stream (Atlantic Ocean) off Bahama Islands, 5 m depth. (b) San Vicente Reservoir, San Diego, California, USA, 1 m depth.

markedly influenced by the variation in the ratio of $b_b$ to $a$ across the spectrum. In clear oceanic waters, for example, $R$ can be as high as 10% at the blue (400 nm) end of the photosynthetic spectrum where pure water absorbs weakly but backscatters relatively strongly (see §4.3), and as low as 0.1% at the red (700 nm) end where water absorbs strongly.[333] Fig. 6.7 shows the spectral distributions of upward irradiance and irradiance reflectance in a clear oceanic water and in an inland impoundment. The upwelling flux in the oceanic water consists mainly of blue light in the 400–500 nm band. In productive oceanic waters with high levels of phytoplankton, the photosynthetic pigments absorb much of the upwelling blue light and so the peak of the upwelling flux is shifted to 565–570 nm in the green.[333] There is also a peak at 685 nm due to fluorescence emission by phytoplankton chlorophyll. In the inland water (Fig. 6.7b), yellow substances and phytoplankton absorb most of the blue light and a broad band, peaking at about 580 nm, with most of the quanta occurring between 480 and 650 nm, is observed: the chlorophyll fluorescence emission at about 680 nm can be seen in this curve.

In the oceanic waters studied by Tyler & Smith (1970), irradiance reflectance for total PAR at 5 m depth varied from about 2 to 5%. For slightly to very turbid inland water bodies in southeast Australia, irradiance reflectance values for PAR just below the surface were usually between 4 and 10%, but values as low as 2% and as high as 19% were observed, the higher values being associated with higher turbidity.[255, 257] The reflectance values increased somewhat with depth (see §6.6), the maximum value observed so far being about 24%. In inland waters with low scattering, but intense colour due to high concentrations of soluble yellow substances, reflectance of PAR can be very low. In a series of lakes of this type in Tasmania, irradiance reflectance for PAR just below the surface ranged from about 1.2% down to 0.14%.[463]

In water bodies which are sufficiently shallow for significant light to reach the bottom, then unless the bottom is very dark in colour there is an increase in $E_u$ near the bottom due to reflection from it.

## 6.5    Scalar irradiance

For randomly oriented phytoplankton cells, photons are equally useful in photosynthesis regardless of the direction from which they come. Scalar irradiance, $E_0$, is therefore the best all-round measure of the availability of light for photosynthesis at a given depth. For monochromatic light scalar irradiance, like downward irradiance, diminishes with depth in an approximately exponential manner, as the linear graph (Fig. 6.8) of log $E_0$ against depth for green light in the Danish Sound[189] shows. When scalar irradiance is measured with a quanta meter which responds to the whole photosynthetic waveband, the variation of log $E_0$ with depth in fairly clear water is biphasic, the rate of attenuation being higher near the surface than lower down. This can be seen in the data for Burrinjuck Dam, Australia, in Fig. 6.8. The explanation is the same as that for the analogous phenomenon observed with downward irradiance (Fig. 6.5): the more strongly absorbed wavelengths are being removed in the upper layer, leaving the weakly absorbed wavelengths to penetrate lower down.

Although scalar irradiance is the best parameter in terms of which to express the availability of light for photosynthesis, the most commonly measured parameter is downward irradiance, $E_d$. It is of interest to examine the relation between them. As might be expected, since scalar irradiance includes both the upwelling and the downwelling light, and since it represents all angular directions equally (whereas downward irradiance, in accordance with the cosine law, is progressively less affected by light flux as its zenith angle increases), at any point in the water the value of scalar irradiance is always higher than that of downward irradiance.

From the radiance distribution data for Lake Pend Oreille, USA,[376, 511] it is possible to calculate that in the 480 nm waveband the ratio $E_0/E_d$ is 1.30–1.35 for various depths within the euphotic zone. The higher the ratio of scattering to absorption, the more diffuse the underwater light field

Fig. 6.8. Attenuation of scalar irradiance with depth. The data for Burrinjuck Dam (an Australian inland impoundment) were obtained with a scalar quantum irradiance sensor measuring the full 400–700 nm waveband (Kirk, unpublished measurements). The data for the Danish Sound (Baltic Sea) are measurements by Højerslev (1975) at 532 nm.

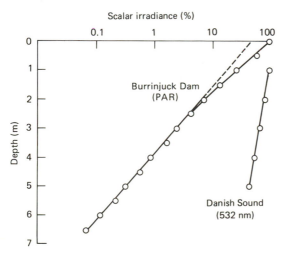

becomes, and the greater the difference between $E_0$ and $E_d$. Fig. 6.9 shows the way $E_0/E_d$ at the mid-point of the euphotic zone ($z_m$) increases as the ratio of the scattering coefficient to the absorption coefficient increases: $E_0/E_d$ is equal to 1.5, 1.75 and 2.0 for $b/a$ values of 6, 10 and 18, respectively. Since the absorption coefficient of natural waters can vary markedly with wavelength, the ratios of scalar to downward irradiance will vary across the spectrum. Taking the whole photosynthetic waveband, for clear oceanic waters the average $b/a$ ratio is low enough to give rise to $E_0/E_d$ values of up to about 1.2. For typical inland and some coastal waters, however, with $b/a$ values in the region of 4 to 10, or even up to 20 or 30 in the most extreme cases, we might commonly expect $E_0/E_d$ to be in the region of 1.4–1.8, rising to 2.0–2.5 in the very turbid waters.

Thus, if it is wished to determine the absolute amount of light available

Fig. 6.9. Ratio of scalar to downward irradiance at the mid-point of the euphotic zone as a function of the ratio of scattering to absorption coefficients. Data obtained by Monte Carlo calculation for vertically incident monochromatic light using the method of Kirk (1981*a*, 1981*c*).

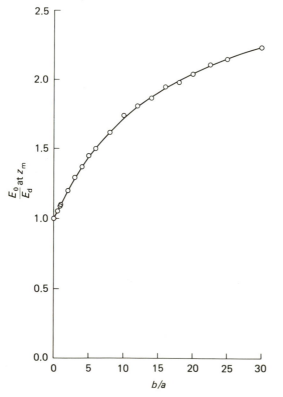

for photosynthesis at a given depth in a water body, a measurement of downward irradiance may seriously underestimate this, particularly in turbid waters. On the other hand, the vertical attenuation coefficient, $K_d$, for downward irradiance is close in value to the vertical attenuation coefficient, $K_0$, for scalar irradiance. Monte Carlo calculations (Kirk, unpublished) show that for media with $b/a$ ranging from 0.3 to 30, $K_d/K_0$ varies only between about 1.01 and 1.06. The measured value of $K_d$ can therefore be taken as a reasonable estimate of the value of $K_0$, and used to predict the attenuation of scalar irradiance with depth.

It would be useful to have, for any given water body, a measure of the total amount of light available for photosynthesis at a given instant, throughout the whole water column, 1 m square. We shall give this parameter the symbol, $Q_t$: it has the units joules or quanta. Making the simplifying approximate assumption that $K_0$ does not vary with depth it can be shown that

$$Q_t = \frac{E_0(0)}{cK_0} \tag{6.5}$$

where $E_0(0)$ is the scalar irradiance just below the surface and $c$ is the speed of light in the medium. Strictly speaking, this equation applies only to monochromatic light (for which the variation of $K_0$ with depth is small) but may be regarded as very approximately applicable to PAR. Thus if one water body has a $K_0$ (or $K_d$, since $K_0 \simeq K_d$) value twice that in another water body, it has about half the total amount of light available for photosynthesis.

## 6.6 Angular distribution of the underwater light field

As sunlight penetrates a water body, its angular distribution begins to change – to become less directional, more diffuse – immediately the light penetrates the surface, as a result of scattering of the photons. The greater the depth, the greater the proportion of photons which have been scattered at least once. The angular distribution produced is not, however, a function of scattering alone: the less vertically a photon is travelling, the greater its pathlength in traversing a given depth, and the greater the probability of its being absorbed within that depth. Thus the more obliquely travelling photons are more intensely removed by absorption and this effect prevents the establishment of a completely isotropic field. The resultant angular distribution is determined by this interaction between the absorption and the scattering processes. Eventually the angular distribution of light intensity takes on a fixed form – referred to as the *asymptotic radiance distribution* – which is symmetrical about the vertical axis and whose shape is determined only by the values of the absorption coefficient, the scattering coefficient, and the volume scattering function.[103, 214, 376, 385, 508] That such an equilibrium radiance distribution would be established at great depth was predicted on theoretical grounds independently by Whitney (1941) in the USA and Poole (1945) in Ireland, and a rigorous proof of its existence was given by Preisendorfer (1959). Early measurements by Jerlov & Liljequist (1938) in the Baltic Sea showed some movement of the radiance distribution towards a symmetrical state with increasing depth. The particularly accurate measurements of Tyler (1960) down to 66 m in Lake Pend Oreille, USA, demonstrated the very close approach of the radiance distribution to the predicted symmetrical state at this near asymptotic depth.

The progression of the angular structure of the underwater light field towards the asymptotic state can be seen from the change in the radiance distribution with increasing depth. This is illustrated in Fig. 6.10 which is

Fig. 6.10. Radiance distribution of underwater light field at different depths. Plotted from measurements at 480 nm made by Tyler (1960) in Lake Pend Oreille, USA, with solar altitude 56.6°, scattering coefficient 0.285 m⁻¹ and absorption coefficient 0.117 m⁻¹. (*a*) Radiance distribution in the plane of the Sun. (*b*) Radiance distribution in a plane nearly at right angles to the Sun.

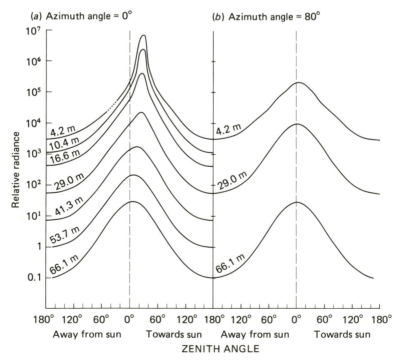

based on the measurements of Tyler (1960) in Lake Pend Oreille. Fig. 6.10*a* is for radiance at different vertical angles in the plane of the Sun. Near the surface the light field is highly directional with most of the flux coming from the approximate direction of the Sun. With increasing depth the peak in the radiance distribution becomes broader as its centre shifts towards $\theta = 0°$. The final, asymptotic, radiance distribution would be symmetrical about the zenith. It is noticeable that in the plane of the Sun this final state had not quite been reached even at 66 m. Fig. 6.10*b* shows that in a plane almost at right angles to that of the Sun, the radiance distribution is nearly symmetrical about the vertical even near the surface. At intermediate azimuth angles, radiance distributions intermediate between those in Fig. 6.10*a* and *b* exist.

Fig. 6.11 shows the near-asymptotic radiance distribution in Lake Pend

Fig. 6.11. Near-asymptotic radiance distribution in Lake Pend Oreille, USA, plotted as a polar diagram. The data are the same as those in the lowest curve of Fig. 6.10*b*.

Oreille in the form of a polar diagram. If in a natural water the ratio of scattering to absorption increases to high levels, then the shape of such a polar diagram of the asymptotic radiance distribution tends towards a circle (i.e. the light field approaches the completely diffuse state). If, however, scattering decreases to low values relative to absorption, then the polar diagram takes on the form of a narrow, downward pencil.[375]

In the context of photosynthesis, what is significant about a radiance distribution is its relevance to the rapidity of attenuation of light intensity with depth. The azimuthal distribution of radiance at each vertical angle has no bearing on this. It is therefore legitimate and, in the interests of simplicity, advantageous to express the angular distribution of the light field at any depth in terms of the radiance averaged over all azimuth angles at each vertical angle. Fig. 6.12 shows a series of such radiance distributions at increasing depth, calculated by the Monte Carlo procedure, for PAR in an Australian lake. The much more rapid approach to a final symmetrical distribution in this case is partly the result of averaging over all azimuth angles, and partly because of the increased scattering relative to absorption in this lake relative to Lake Pend Oreille.

An even simpler way of expressing the angular distribution of the light field is in terms of the three average cosines (see §1.3) and the irradiance reflectance. In Fig. 6.13 the values of these parameters, obtained by Monte

Fig. 6.12. Radiance distribution of underwater light field averaged over all azimuth angles in a turbid lake. Radiance values for total PAR in Lake Burley Griffin, Australia, were obtained by Monte Carlo calculation using the method of Kirk (1981a, 1981c) on the basis of the measured absorption and scattering properties (absorption spectrum – Fig. 3.9b; scattering coefficient – 15.0 m⁻¹; solar elevation – 32°).

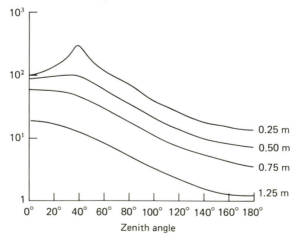

Carlo calculation[259, 261] are plotted against the optical depth ($\zeta = K_d z$) in water in which $b/a = 5.0$. The average cosine and the average downward cosine initially both diminish sharply in value but then begin to level off as the angular distribution of the light approaches the asymptotic radiance distribution. There is no significant further change in the angular distribution beyond the depth ($z_{eu}, \zeta = 4.6$) at which irradiance has been reduced to 1% of the subsurface value, and indeed most of the change takes place before $z_m$ ($\zeta = 2.3$) is reached.

It will be noted that for vertically incident light $\bar{\mu}_d$ starts (at $z = 0$) from a value of slightly less than 1.0. This is because the downwelling light just below the surface includes not only those photons which have just passed down through the surface (and these do have $\bar{\mu}_d = 1.0$), but also those photons which were part of the upwelling stream and which have just been reflected downwards from the surface. These latter photons have a $\bar{\mu}_d$ much

Fig. 6.13. Variation of average cosines for downwelling ($\bar{\mu}_d$), upwelling ($\bar{\mu}_u$) and total ($\bar{\mu}$) flux, and irradiance reflectance ($R$) with optical depth ($\zeta = K_d z$) in water with $b/a = 5$. Data obtained by Monte Carlo calculation.[259] (Vertically incident light ———. Light incident at 45° ------.)

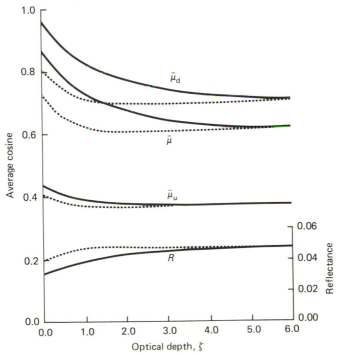

less than 1.0 and so, although they constitute only a small proportion of the total, they bring the average value of $\bar{\mu}_d$ significantly below 1.0.

For light incident on the surface at angles other than the vertical, the behaviour is much the same as for perpendicularly incident light except that $\bar{\mu}$ and $\bar{\mu}_d$ just below the surface have lower values, and the light field reaches its asymptotic state at shallower depths: compare the curves for $\theta = 0°$ and $\theta = 45°$. In the case of a non-perpendicular incident beam it should be realized that although the angular distribution averaged over all azimuths has settled down to almost its final form at $z_{eu}$, the downward radiance distribution in the vertical plane of the Sun is still significantly different from the asymptotic distribution: this can be seen in Fig. 6.10a in the curves for 29 m ($z_{eu} \simeq 28$ m).

The angular distribution of the upwelling stream acquires its final form a very short distance below the surface. In Lake Pend Oreille, USA, $\bar{\mu}_u$ varied only between 0.37 and 0.34 at all depths between 4.2 and 66.1 m.[514] Monte Carlo calculations[259] indicate that for waters with $b/a$ values ranging from 0.1 to 20, $\bar{\mu}_u$ is typically in the range 0.35 to 0.42 throughout the euphotic zone. That the upwelling stream should have these characteristics is not surprising since it consists predominantly of backscattered light: backscattering does not vary strongly with angle and so rather similar upward radiance distributions are produced whatever the downward radiance distribution.

The radiance distribution of the upwelling light stream near the surface in Lake Pend Oreille is shown in the form of a polar diagram in Fig. 6.14. If, as is sometimes assumed (see §6.4), the upwelling flux has the same radiance distribution as that above a Lambertian reflector (same radiance at all angles), all the radiance vectors in Fig. 6.14 would be the same length. It is clear from the data that the upward radiance distribution is far from Lambertian.

As might be expected from the changes in angular distribution with depth, irradiance reflectance increases with depth but levels off at a final value concurrently with the settling down of $\bar{\mu}$ and $\bar{\mu}_d$ to their final values

Fig. 6.14. Vector diagram of radiance distribution of upwelling light at 4.24 m depth in Lake Pend Oreille, USA. Radiance measurements of Tyler (1960) averaged over all azimuth angles. Radiance vectors are at nadir angle intervals of 10°.

(Fig. 6.13). Thus, in natural water bodies the depth at which the asymptotic radiance distribution (averaged over all azimuth angles) is established, can be found by determining at what depth the value of irradiance reflectance ceases to increase.

### 6.7 Dependence of properties of the field on optical properties of the medium

We have noted earlier that the diminution of light intensity with depth is due to a combination of absorption and scattering processes, and also that the angular distribution of light intensity is determined by interaction between absorption and scattering. We shall now examine the quantitative relations that exist between the inherent absorption and scattering properties of the medium and the angular structure and vertical attenuation of irradiance in the light field that is established.

*Angular structure*

The shape of the volume scattering function, as defined by $\tilde{\beta}(\theta)$, the normalized volume scattering function (§1.4), can have a major influence on the character of the light field. In the majority of natural waters, however, scattering is particle-dominated to such an extent that $\tilde{\beta}(\theta)$ curves are rather similar in shape from one water to another. We shall not, therefore, consider the effects of variation in $\tilde{\beta}(\theta)$, and shall assume that the waters under consideration have a typical particle-dominated volume scattering function, similar to that measured by Petzold in San Diego harbour (Fig. 4.9).

To give the conclusions as much generality as possible, our analysis will be in terms of optical depth ($\zeta = K_d z$) rather than actual depth. This makes it permissible to confine our attention to the ratio of scattering, to absorption coefficients, $b/a$, rather than to their absolute values. A convenient optical depth at which to study the effects of changes in $b/a$ is $z_m$, the mid-point of the euphotic zone ($\zeta = 2.3$). Fig. 6.15 shows the total and downward average cosines, and reflectance at $z_m$, as a function of $b/a$, determined by Monte Carlo calculation.[259] As scattering increases relative to absorption, so the underwater light field at this depth becomes less vertical, more diffuse, as shown by the decrease in $\bar{\mu}$ and $\bar{\mu}_d$. Reflectance increases almost linearly with $b/a$ over much of the range, but the curve as a whole has a slightly sigmoid character. If the zenith angle of the incident light is changed from 0 to 45°, it makes very little difference to the reflectance, and not much to the average cosine except at low values of $b/a$.

In addition to delineating the quantitative dependence of the angular structure of the light field on $b/a$, Fig. 6.15 illustrates a useful general conclusion. Namely, that for water with a given $\tilde{\beta}(\theta)$, the relations between $\bar{\mu}$, $\bar{\mu}_d$, $\bar{\mu}_u$, $R$ at a specified optical depth, and $b/a$, are fixed at any given angle of incident light, and indeed are largely independent of the angle of incident light. The fixed nature of the relation between $\bar{\mu}$, $R$ and $b/a$ provides the basis for a method of estimating the absolute values of both $b$ and $a$ for any real water body, using only measurements of underwater irradiance (see §4.2).

### Vertical attenuation of irradiance

In the absence of scattering ($b = 0$), $K_d$, the vertical attenuation coefficient for downward irradiance, is determined only by the absorption

Fig. 6.15. Average cosines and irradiance reflectance at mid-point ($z_m$) of euphotic zone as a function of $b/a$. Data obtained by Monte Carlo calculation.[259] (Vertically incident light ——. Light incident at 45° ------.)

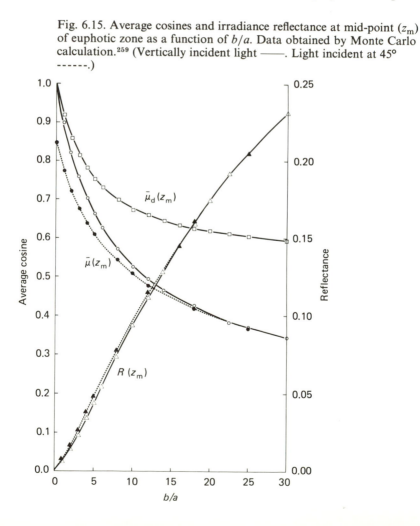

coefficient and the zenith angle, $\theta$, of the light beam within the water, in accordance with $K_d = a/\cos\theta$. In a scattering medium $K_d$ is increased, partly because of the altered angular distribution of the downwelling light, and partly because of upward scattering of the downwelling stream. Concurrently with the progressive change in angular distribution with depth, so the value of $K_d$ increases with depth, levelling off at that optical depth at which the asymptotic radiance becomes established. Fig. 6.16 shows $K_d$ as a function of optical depth for vertically incident light in a medium for which $b/a = 5.0$.

While $K_d$ always increases with depth when the incident light is parallel, it is possible that the converse might be true if the incident light was diffuse, e.g. from an overcast sky. If, as could be the case, especially in water with a low value of $b/a$, the final asymptotic radiance distribution was more vertical than the initial radiance distribution, then $K_d$ would decrease with depth.

The relation between $K_d$ and the absorption and scattering coefficients may, for a given incident light field and a given $\tilde{\beta}(\theta)$, be expressed with complete generality by expressing $K_d/a$ as a function of $b/a$. The absolute values of $a$, $b$ and $K_d$ do not have to be specified: it is the ratios which are important. A certain ratio of $b$ to $a$ gives rise to a specific ratio of $K_d$ to $a$, regardless of the actual value of any of the coefficients. The calculated

Fig. 6.16. Vertical attenuation coefficient for downward irradiance as a function of optical depth for $b/a = 5$. Data obtained by Monte Carlo calculation for vertically incident light.[259]

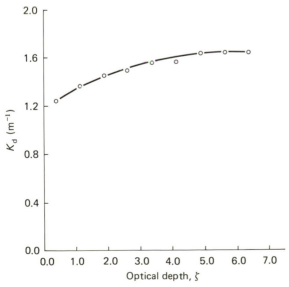

value of $K_d/a$ at $z_m$ for vertically incident light is plotted against $b/a$ in Fig. 6.17. $K_d$ starts off equal to $a$ when $b = 0$, and rises progressively as $b$ increases, in a linear manner to begin with, but curving over at high ratios of $b$ to $a$. The values of $K_d(z_m)/a$ over the whole range, up to $b/a = 30$, were found[259] to fit very closely to the curve specified by the equation

$$K_d(z_m)/a = (1+0.256\,b/a)^{\frac{1}{2}} \tag{6.6}$$

or

$$K_d(z_m) = (a^2+0.256\,ab)^{\frac{1}{2}} \tag{6.7}$$

As an indication of the extent to which scattering intensifies vertical attenuation of light it may be noted from Fig. 6.17 that $b/a$ ratios of about 5 and 12 increase $K_d$ by 50 and 100%, respectively. Comparable effects can be shown for real water bodies. In Table 6.2, data which show the effect of scattering on attenuation of the whole photosynthetic waveband in four Australian inland waters of varying turbidity are presented. The table compares the values of $K_d$ for PAR calculated from the total absorption spectrum of the water, assuming no scattering, with those actually measured within the water with an irradiance (PAR) meter. The ratio of the observed to the calculated value of $K_d$ is a measure of the extent to which scattering intensifies light attenuation. The increase in $K_d$ (PAR) due to scattering ranged from 16% in Burrinjuck Dam, which was rather clear

Fig. 6.17. Ratio of vertical attenuation coefficient for downward irradiance (at $z_m$) to absorption coefficient as a function of $b/a$. Data obtained by Monte Carlo calculation for vertically incident light.[259]

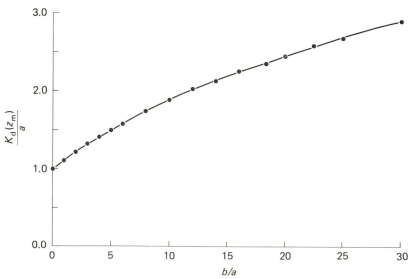

at the time, to more than three-fold in the very turbid water of Lake George.

In our consideration of the ways in which scattering affects attenuation we can now make use of eqn 1.53, derived from radiative transfer theory by Preisendorfer (1961), which expresses the vertical attenuation coefficient for downward irradiance as a function of the diffuse absorption and backscattering coefficients.

$$K_d(z) = a_d(z) + b_{bd}(z) - b_{bu}(z) R(z)$$

Although this relation is not, in view of the difficulty of measuring the diffuse coefficients, of everyday practical use, it is conceptually valuable in helping us to understand the nature and relative importance of the radiation transfer mechanisms underlying the attenuation process. It tells us that attenuation of downward irradiance is the result of three different processes, represented by the three different terms on the right-hand side of the equation. Absorption from the downwelling stream is accounted for by the diffuse absorption coefficient $a_d(z)$ which is equal to $a/\bar{\mu}_d(z)$, and so must increase as scattering causes $\bar{\mu}_d$ to decrease. Light is also removed from the downwelling stream by upward scattering of the photons: this process is represented by the diffuse backscattering coefficient for the downwelling stream, $b_{bd}(z)$. Opposing these two processes is the downward scattering of photons from the upwelling stream, which acts to increase the downward flux: this is represented by the term, $-b_{bu}(z) R(z)$, where $b_{bu}(z)$ is the diffuse backscattering coefficient for the upwelling stream and $R(z)$ is irradiance reflectance.

For any given values of $a$ and $b$, and having specified $\tilde{\beta}(\theta)$ and the incident light field, then if the radiance distribution at depth $z$ m is

Table 6.2. *Effects of light scattering on vertical attenuation coefficient ($K_d$) for irradiance of PAR (400–700 nm), in four water bodies of differing turbidity, in southeast Australia. $K_d$ values calculated from the measured absorption spectra are compared with those obtained* in situ *with a quanta meter*[259]

| Water body | Turbidity (NTU) | $K_d$ (calculated) (m$^{-1}$) | $K_d$ (observed) (m$^{-1}$) | Effect of scattering $\dfrac{K_d(\text{obs.})}{K_d(\text{calc.})}$ |
|---|---|---|---|---|
| Burrinjuck Dam | 1.8 | 0.775 | 0.90 | 1.16 |
| L. Ginninderra | 4.6 | 0.547 | 0.90 | 1.65 |
| L. Burley Griffin | 17.4 | 1.333 | 2.43 | 1.82 |
| L. George | 49.0 | 1.742 | 5.67 | 3.25 |

calculated by a Monte Carlo, or other, procedure, it is possible to calculate the diffuse optical properties at that depth.[257] Thus all the terms on the right-hand side of eqn 1.53 may be determined for a series of values of $b$ and $a$, and so the relative importance of the three different processes can be evaluated for different types of water.

Fig. 6.18 shows the change in value of each of these three terms, compared to the value of $K_d$, at $z_m$, as the ratio of scattering to absorption coefficients increases from 0.0 to 30.0. At low values of scattering, up to $b/a$ ratios of about 3, attenuation is almost entirely due to absorption, and furthermore the scattering-induced increase in attenuation is mainly a consequence of the increase in absorption resulting from the changed angular distribution of the downwelling flux. Upward scattering of the downwelling stream contributes only slightly to attenuation.

When $b/a$ has risen to about 7, upward scattering of the downwelling stream accounts for about 25% of all the attenuation and contributes as much to the increase in attenuation as does the increase in absorption of

Fig. 6.18. Changes in the different components of the attenuation processes as a function of $b/a$. The calculated values of the three right-hand terms in eqn 1.53 are plotted alongside the corresponding values of $K_d$ at $z_m$ (data from Kirk, 1981$a$).

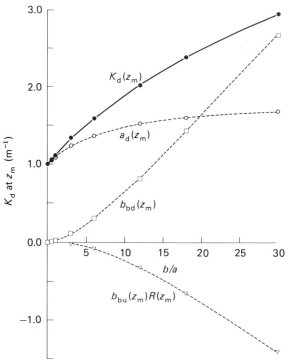

the downwelling light. At $b/a$ values of about 11 and 20, upward scattering of the downwelling light accounts for 0.5 and 1.0 times as much attenuation, respectively, as does absorption, and at higher $b/a$ values it becomes the major mechanism for attenuation. The downward scattering of the upwelling stream, which acts to diminish attenuation, has little effect over the lower part of the range of $b/a$ values studied, but becomes significant from about $b/a = 6$ onwards, and counteracts a large part of the attenuation due to the other two processes at $b/a$ values in the range 12 to 30.

## 6.8     Partial vertical attenuation coefficients

In any medium containing a number of absorbing and scattering components, the contributions of the different components to any one of the absorption and scattering coefficients (normal, diffuse, forward, back) are additive, e.g. if there are $n$ components, then

$$a_{\text{total}} = a_1 + a_2 + \ldots + a_n$$

and similar relations hold for $b$, $a_d(z)$, $b_{bd}(z)$ etc. Consequently, for an aquatic medium containing $n$ absorbing/scattering components we can expand eqn 1.53 to

$$K_d(z) = a_d(z)_1 + a_d(z)_2 + \ldots + a_d(z)_n$$
$$+ b_{bd}(z)_1 + b_{bd}(z)_2 + \ldots + b_{bd}(z)_n$$
$$- b_{bu}(z)_1 R - b_{bu}(z)_2 R - \ldots - b_{bu}(z)_n R$$

Rearranging, we obtain

$$K_d(z) = [a_d(z)_1 + b_{bd}(z)_1 - b_{bu}(z)_1 R]$$
$$+ [a_d(z)_2 + b_{bd}(z)_2 - b_{bu}(z)_2 R]$$
$$+ \ldots + [a_d(z)_n + b_{bd}(z)_n - b_{bd}(z)_n R]$$

which can be written in the form

$$K_d(z) = K_d(z)_1 + K_d(z)_2 + \ldots + K_d(z)_n \qquad (6.8)$$

Thus the vertical attenuation coefficient for downward irradiance in a natural water can be partitioned into a set of partial vertical attenuation coefficients, each corresponding to a different component of the medium. The partial vertical attenuation coefficient for the $i$th component is given by

$$K_d(z)_i = a_d(z)_i + b_{bd}(z)_i - b_{bu}(z)_i R \qquad (6.9)$$

It is important to remember that although the contributions of the different components of the medium to total attenuation of irradiance can be simply added together in accordance with eqn 6.8, the nature of their contributions to attenuation can vary markedly from one component to another. Consider, for example, a water in which component $j$ is dissolved yellow colour, and component $j+1$ consists of suspended mineral particles, scattering intensely but with little intrinsic colour. For the $j$th component, $K_d(z)_j$ will consist mainly of the absorption term, $a_d(z)_j$, in eqn 6.9, whereas for the $(j+1)$th component, $K_d(z)_{j+1}$ will consist mainly of the backscattering term, $b_{bd}(z)_{j+1}$.

It is also important to remember that the value of $K_d(z)_i$, and consequently the contribution of the $i$th component to $K_d(z)$, is not a linear function of the concentration of that component. Any substantial change in the absorption and/or scattering characteristics of the medium resulting from a change in concentration of one of the components will inevitably affect the radiance distribution at depth $z$. The absorption and scattering coefficients on the right-hand side of eqn 6.9, being quasi-inherent rather than inherent optical properties (see §1.5) are functions of the radiance distribution as well as of the concentration of the component in question. Consequently, although their values will increase with the concentration of the component, the increase will not be linear with concentration. Furthermore, $R$, being a function of the radiance distribution, will also change as the concentration of any component changes. In short, $K_d(z)_i$ and $K_d(z)$ will certainly increase as the amount of any of the components of the medium increases, but except for small increments in concentration, the increases will not be linearly related to the increase in concentration.

# 7

.................................................................................

# Remote sensing of the aquatic environment

In any water body some of the light which penetrates the water is caused, by scattering within the water, to pass up through the surface again. Of this emergent flux, 90% originates within the depth (equal to $1/K_d$) in which downward irradiance falls to 37% ($1/e$) of the subsurface value.[160] It can be regarded as a sample derived from the underwater light field, and so by studying it with appropriate detection instruments above the surface, information about that field and therefore about the optically significant components of the medium, can be obtained. There is not much point in having detection instruments just above the surface: they would be more useful below. If, however, this emerging radiant flux can be studied by remote sensing instruments, located a considerable distance above the surface, in an aeroplane or space satellite, then the considerable advantage is gained that information about the underwater environment over a large area can be obtained in a short time. This makes it possible to acquire a synoptic view of a large aquatic ecosystem, with a fraction of the time and effort that would be involved in carrying out measurements over the same area from a surface vessel.

It will readily be appreciated, however, that a price must be paid. Measurements of the emergent flux, from a great distance, cannot be as accurate, or yield as much information, as measurements carried out within the water itself. We shall now consider the kinds of measurements that can be made, the correction procedures that must be carried out, and the nature of the information that may be obtained. We shall then go on to examine some of the studies that have been carried out so far. The subject up to 1972, with particular reference to the underlying optical phenomena, was reviewed by Austin (1974a).

## 7.1    The upward flux and its measurement

A photometer in an aeroplane or satellite, pointed down at the ocean or any other water body can receive light originating in four different ways: by scattering from below the surface, by reflection of skylight at the surface, by reflection of the direct solar beam at the surface, and by scattering within the atmosphere (Fig. 7.1). Only the first of these four light fluxes, which we shall refer to as the emergent flux, contains information about the underwater light field and the composition of the aquatic medium. The essential problem, therefore, is to quantify the emergent flux in the presence of the other light fluxes.

The photometer used to measure the light flux above the water surface could in principle be either an irradiance meter receiving all the upwelling light through an angle of 180° with its cosine collector, or could be a radiance meter with a narrow angle of acceptance. In practice the sensor is usually a radiance meter since the irradiance meter has the serious disadvantages that it has to receive all the upwelling light including the reflected solar beam, and also that it cannot be directed at specific areas on the surface.

To avoid receiving the surface-reflected solar beam, the radiance meter can be directed at a part of the surface well outside the solar glitter pattern.

Fig. 7.1. The different origins of the light received by a remote sensor above a water body.

A more difficult problem is that of accounting for that part of the measured radiance which originates within the atmosphere. About 80–90% of the radiance received by a satellite-borne radiance meter, directed at the ocean surface outside the Sun glitter pattern, originates in atmospheric Rayleigh (air molecule) and aerosol (particle) scattering. Ways of correcting for this flux are considered in a later section. This problem is largely eliminated when measurements are taken from low-flying aircraft (see Fig. 7.6a).

Skylight also originates in atmospheric scattering. Correction for the contribution of reflected skylight to the measured upwelling flux can be lumped in with the total atmospheric scattering correction.[157] An alternative procedure is to carry out the measurements at an angle of 53° to the vertical.[80] Light reflected from a plane surface at this angle (Brewster's angle) is polarized and can be eliminated by placing a polarizing filter oriented at right angles to the major axis of polarization over the aperture of the radiance meter. A consequence of viewing at this angle, however, is that the effective air mass through which the flux from within the water must travel is increased by a factor of 1.67:[16] this would be a serious disadvantage in the case of a satellite or high-flying aircraft, but not in the case of a low-flying aircraft.

### Measurement systems – general considerations

While information about natural waters can be obtained from measurements of upward radiance in just one waveband, the possibility of estimating specific components of the medium is greatly enhanced if radiance values in more than one spectral region can be measured. Some remote sensing optical systems used in low-flying aircraft have used spectroradiometers which determine the spectral distribution of radiance throughout the visible, or visible/near infrared spectrum.[11, 80, 329, 341] Spectral radiance meters developed for satellites or high-altitude aircraft measure radiance simultaneously in a finite number of spectral bands obtained with interference filters, or a diffraction grating (a detailed account of some satellite optical systems may be found in the book by Slater, 1980). In this way simultaneous data for all the wavebands is obtained for each unit element of the Earth's surface (see later) viewed from the satellite or aircraft.

Remote sensing photometers used in low-altitude aircraft normally have a fixed direction of view. They measure upward radiance at a series of points along a linear track determined by the flight path of the aircraft. For two-dimensional mapping, the aircraft must traverse the area of interest many times.

With increasing altitude the area that it is practicable to 'view' increases.

To take advantage of this, remote sensing from satellites and high-altitude aircraft such as the U-2 uses spatially scanning photometers. These incorporate a rotating or oscillating mirror which ensures that the direction of viewing is continually moving from one side to the other at right angles to the satellite/aircraft trajectory. The part of the Earth's surface which is remotely sensed is thus not just a line, but a broad swath which might be of the order of 25 km wide in the case of a U-2, or hundreds of kilometres wide in the case of a satellite.

The scene reflected in the mirror is viewed through a telescope, with a very small acceptance angle (0.005–0.05°) so that at any given instant only a small unit element of the Earth's surface is in view. The size of this element depends on the altitude as well as on the power of the telescope. For the kinds of scanning photometers developed for oceanic studies in the visible region, the element of area might typically be about 60 m square when viewed from a U-2 and about 1 km square when seen from satellite altitude.

As the mirror rotates, the photometer scans from one side of the swath to the other. A series of contiguous but non-overlapping, square image elements, corresponding to a linear sequence of elements of area at the surface, extending across the swath, is recorded. By the time the next side-to-side scan begins the satellite or aircraft has moved on so that the next left-to-right sequence of image elements adjoins, but does not overlap, the previous sequence. The accumulated set of image elements constitutes a mosaic corresponding to that strip of the Earth's surface defined by the path of the satellite/aircraft and its maximum lateral angle of view.

For each image element, i.e. for each element of surface, the photometer has recorded a set of radiance values corresponding to the number of spectral bands with which it operates. This information, after being relayed to a receiving station in the case of a satellite, is stored on tape. The data can then be used, after appropriate computer processing, to prepare a map of any part of the Earth's surface that lay beneath the satellite's orbit, or aircraft's path. The continuous swath covered by a satellite is broken down for mapping purposes into individual sections which might be some hundreds of kilometres long.

Each small element of the map, referred to as a *pixel*, corresponds to one of the square unit elements of surface viewed by the scanning photometer. The intensity, or the colour within a selected colour scale, of each pixel in the final map displayed on the computer screen can be directly related to the radiance value in any one of the spectral bands, or can be determined by the value of some parameter (e.g. chlorophyll concentration) derived by calculation from more than one of the radiance values.

### Low altitude (100–500 m) systems

We consider here instruments which measure radiance only along a linear path, i.e. which do not simultaneously scan from side to side.

If a spectroradiometer is used to measure radiance, it should complete its spectral scan quickly to minimize the effects of changes in the emergent flux from point to point along the aircraft's path. Clarke, Ewing & Lorenzen (1970) used a spectroradiometer with a $3° \times 0.5°$ field of view which scanned from 400 to 700 nm in 1.2 s. The instrument was operated at the Brewster angle in conjunction with a polarizer, to eliminate reflected skylight. Neville & Gower (1977) used a spectroradiometer designed by Walker *et al.* (1974) in which the spectrum produced by a diffraction grating was distributed over an array of 256 silicon diodes, each of which continuously detected the radiant flux in its own narrow region of the spectrum. It was possible to read out the complete spectrum, 380–1065 nm, at 2 s intervals.[56] This instrument was also operated at the Brewster angle. Miller *et al.* (1977) used a spectroradiometer which scanned simultaneously across four spectral bands: 440–460 nm, 550–590 nm, 680–700 nm and 720–750 nm. This took 4 s to complete a scan.

Viollier, Deschamps & Lecomte (1978) used interference filters to carry out virtually simultaneous measurements at four wavelengths, 466, 525, 550 and 600 nm, in a study of remote sensing under cloudy conditions. An unusual feature of their system was that they used irradiance collectors for measuring both the upward and downward light fluxes. Under cloudy skies (so that Sun glint was not a problem) they found that measuring the upward irradiance gave results as good as those obtained with measurements of upward radiance.

Arvesen, Millard & Weaver (1973) developed a differential radiometer which carried out simultaneous measurements of upward radiance at 443 and 525 nm, and continuously compared one with the other. This had the advantage that changes in incident light intensity or variations in surface water roughness had similar effects on the flux in both wavebands and so were automatically corrected. Changes in phytoplankton concentration, however, which would be expected specifically to affect the ratio of upward flux in these two wavebands (see §7.5), markedly affected the signal.

### Medium/high altitude (2–20 km) systems

A spatially scanning photometer – the Ocean Colour Scanner (OCS) – was developed by the NASA Goddard Space Flight Centre for remote sensing of ocean colour from a U-2 aircraft at altitudes of 18 to 20 km.[194, 248] This has 10 spectral bands, ranging from 433 to 772 nm, with bandwidths of about 20 nm. It scans by means of a rotating mirror at 45°

and uses a telescope with a field of view of ~ 0.2°, corresponding, at the U-2 operating altitude of 19.8 km, to a unit element of area at sea level of about 75 m square. The swath covered is 25 km wide.

The M2S Multispectral Scanner is designed for use at intermediate altitude. It has 10 spectral bands covering the range 380 to 1060 nm, bandwidths being about 40 nm. From an altitude of 3 km, the swath width at the surface is 8.5 km and the unit element of area viewed is 8 m square.[225]

### Satellite systems

Three Landsat satellites[428] have been placed in orbit so far. All contain the MSS Multispectral Scanner with spectral bands 4 (500–600 nm), 5 (600–700 nm), 6 (700–800 nm) and 7 (800–1100 nm). The Landsat optical system was designed primarily for remote sensing of terrestrial

Fig. 7.2. The Nimbus-7 satellite with its various sensors (NASA-PHOTO).

regions. Its sensitivity is inadequate to detect the subtle variations in reflectance that occur in the aqueous parts of the globe, and in particular it has no spectral band in the crucially important blue (400–500 nm) region of the spectrum. A mirror oscillating through an angle of 2.9° causes the direction of view to scan from side to side. From its altitude of 920 km it covers a swath 185 km wide, and the unit element of area viewed is about 80 m square.

In October 1978 the Nimbus-7 satellite (Fig. 7.2) was launched, carrying the Coastal Zone Colour Scanner (CZCS) which was designed specifically for remote sensing of the aquatic environment: its optical system is shown in Fig. 7.3. The CZCS has four bands in the visible region, each 20 nm wide, centred on 443, 520, 550 and 670 nm, one band in the near infrared (700–800 nm) and another band in the infrared (10.5–12.5 $\mu$m) for temperature sensing. To facilitate avoidance of Sun glitter the sensor can be tilted so that it scans at up to 20° from the nadir, ahead of or behind the spacecraft. It has a mirror at 45° rotating at 8.1 revolutions per second, at which is directed a telescope which gives an instantaneous field of view of 865 $\mu$radians (0.05°) corresponding, at the satellite altitude of 955 km, to a unit element of area 825 m square at the surface.[193, 194] It covers a continuous swath 1636 km wide, which for mapping purposes is broken down into individual sections which might typically be 700 km along the spacecraft's track. The radiometric sensitivity of the CZCS is 60-fold

Fig. 7.3. Optical system of the Coastal Zone Colour Scanner in the Nimbus-7 satellite (by permission, from Hovis, 1978).

CZCS Optical Arrangement

higher than that of the MSS on Landsat. A simple threshold, using the high reflectance of clouds and land in the 700–800 nm waveband, is used to distinguish these area from open water: it is only to these latter picture elements that the atmospheric correction procedure described later is applied. The scanning geometry of the CZCS is shown in Fig. 7.4.

Although the CZCS represents a major advance in the technology of remote sensing of the aquatic environment, the unit element of area it resolves (825 m square) is rather large for studies on inland water bodies or smaller coastal features such as estuaries and small bays. Design studies are being carried out for possible future spaceborne ocean scanners with a spatial resolution of the order of 100 m.[248]

The latest Landsat satellite (Landsat D, launched 1982, altitude 705 km) carries, as well as an MSS scanner, a new scanner known as the Thematic Mapper, which has a band in the blue (450–520 nm) as well as bands in the green (520–600 nm), red (630–690 nm) and near-infrared (760–900 nm) wavelengths plus three infrared bands. The unit element of area viewed is 30 m square. Although primarily designed for prediction of crop production, the Thematic Mapper may prove also to be of value for remote sensing of the aquatic environment.

Fig. 7.4. Scanning geometry of the Coastal Zone Colour Scanner in the Nimbus-7 satellite (adapted from Hovis, 1978).

## 7.2     The emergent flux

The particular light flux which is of the greatest interest in the present context is the upwelling light flux just below the surface. However, the flux which, after due correction, is remotely sensed is the emergent flux – that part of the upwelling flux which succeeds in passing up through the surface. How are the two related?

Although about half the total upwelling light flux is reflected downwards again at the water/air interface, this does not represent a serious loss, since it is mainly the flux at larger angles which undergoes reflection. Remote sensing normally involves measurement of radiance at a specific angle rather than total upward irradiance. In the case of satellites this angle is fairly close to the nadir, and even in the case of aircraft is not likely to be more than 53°. Thus remote sensing is normally concerned with that upwelling flux which, below the surface, is at angles of 0–37° to the vertical, and under calm conditions only 2–4% of this is reflected at the water/air boundary. Reflection in this angular range is somewhat increased under rough conditions: at the smaller angles losses remain insignificant but can rise to 15–20% for angles at the upper end of this range.[16]

As the light passes through the water/air boundary it undergoes refraction which, in accordance with Snell's Law (§2.5) increases its angle to the vertical. A further consequence of refraction is that the flux contained within a small solid angle, $d\omega$, below the surface spreads out to a larger solid angle, $n^2 d\omega$ (where $n$ is the refractive index), above the surface. Because of this effect, the value of emergent radiance at any given angle is about 55% of the corresponding subsurface radiance from which it is derived. Combining this effect with the much smaller effect of internal reflection, Austin (1980) proposes a factor of 0.544 for relating radiance just above the surface, $L_w(\theta', \phi)$, to the corresponding radiance just below the surface, $L_u(\theta, \phi)$

$$L_w(\theta', \phi) = 0.544\, L_u(\theta, \phi) \tag{7.1}$$

$\theta$ being the nadir angle within the water and $\theta'$ the angle in air after refraction at the surface.

If $L_w(\theta', \phi)$ can be determined from remote sensing measurements, then it can be multiplied by 1.84 to give $L_u(\theta, \phi)$, and thus provide information about the underwater light field. Upward radiance within the water at nadir angles up to about 30° varies only slightly with azimuth angle (see §6.6). Thus, whatever the azimuth angle of observation, provided the nadir angle of observation is not more than about 42°, as will usually be the case, the estimated subsurface radiance value, $L_u(\theta, \phi)$ can be taken as the value at that subsurface nadir angle averaged over all azimuth angles, $L_u(\theta)$.

Furthermore, $L_u(\theta)$ does not change much with nadir angle in the range 0 to 30°, and so the value of $L_u(\theta)$ can be taken as an estimate of the radiance, $L_u$, in the vertically upward direction.

### 7.3     Correction for atmospheric scattering and solar elevation

We have already noted that about 80% of the upward radiance measured by a photometer in a satellite, or in a very high altitude ($\sim$ 20 km) aircraft, above the ocean originates by scattering of the solar beam by air molecules and by aerosol particles (dust, water droplets, salt, etc.) within the atmosphere – this is known as the *path radiance*. To arrive at a value of $L_w$ from the measured radiance, the path radiance must be removed, and furthermore the attenuation of the emergent flux during its passage through the turbid atmosphere to the sensor must be allowed for.

As discussed in a later section, it is the ratio of the emergent flux to the incident downward flux, rather than the absolute value of emergent flux alone, which is directly related to the optical properties of the aquatic medium. It is therefore desirable to have an estimate of the downward irradiance at the surface of the water, and this will be determined largely by solar elevation. In addition the contribution of atmospheric scattering to the measured upward radiance is a function of incident irradiance at the top of the atmosphere, which is also a function of solar elevation. When using satellite/high-altitude radiance data to provide information about aquatic systems, this dual effect of solar elevation must be taken into account.

#### *Landsat*

In the case of Landsat studies on aquatic systems, no generally applicable method of correcting for these atmospheric effects appears to have been devised. The use of Landsat for remote sensing of water composition has hitherto involved correlating increases in radiance in particular wavebands with increases in concentration of particular types of suspended particles, the increased radiance being due to increased backscattering. In some cases correlations between the uncorrected radiances and concentration of a particular component have been used. The problem in such a case is that on another occasion, when atmospheric conditions have changed, the previously derived relation will no longer apply. In studies on Lake Superior, Canada, for example, the curve of measured radiance against turbidity shifted up and down and changed somewhat in slope on different days, in accordance with changes in atmospheric conditions.[416]

If some simultaneous sea/lake measurements of the concentration of particles are available, then from the measured values of radiance at the corresponding points in the satellite image it may be possible to determine which particular calibration curve to use (if a set of such curves, for differing atmospheric conditions, has previously been obtained). Alternatively, the simultaneous sea/lake measurements may provide enough data for a calibration curve. Once such a curve is established, then, subject to the provision that the atmospheric conditions vary little over the area of interest, a complete map of the water parameter in question can be derived from the satellite data.

The above approach, even when a simultaneously determined calibration curve is available, still makes use of the total uncorrected radiance values. It seems reasonable to suppose that the greater the extent to which extraneous contributions (other than the true emergent flux) can be removed, the better the correlation between radiance and water composition should be. Munday & Alföldi (1979) in their study on the Bay of Fundy used simultaneous sea data on suspended solids to extrapolate radiance to zero suspended solids. They took the radiance value obtained in this way as an estimate of the atmospheric plus surface-reflected contribution, and subtracted it from the measured radiance values to give residual radiance values which were a better guide to the concentration of suspended solids.

In the absence of simultaneous sea/lake data then it may be possible to identify a part of the image where the suspended particle concentration is low. The radiance from this location can be used as a baseline value which will include both the radiance contributed by scattering within the atmosphere and the background radiance due to backscattering within water with a small particulate component. Subtraction of this from the observed radiance values in other parts of the image gives residual radiance values which can be used to provide information about suspended particles. The essential problem is how to identify a part of the image where particle concentration is low. Sydor (1980) collected data for the radiance obtained in three Landsat wavebands under clear atmospheric conditions at different times of the year, from clear-water areas of Lake Superior. Any part of a subsequent image of the lake which had a radiance value close to that previously obtained for clear water at that time of the year could then be assumed to have less than a certain level of suspended solids, and so could be used as the baseline value to correct radiance values elsewhere in the image. This procedure, however, only works for clear atmospheric conditions. No way of estimating atmospheric turbidity from Landsat data alone appears yet to have been devised.

The effects of solar altitude on the remotely sensed radiance – through its influences on surface irradiance and path radiance – vary approximately in proportion to the cosine of the zenith angle of the Sun. A first-order correction for the effects of solar elevation may therefore be made by dividing the measured radiance by the cosine of the solar zenith angle.[1, 274, 337] By, in effect, normalizing radiance values to a standard downward irradiance, a parameter proportional to reflectance is obtained and this, as discussed elsewhere, is preferable for relating to the optical properties of the medium.

### Nimbus-7

Remote sensing of water composition with Landsat involves relating *increased* radiance values to the concentration of different kinds of suspended particles, and useful results can be obtained by subtracting comparatively crude estimates of path radiance from the observed radiance values. The Coastal Zone Colour Scanner (CZCS) carried by the Nimbus-7 satellite, in contrast, is required to detect decreases in radiance in some wavebands as well as increases in others, and a much better estimate of the atmospheric scattering contribution to the measured radiance is needed. A theoretical basis for carrying this out, taking advantage of the fact that radiance in the red/near infrared (strongly absorbed by water) is nearly all due to atmospheric scattering,[86] was developed by Gordon (1978).

A procedure for removing the atmospheric scattering contribution to the CZCS measurements, described by Gordon, Clark, Mueller & Hovis (1980), and in a modified form by Smith & Wilson (1980), takes as its starting point the proposition that the total radiance, $L^\lambda$, observed by the sensor at wavelength $\lambda$, is the sum of the radiance due to Rayleigh scattering ($L_R^\lambda$), that due to aerosol scattering ($L_A^\lambda$), and that backscattered out of the ocean ($L_W^\lambda$) attenuated by the atmospheric transmittance factor $t^\lambda$ during its passage from the surface to the satellite

$$L^\lambda = L_R^\lambda + L_A^\lambda + t^\lambda L_W^\lambda \qquad (7.2)$$

$L_R^\lambda$ includes the Rayleigh component of the skylight reflected towards the sensor by the surface, as well as that backscattered from the direct solar beam. $L_A^\lambda$, similarly, includes the aerosol component of the reflected skylight as well as that light scattered back by aerosol particles towards the sensor from the solar beam.

The correction procedure goes on to take advantage of the fact that water absorbs light strongly in the red/near infrared region, and so essentially all the radiance measured at this end of the spectrum can be attributed to scattering within the atmosphere. For oceanic waters, which are usually very clear, the 670 nm band may be used. In more turbid waters, however, such as those of inland water bodies or near the coast, the concentration of particulate matter can be high enough to backscatter significant amounts of 670 nm light out of the water[68] and in these cases a more strongly absorbed, near infrared waveband would be preferable.

For the 670 nm radiance, since $L_W^{670} \simeq 0$, eqn 7.2 becomes

$$L^{670} = L_R^{670} + L_A^{670} \qquad (7.3)$$

Since the contribution to scattering by air molecules is essentially constant, the radiance at the sensor's viewing angle due to Rayleigh scattering is a known function of solar elevation and wavelength, and so $L_R^{670}$ and $L_R^\lambda$, can be calculated and are the same for all pixels. The value of $L_A^{670}$ is then obtained from the measured $L^{670}$, using eqn 7.3.

The approximate assumption is now made that the radiance due to aerosol scattering at wavelength $\lambda$, while not identical to that at 670 nm, is at least proportional to it, i.e. as $L_A^{670}$ varies with atmospheric turbidity from one part of the scene to another, so $L_A^\lambda$ will vary in parallel with it. The ratio, $L_A^\lambda / L_A^{670}$, we shall refer to as $\varepsilon(\lambda, 670)$. Substituting for $L_A^\lambda$ in eqn 7.2, and rearranging, we obtain

$$L_W^\lambda = (1/t^\lambda)[L^\lambda - L_R^\lambda - \varepsilon(\lambda, 670)\, L_A^{670}] \qquad (7.4)$$

an equation for the emergent radiance at wavelength $\lambda$, with the remaining unknowns, $t^\lambda$ and $\varepsilon(\lambda, 670)$.

These are best determined by carrying out sea measurements at one or a few locations simultaneously with the satellite overpass. Atmospheric transmittance at wavelength $\lambda$ can be determined with a suitable Sun photometer, and $L_W^\lambda$ can be measured with a radiance meter. Use of these values in eqn 7.4 in conjunction with the satellite radiance data and the calculated Rayleigh radiance will enable the value of $\varepsilon(\lambda, 670)$ to be determined. This factor, in conjunction with the value of $L_A^{670}$ determined for each pixel, enables the variation in aerosol-generated radiance at wavelength $\lambda$, from place to place, to be taken into account.

The variation in atmospheric transmittance throughout the area covered by the satellite image still has to be allowed for. The diffuse transmittance of the atmosphere for wavelength $\lambda$, is given by

$$t^\lambda = e^{-\tau_{tot}^\lambda/\mu}$$

where $\tau_{tot}^\lambda$ is the total optical thickness of the atmosphere at wavelength $\lambda$, and $\mu$ is the cosine of the nadir viewing angle of the CZCS. The total optical thickness is given by

$$\tau_{tot}^\lambda = \tau_R^\lambda + \tau_{Oz}^\lambda + \tau_A^\lambda$$

where $\tau_R^\lambda$, $\tau_{Oz}^\lambda$ and $\tau_A^\lambda$ are the optical thicknesses due to Rayleigh scattering, ozone absorption, and aerosol scattering, respectively. The first two of these can be calculated and are the same for all pixels. Thus to determine the variation in $t^\lambda$ from place to place we must determine the variation in $\tau_A^\lambda$.

The radiance generated by aerosol scattering at any wavelength is approximately proportional to the aerosol optical thickness at that wavelength.[154] If sea measurements with a Sun photometer provide a value of $\tau_A^{670}$ for a particular pixel, then from the corresponding satellite-derived value for $L_A^{670}$, the constant of proportionality relating $\tau_A^{670}$ to $L_A^{670}$ can be determined. Using this, the value of $\tau_A^{670}$ can be determined for all the pixels.

The aerosol optical thickness at wavelength $\lambda$ is related to that at 670 nm by

$$\frac{\tau^{670}}{\tau^\lambda} = \left(\frac{670}{\lambda}\right)^{-\alpha}$$

where $\alpha$ is known as the Angstrom exponent. The value of $\alpha$ is determined by the kind of aerosol particles present. It does not vary markedly for oceanic atmospheres[524] and could reasonably be assumed to be constant over a particular oceanic scene viewed by the satellite. Sea measurements at one location with a Sun photometer at two wavelengths will provide a value for $\alpha$, by means of which the value of $\tau_A^\lambda$ for each pixel can be calculated from the value of $\tau_A^{670}$. Thus, $t^\lambda$ for each pixel can now be determined. In the absence of any sea level measurements of atmospheric transmittance, the measured radiance in the red/near infrared alone, since it is composed of a known constant Rayleigh component and a variable aerosol component, can be used to provide a value for the aerosol optical thickness in this waveband using empirical and/or calculated relations.[248] The Angstrom exponent, $\alpha$, can be assigned a suitable value typical of a marine atmosphere, to permit calculation of aerosol optical thickness at $\lambda$ from the long-wavelength value.

If sea values of $L_W^\lambda$, as well as atmospheric transmittance measurements, are not available, then an alternative procedure is to identify from the satellite data a part of the image where the concentration of phytoplankton and other components is very small and then assign to this pixel a value of $L_W^\lambda$ typical for a clear oceanic water at the existing solar elevation. This approach was found to work reasonably well for the Gulf of Mexico and other regions in the Atlantic Ocean.[158, 159]

With or without the assistance of sea data, all the factors on the right-hand side of eqn 7.4 can be determined, and the value of $L_W^\lambda$, the emergent radiance at wavelength $\lambda$, can be obtained for every pixel in the satellite image. If desired this can be converted to the subsurface upward radiance, $L_u^\lambda$, by dividing by 0.544 (§7.2).

Up to now, use of the CZCS data for mapping phytoplankton distribution has made use of the ratio of $L_W$ values at pairs of wavelengths (see §7.4), rather than radiance reflectance values. If it is wished to determine the radiance reflectance (upward radiance divided by downward irradiance) at or below the surface, this can be calculated for the appropriate solar elevation, using the atmospheric transmission information derived from the satellite's own measurements, together with reasonable assumptions about Rayleigh scattering.[524]

### High-altitude aircraft

Some remote sensing studies on water composition carried out from high-altitude aircraft have not attempted to correct for the contribution of atmospheric scattering to the measured radiance. They have relied instead on multiregression analysis to detect the correlation between total radiances in various wavebands, and the changing water composition parameters.[223, 224]

Kim, McClain & Hart (1979) used a correction procedure analogous to that described above for the Nimbus-7 measurements, to correct their data for Californian coastal waters obtained with the Ocean Colour Scanner from a U-2 aircraft at 19.8 km altitude. They assumed that all the radiance measured at 778 nm was due to atmospheric scattering. From the 778 nm radiance, using a mathematical model, they calculated, for all the other wavebands used by their scanner, the expected contribution of atmospheric scattering to observed radiance, and these values were subtracted from the observed values.

### 7.4    Remote sensing of optical properties – theoretical approach

While values for upward radiance are of some interest in themselves, their main significance lies in what they can tell us about the optical properties of the medium, and thus about the concentrations of the optically significant constituents in the water. A possible route by which such information may be acquired makes use of the approximate relations, derived from theoretical modelling of the underwater light field (§6.4, eqns 6.3 and 6.4), that exist between radiance reflectance ($L_u/E_d$), or irradiance reflectance ($E_u/E_d$) and the ratio of backscattering coefficient to absorption coefficient.

Whichever relation is used, it is necessary to have a value for the downward irradiance, $E_d$, just below the surface. Knowing the solar elevation, making reasonable assumptions about Rayleigh scattering, and assigning a plausible value to aerosol scattering (based, if possible on the remote sensor's own measurements as described previously) the atmospheric transmittance may be calculated[524] and so the downward irradiance on the surface determined. From this, making due allowance for reflection at the surface, a value for $E_d$ could be obtained. In the case of remote sensing from low-flying aircraft, downward irradiance can be measured on the aircraft and assumed to be equal to that just above the water surface.

$L_u$ is obtained by multiplying $L_w$ by 1.84 (see §7.2). $E_u$ is obtained by multiplying $L_u$ by 5, or some slightly different factor depending on the solar elevation (see §6.4). If, as could be the case in low-altitude studies, the remote sensor measures irradiance, the value of $E_u$ may be obtained by dividing the observed value of upward irradiance above the surface by 0.52 to allow for the approximately 48% reflection of diffuse upwelling irradiance at the water/air boundary.[16]

Using the appropriate equation relating $L_u/E_d$ or $E_u/E_d$ to $b_b/a$, the ratio of the backscattering coefficient to the absorption coefficient could be estimated. If, on the basis of other information, a plausible value could be selected for $b_b$, then an approximate value for $a$ could be arrived at. Alternatively, if measurements are made at three wavelengths, and reasonable assumptions are made about the manner in which $b_b$ and the values of $a$ for the different absorbing components (water, phytoplankton, non-living colour), vary with wavelength, then the actual concentrations of phytoplankton and non-living pigmented material can be estimated: the theoretical basis for such calculations is given by Morel (1980). Ways of calculating phytoplankton concentration from above-surface reflectance measurements have also been described by Pelevin (1978) and Kopelevich & Mezhericher (1979).

Methods of remotely sensing phytoplankton concentration are commonly based on the fact that the photosynthetic pigments absorb strongly in the blue region but only weakly in the green. Viollier, Deschamps & Lecomte (1978) have used eqn 6.3, together with the assumption that the particle scattering coefficient ($b_p$) varies inversely with wavelength, to calculate a nomogram from which, given values of ($R_{466} - R_{525}$) and ($R_{550} - R_{600}$), approximate values for phytoplankton chlorophyll and $b_p$ may be read out. $R_\lambda$ is the remotely sensed subsurface irradiance reflectance at wavelength $\lambda$. This is only applicable to waters (such as some oceanic waters) in which the phytoplankton is the only significant light-absorbing component, other than water itself. Viollier, Tanré & Deschamps (1980) have proposed a further simplification of this treatment for any water in which the particle scattering coefficient could be considered to covary linearly with phytoplankton concentration. They present a calculated curve of ($R_{525} - R_{466}$) against the logarithm of phytoplankton chlorophyll concentration which seems to agree quite well with an experimentally measured relation between these quantities determined at locations in the North Sea and the Atlantic Ocean.

Jain & Miller (1976) used a simplified model of the underwater light field to develop a computer calculation method, involving an iterative optimization procedure, by means of which, given a set of reflectance values at a series of wavelengths, the phytoplankton chlorophyll concentration and the scattering coefficient can be determined. The method seems to be designed for waters in which phytoplankton are the major light absorbers apart from water itself. The values of chlorophyll concentration obtained in this way from low-altitude aircraft measurements in certain Canadian lakes correlated quite well with values determined by analysis of water samples.[201, 316]

## 7.5    Remote sensing of optical properties – empirical approach

Most remote sensing studies of water composition have made use, not of the *a priori* deductive method outlined above, but of an empirical approach in which relations have been sought between remotely measured radiances in particular wavebands, or functions of one or more radiances, and the concentrations of specific components such as phytoplankton and suspended solids determined by *in situ* measurements within water bodies. Once such relations are established it should in principle be possible to use them to determine the optical properties and composition of water bodies from remote sensing data alone.

### Suspended solids

One of the water quality parameters which has been most extensively studied in this way is the concentration of total suspended solids (TSS). Increases in the amount of suspended particles will, at wavelengths where the particles do not absorb strongly, increase the backscattering coefficient of the water more than the absorption coefficient, and so, in accordance with eqns 6.3 and 6.4, increase the emergent flux.

In the case of reservoirs in Mississippi, it was found that the irradiance reflectance measured above the water surface increased with the concentration of suspended (inorganic) sediment at all wavelengths from 450 to 900 nm (Fig. 7.5).[401] The best, linear, correlation between reflectance and suspended solids concentration was found between 700 and 800 nm. Correlations between suspended solids or nephelometric turbidity (proportional to suspended solids) and radiances in particular wavebands have been obtained with the Landsat satellites. Klemas, Borchardt & Treasure (1973) found that in Delaware Bay the distribution of MSS 600–700 nm radiance corresponded best with sediment load in the upper 1 m of the water; the 500–600 nm radiance was more subject to interference from atmospheric haze while the 700–800 nm waveband did not penetrate sufficiently into the water column. TSS corresponded best with an exponential, rather than linear, function of radiance in this system.[265] Suspended solids values for the Bay of Fundy, Nova Scotia, Canada, correlated with particular functions of upward radiance in the 500–600, 600–700 and 700–800 nm bands, but the latter two bands gave the highest correlation coefficients.[337] In the case of Lake Ontario, Canada, a good linear relation was observed between radiance reflectance in the 600–700 nm band and measured turbidity.[67] For San Francisco Bay, Khorram (1981*b*) developed regression equations relating suspended solids to a function of the radiance values in Landsat bands 5, 6 and 7, and relating turbidity to

Fig. 7.5. Intensity and spectral distribution of upwelling radiation measured 20–50 cm above the surface of fresh-water bodies containing various levels of suspended solids (after Ritchie, Schiebe & McHenry, 1976). The concentration of suspended solids in mg l$^{-1}$ is indicated next to each curve.

a function of the radiance values in bands 4, 5 and 6. Correlation was better in the case of turbidity ($r = 0.907$) than in the case of suspended solids ($r = 0.674$).

If it is assumed that $R$ is proportional, not simply to $b_b/a$, but to $b_b/(a+b_b)$ (see §6.4), then a linear relation between reflectance and suspended solids ($b_b$ being proportional to TSS) is not to be expected. Munday & Alföldi (1979) point out that in this case the reciprocal of the radiance from within the water should be linearly related to $1/TSS$ but with a non-zero intercept on the $1/L_W$ axis. Their data (Landsat 600–700 nm and 700–800 nm bands) for the Bay of Fundy appear to conform better to a relation of this type than to one based on proportionality of $R$ to $b_b/a$, which would require the graphs to pass through the origin.

It is important to note that since reflectance varies inversely with $a$, at the same time as it increases with $b_b$, a reproducible linear relation between emergent radiance and TSS, or between $1/L_W$ and $1/TSS$, is only to be expected if the total absorption coefficient of the aquatic medium remains constant while the concentration of suspended solids varies. This need not be the case, and variations in absorption coefficient from place to place or over a period of time, within the water body under study, would affect the reproducibility of any empirically derived relation between $L_W$ and TSS, and so affect the accuracy with which the concentration of suspended solids can be determined from remotely sensed data. In the red and near infrared wavebands, however, absorption is mainly due to water itself, and so fluctuations in absorption should not be large.

Johnson (1978, 1980) has studied the feasibility of mapping the concentration of suspended solids from an aircraft using spectral scanners operating with 10 visible/infrared wavebands. A multiple regression analysis procedure is applied to the data in which correlations, either positive or negative, are sought between the water property of interest, e.g. TSS, and the radiance values in any of the bands. Regression equations relating suspended solids concentration to radiances in one or more bands have been developed, which have correlation coefficients of 0.90 to 0.96.

The spectral distribution of light reflected by suspended inanimate particulate matter varies in accordance with its own absorption properties, which are determined by its chemical nature. Sydor (1980), on the basis of his measurements of the spectral distribution of light scattered upwards from suspensions of particular types (red clay, mine tailings) has concluded that evidence on the identity as well as concentration of the suspended solids may be derived using the 500–600, 600–700 and 700–800 nm bands of Landsat.

*Phytoplankton – increased reflectance*

One empirical approach to the remote sensing of phytoplankton is to treat it as a special case of suspended solids and to use the increased reflectance in the near infrared associated with the increased biomass. In the Landsat MSS red (600–700 nm) band, high concentrations of algae are associated with decreased radiance because of the chlorophyll absorption in this region. Strong (1974) observed in the case of Landsat pictures of an algal bloom on Utah Lake, USA, that there was a contrast reversal in going from the red to the infrared (700–800 nm or 800–1100 nm) bands: areas dark in the red image showed up as light against a darker background in the infrared images. Bukata, Harris & Bruton (1974) found that the reflected radiance in the Landsat 700–800 nm band correlated well with the logarithm of the chlorophyll concentration in Lake Ontario, Canada, and some nearby eutrophic water bodies. The sensitivity of this method is, however, comparatively low ($> 10$ mg chlorophyll $a$ m$^{-3}$) and it is perhaps most applicable to highly productive inland waters which develop algal blooms. Suspended inorganic particles can also increase reflectance in the 700–800 nm band. A comparison of the distribution of 600–700 nm waveband reflectance (which correlates better with inorganic turbidity than with biomass) with that of 700–800 nm reflectance is necessary to ensure that suspended sediments are not mistaken for phytoplankton.[66]

*Phytoplankton – increased absorption*

The best approach to the remote sensing of phytoplankton is undoubtedly that which makes use of the increased absorption, and therefore reduced radiance reflectance, at the blue end of the spectrum caused by the photosynthetic pigments. Clarke, Ewing & Lorenzen (1970) measured from a low-flying aircraft, the spectral distribution of the emergent flux in different regions of the Northwest Atlantic Ocean with chlorophyll $a$ concentrations varying from $< 0.1$ to 3.0 mg m$^{-3}$. The general tendency (Fig. 7.6$b$) was that as the phytoplankton concentration increased, reflectance decreased in the blue (400–515 nm) and increased in the green (515–600 nm). The increased reflectance in the green may be attributed to the fact that phytoplankton, being refractive particles, increase scattering at all wavelengths, but in this spectral region absorb only weakly. Empirical methods of remotely estimating phytoplankton concentration generally take advantage of both the decreased reflectance in the blue and the increased reflectance in the green, by working in terms of the ratios or differences between reflectances in these two wavebands.

Morel & Prieur (1977) on the basis of their measurements of the spectral distribution of the upwelling and downwelling light flux in various oceanic waters found that the ratio of underwater irradiance reflectance (related in the manner discussed above to the reflectance estimated by remote sensing above the water) at 440 nm to that at 560 nm decreases as the phytoplankton pigment content increases. If data for waters with high scattering coefficients due to suspended sediments are omitted then an approximately linear relation is obtained between the logarithm of $R_{440}/R_{560}$ and the logarithm of the chlorophyll plus phaeopigment (chlorophyll breakdown products) concentration over the range 0.02 to 20 mg total pigment m$^{-3}$.[329]

Gordon & Clark (1980) have used the effect of phytoplankton on the

Fig. 7.6. Spectral distribution of reflectance above the Atlantic ocean off New England, USA (after Clarke *et al.*, 1970). Measurements were made at a zenith angle of 53°, using a polarizer to eliminate reflected skylight. (*a*) Dependence of reflectance on altitude: measurements were made at five altitudes over the same location (1° latitude east of Cape Cod). The large, and increasing, contribution of atmospheric scattering to measured reflectance at altitudes above 305 m is readily apparent. (*b*) Dependence of reflectance on phytoplankton content. Measurements were made at low altitude (305 m). The phytoplankton chlorophyll level (mg chl *a* m$^{-3}$) is indicated above each curve. The three curves are for water on the northern edge of the Gulf Stream (0.3 mg chl *a* m$^{-3}$), on the edge of the continental shelf (0.6 mg chl *a* m$^{-3}$), and over the productive Georges Shoals (3.0 mg chl *a* m$^{-3}$).

spectral distribution of the upwelling flux to develop an algorithm by means of which $L_W$ values obtained from the Coastal Zone Colour Scanner in the Nimbus-7 satellite can be used to calculate phytoplankton pigment concentration. They measured the upward nadir radiance within the water (which as we saw earlier is directly related to the emergent flux) at 440, 520 and 550 nm in marine waters with chlorophyll $a$ (plus phaeopigment) concentrations ranging from 0.07 to 78 mg m$^{-3}$. The data conformed reasonably well to relations of the type

$$\log_{10} C = \log_{10} a + b \log_{10} R_i \qquad (7.5)$$

where $C$ is the chlorophyll $a$ (plus phaeopigment) concentration in mg m$^{-3}$, and $R_i$ is the 440/550, 440/520 or 520/550 ratio of radiances. In the case of the 440/550 nm pair, for example, $\log_{10} a$ is $-0.297$ and $b$ is $-1.269$, i.e. as phytoplankton concentration increases so the ratio of upwelling radiance at 440 nm to that at 550 nm decreases. By substitution of a pair of values of upwelling radiance into the appropriate form of eqn 7.5 the chlorophyll (plus phaeopigment) concentration can be calculated. For chlorophyll concentrations below 0.6 mg m$^{-3}$ the radiances at 443 and 550 nm are most useful. At higher concentrations, however, $L_W^{443}$ becomes too small for accurate determination by the CZCS, and the radiances at 520 and 550 nm are used instead. Due to unavoidable uncertainties in the atmospheric correction it seems that this method should work reasonably well for pigment concentrations in the range 0.05 to 10.0 mg m$^{-3}$, but not above.[157] Approximate agreement was found between the variation of chlorophyll concentration determined by surface sampling along a ship course of several hundred km in the Gulf of Mexico with that calculated for the same course from CZCS data.[159]

Johnson (1978) used multiple regression analysis to relate radiances in various visible wavebands obtained with airborne M2S and OCS Multi-spectral Scanners to chlorophyll $a$ concentrations at an estuarine and a coastal site on the American Atlantic seaboard. For the estuarine site, a linear regression equation relating chlorophyll negatively to the 440–490 nm radiance, and positively to the 620–660 nm and 700–740 nm radiances, was developed. The correlation coefficient was 0.96 over the range 1.6 to 19.5 mg chlorophyll m$^{-3}$. For the coastal site, the linear regression equation related chlorophyll $a$ negatively to the 499–519 nm radiance and positively to the 610–630 nm radiance, with a correlation coefficient of 0.83. The difference between the two equations is in part due to the fact that the coastal site was overflown at a much higher altitude (19.7 km) than the estuarine site (2.4 km), and no correction for atmospheric effects was made.

Khorram (1981*a*), using the OCS in a U-2 at 19.8 km over San Francisco Bay, found a regression equation relating phytoplankton chlorophyll inversely to the 655–679 nm radiance (this presumably being reduced due to absorption by the chlorophyll absorption peak in the red) and directly to the 766–790 nm radiance (reflectance in the near infrared being increased by phytoplankton). No atmospheric correction was made. The correlation coefficient was 0.934.

Kim *et al.* (1979) used the OCS Multispectral Scanner operating from a high-altitude aircraft (a U-2 at 19.8 km) off the California coast. The observed radiances were corrected for the atmospheric contribution. It was found that various combinations of radiances could yield information about the chlorophyll distribution: for example, the ratio of radiance at 472 nm to that at 548 nm was negatively correlated with chlorophyll concentration.

Borstad, Brown & Gower (1980) eliminated atmospheric effects by flying at a height of 100 m and minimized sky reflection by taking measurements of radiance at the Brewster angle, with a polarizing filter over the aperture of their scanning spectrometer. For the relatively eutrophic British Columbian coastal waters they obtained an approximately linear relation between the ratio of radiance reflectance at 560 nm to that at 440 nm and the chlorophyll *a* concentration measured at 3 m depth, over the range 2 to 13 mg m$^{-3}$.

Viollier *et al.* (1980) carried out low-altitude airborne measurements of reflectance at 466 and 525 nm, together with measurements of chlorophyll, in the tropical East Atlantic Ocean and in the North Sea. They obtained an approximately linear relation between ($R_{525} - R_{466}$) and the logarithm of chlorophyll concentration over the range 0.2 to 3 mg m$^{-3}$.

The validity of any relation, such as those described above, between the ratio of radiances or radiance reflectances in particular pairs of wavebands and the phytoplankton pigment concentration is entirely dependent on there being no substantial change in the background optical properties of the medium. The relations are likely to be rendered useless, for example, by a significant increase in dissolved yellow substances whose absorption in the blue would be compounded with that due to the photosynthetic pigments, or in suspended sediments which would increase reflectance at all wavelengths. Remote sensing of phytoplankton distribution by these procedures should be most applicable in oceanic waters whose optical properties are not greatly influenced by river discharge or shallow sediments. Relations applicable to yellow and/or turbid waters have yet to be developed.

### Phytoplankton – fluorescence

The methods for remote sensing of phytoplankton described above make use of the fact that algal cells absorb light in a certain region of the spectrum. However, algal cells also emit light: about 1 % of the light a photosynthesizing cell absorbs is re-emitted as fluorescence, with a peak at about 685 nm. This fluorescence is too feeble to be detected in the downwelling light stream within the water but can show up as a distinct peak in the spectral distribution of the upwelling stream or in the curve of apparent reflectance against wavelength (see Fig. 6.7).[333, 516]

This peak can also be detected in the spectral distribution of the emergent flux. Neville & Gower (1977) showed that in the radiance reflectance (upward radiance/downward irradiance) spectra of productive British Columbian coastal waters obtained at the Brewster angle from a low-flying aircraft, a distinct peak at 685 nm was present, the height of which was proportional to chlorophyll concentration in the upper few metres of water. In a coastal inlet the height of the fluorescence peak above the baseline (measured above the water, but from a boat) was linearly correlated ($r^2 = 0.85$) with a weighted average (allowing for vertical distribution) chlorophyll $a$ concentration over the range 1 to 20 mg m$^{-3}$).[163] On the basis of the observed height of the peak it seems unlikely that the results would be of acceptable accuracy below about 1 mg chlorophyll $a$ m$^{-3}$.

Phytoplankton chlorophyll can be estimated using the fluorescence excited by light from an airborne laser rather than by sunlight, and the method has certain advantages. A key problem in the use of remotely sensed chlorophyll fluorescence as an indication of phytoplankton biomass is that the proportion of the fluorescent light which succeeds in passing up to the sensor, as well as the proportion of the exciting light (solar or laser) which succeeds in reaching the algae, depends on the optical properties of the water. Two different water bodies with the same phytoplankton content, but different attenuation properties, could give quite different fluorescence signals. To correct for the effects of attenuation by the aquatic medium, Hoge & Swift (1981) and Bristow, Nielsen, Bundy & Furtek (1981) have made use of the laser-induced Raman emission of water.

When water molecules scatter light, most of the scattered light undergoes no change in wavelength. A small proportion of the scattered photons, however, when they interact with the scattering molecule, lose or gain a small amount of energy corresponding to a vibrational or rotational energy transition within the molecule, and so after scattering are shifted in wavelength. These appear in the scattered light as emission bands at wavelengths other than that of the exciting light, and are referred to as Raman emission lines, after the Indian physicist who discovered this phenomenon. A particularly strong Raman emission in the case of liquid water arises from the O—H vibrational stretching mode: this shows up as an emission band roughly 100 nm on the long-wavelength side of the exciting wavelength (Fig. 7.7). Since the water content of the aquatic medium is essentially constant, the intensity of this Raman emission, when remotely sensed from above the water is, for a given exciting light source, determined entirely by the light-attenuating properties of the water, low Raman signals indicating high attenuation, and *vice versa*.

Thus to correct for variation in the optical character of the water, the measured chlorophyll fluorescence intensity at each station is divided by the corresponding water Raman emission intensity. Chlorophyll fluorescence values measured from the air, and normalized in this way, have been found to correlate closely with chlorophyll contents determined on water samples.[60]

Bristow *et al.* (1981) used a pumped dye laser, emitting at 470 nm, and exciting Raman emission at 560 nm, operated from a helicopter at 300 m above the water (Fig. 7.8*a*). The light returning from the water was collected by a telescope with a 30 cm diameter Fresnel lens; a beam splitter, interference filters, and separate photomultipliers (Fig. 7.8*b*) being used to

Fig. 7.7. Idealized emission spectrum of natural waters resulting from excitation with a laser at 470 nm (by permission from Bristow, Nielsen, Bundy & Furtek (1981), *Applied Optics*, **20**, 2889–906).

Fig. 7.8. Mode of operation of airborne laser system for detection of chlorophyll fluorescence and water Raman emission. (*a*) Schematic diagram of light fluxes. (*b*) Diagram of laser and optical receiver system (by permission, from Bristow, Nielsen, Bundy & Furtek (1981), *Applied Optics*, **20**, 2889–906).

separately detect the chlorophyll fluorescence and Raman emission. Hoge & Swift (1981) used a Nd:YAG laser emitting at 532 nm, and exciting a water Raman emission at 645 nm: the equipment was flown in a P-3A aircraft at 150 m above the water surface. It is worth noting that the Raman signal itself can be used for mapping the attenuation properties of the water: the reciprocal of the signal is proportional to an attenuation coefficient intermediate in value between $K_d$ and $c$.

### Recognition of water optical type

In marine waters away from the influence of land, the major factor determining the optical character of the medium, apart from water itself, is the content of phytoplankton and their associated degradation products. This means that the reflectance spectra of oceanic waters vary in a roughly systematic way. A family of curves,[19] of progressively changing shape, determined mainly by phytoplankton concentration, is observed. Thus, for any given oceanic water, specification of the ratio of radiances or radiance reflectances at any two wavelengths should, in effect, specify the whole radiance reflectance curve, and therefore the optical character of the water. Austin & Petzold (1981), using measurements of the spectral distribution of the upwelling flux obtained from several different regions of the ocean, found that the ratios of upwelling radiance for one pair of wavelengths correlated well with the ratios for another pair of wavelengths (e.g. $\log[L_u(443)/L_u(670)]$ plotted against $\log[L_u(443)/L_u(550)]$ was a straight line), indicating, in agreement with the above, that if one ratio is known, the shape of the spectral radiance curve is specified.

With a suitable spectral scanner, such as the CZCS on the Nimbus-7 satellite, the ratio of emergent radiant flux in various pairs of wavebands can be remotely sensed, and the variation in optical water type mapped, over large areas. A suitable quantitative parameter in terms of which to indicate the variation in water type must first be selected however. Concomitant with the systematic change in the spectral radiance curve, there is a corresponding progressive change in the curve of vertical attenuation coefficient for downwelling irradiance ($K_d$) as a function of wavelength; thus, the specification of the value of the upwelling radiance ratio at two wavelengths should also approximately specify $K_d$ at any given wavelength. In agreement with this supposition, Austin & Petzold found that for oceanic waters, the ratio of radiances at 443 and 550 nm, and the value of $K_d$ at 490 or 520 nm did indeed vary together (e.g. $\log K_d(490)$ plotted against $\log[L_u(443)/L_u(550)]$ conformed closely to a single curve), and they derived empirical relations by means of which $K_d(490)$ or $K_d(520)$ can be obtained from the ratio of radiances in the CZCS 443 and 550 nm bands. The calculated vertical attenuation coefficient at 490 or 520 nm can thus be used as the quantitative parameter in terms of which variation in oceanic water optical type can be mapped on the basis of satellite measurements. Agreement between the value of $K_d(490)$ estimated from CZCS radiance data and that determined by direct measurement in the water, was found to be good for two stations in the Gulf of Mexico.[21] It is not, however, to be expected that this method would work satisfactorily for waters, such as estuarine and many coastal waters, in which the optical properties are significantly influenced by components other than phytoplankton.

## 7.6    Specific applications of remote sensing to the aquatic environment

Since our concern in this book is with light and its availability for aquatic photosynthesis, we shall consider the remote sensing only of those qualities of the aquatic environment which relate to its optical properties and/or to photosynthetic biomass.

### Total suspended solids

Remote sensing, particularly using Landsat, has been quite commonly used for mapping the distribution of total suspended solids in inland and coastal waters. We shall examine just a few examples.

Klemas *et al.* (1974) used Landsat MSS 600–700 nm radiance values, together with their previously determined empirical relation to study the distribution of TSS in the Delaware River estuary and Delaware Bay. Concentrations varying, through six intervals, from ~ 5.6 ppm to > 211 ppm, were mapped. Turbidity in the system decreased with distance towards the sea: within the bay it was highest around the sides and lowest in the deeper central part (Fig. 7.9). Landsat 600–700 nm radiance values have also been used as a broad indication of the distribution of TSS in western Lake Ontario,[66] in Lake Superior,[416] and off the Georgia coast.[347]

Rouse & Coleman (1976) used Landsat 600–700 nm radiance values to study the distribution of turbidity from the Mississippi River in the Louisiana Bight. They mapped the band 5 radiance (for which they had a calibration curve relating it to TSS) at six different levels of intensity. They were able to show that the size and shape of the discharge plume were a function not only of the amount of fresh water discharged, but also of the speed and direction of the wind.

The limnology of the large impoundment Lake Kainji, Nigeria, is dominated by the alternation of the 'white flood' (sediment-laden water) from the Sokoto River and the 'black flood' (clear water) from the Niger

Fig. 7.9. Distribution of suspended solids concentration in Delaware
River estuary and Delaware Bay derived from Landsat 600–700 nm
radiance measurements (by permission, from Klemas, Bartlett,
Philpot, Rogers & Reed (1974), *Remote Sensing of Environment*, **3**,
153–74).

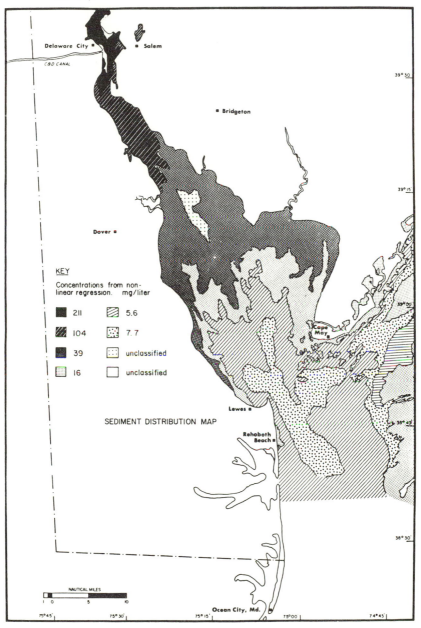

River, which occur at different times of the year. Abiodun & Adeniji (1978) used Landsat 600–700 nm radiance values to follow the movement and mixing of the two types of water in the 1972–74 period. Landsat radiance data, combined with empirical radiance/TSS relations have been used to map the distribution of suspended solids in the Bay of Fundy.[6] Radiance values in bands 4, 5 and 6 of Landsat have been used, in conjunction with appropriate empirical relations, to map the distribution of nephelometric turbidity in an Australian impoundment.[76]

Johnson (1980) used the 540–580 nm and 760–860 nm radiances, measured with the M2S Scanner in an aircraft at 3 km, together with an empirical TSS/radiance relation, to map the distribution of suspended solids resulting from the dumping of sewage sludge in the New York Bight on the Atlantic coast.

A special category of suspended solids is responsible for the phenomenon known as 'whitings'. This is produced by the precipitation of a fine cloud of calcium carbonate particles, possibly resulting from the uptake of $CO_2$ by photosynthesizing phytoplankton. Strong (1978) used Landsat to observe the distribution of 'whitings' in the Great Lakes. The 'whitings' are formed some metres below the surface. They showed up as a strongly increased radiance in the more highly penetrating green (500–600 nm) band, but were hardly detectable in the strongly absorbed red (600–700 nm) band.

### Phytoplankton – aircraft studies

Detection of phytoplankton from aircraft is generally based on the decreased reflectance in the blue, relative to that in the green. Studies carried out from low altitude have used photometers of the non-(spatially)scanning type so that distribution along the aircraft's linear track, rather than over an area, has been determined. Arvesen, Millard & Weaver (1973) used their differential radiometer, operating at 443 and 525 nm from a plane at 150 m altitude, to determine the distribution of phytoplankton within and outside San Francisco Bay and along the Pacific coastline north of San Francisco. Clarke & Ewing (1974) used the ratio of reflectance at 540 nm to that at 460 nm (which they call the colour ratio) measured with an airborne spectrometer, as a broad indication of the phytoplankton content, and hence productivity, of waters in the Caribbean region and Gulf of California.

Viollier *et al.* (1978), using the difference between irradiance reflectance in the green (525 nm) and that in the blue (466 nm) observed from 150 m altitude as an approximate indication of phytoplankton concentration, mapped the changing distribution of phytoplankton during the development of a coastal upwelling in the Gulf of Guinea (tropical East Atlantic). Borstad *et al.* (1980) used both the ratio of radiance reflectance at 560 nm to that at 440 nm, and the chlorophyll fluorescence emission at 685 nm, to map the distribution of phytoplankton along the coast of British Columbia from a plane at 100 m altitude. They concluded that airborne remote sensing is a useful and powerful technique for the detection and measurement of phytoplankton abundance, and can moreover, unlike satellite-based measurements, be used under cloudy conditions.

Moderate- to high-altitude aircraft studies have used spatially scanning photometers to observe phytoplankton distribution within swaths some kilometres in width. Johnson (1978) used the radiances in the 465, 640 and 720 nm bands of the M2S Scanner from an altitude of 2.4 km to map the phytoplankton concentration in the James River estuary, Virginia, USA. He used the 509 and 620 nm radiances measured by the Ocean Colour Scanner from an altitude of 19.7 km to map phytoplankton in the New York Bight: the data showed, for example, the higher concentration of algae in the plume formed by the Hudson River flowing into the Bight. Kim *et al.* (1979) also used the Ocean Colour Scanner, from an altitude of 19.8 km, to map the phytoplankton distribution in Monterey Bay, California, USA, and Khorram (1981*a*) used it from the same altitude to map the phytoplankton distribution in the San Francisco Bay Delta. Yoder *et al.* (1981) used the Ocean Colour Scanner aboard a U-2 to observe a large diatom patch within a Gulf Stream frontal eddy. Use of a scanner at moderate altitude, as in the James River study, provides better spatial resolution, and so is better suited for the study of aquatic systems with complex shapes, such as estuaries.

Hoge & Swift (1981) mapped the chlorophyll distribution in the German Bight (North Sea) from a low-altitude aircraft using the laser-excited chlorophyll fluorescence corrected for variation in water attenuation (which they also mapped) using the Raman emission band.

### *Phytoplankton – satellite studies*

The Landsat MSS optical system can, as we have noted previously, be used to detect high concentrations of phytoplankton by taking advantage of the increased reflectance in the near infrared (700–800 nm) band without a corresponding increase in the red (600–700 nm) band. Bukata & Bruton (1974) used the radiance in the 700–800 nm band as a broad indication of the distribution of phytoplankton in western Lake Ontario, Canada. Strong (1974) used Landsat radiances to map algal blooms in Utah Lake, USA, and in Lake Erie, and also subsequently in Lake Ontario.[474]

With the advent of the Coastal Zone Colour Scanner on the Nimbus-7 satellite it is now possible to remotely sense phytoplankton at the low concentrations that occur in the ocean. The studies carried out so far with this system make use of the decreased ratio of blue to green light in the emergent flux as the algal concentration increases. Hovis *et al.* (1980) and Gordon *et al.* (1980) have been able to characterize the distribution of phytoplankton throughout much of the Gulf of Mexico in this way. The data are presented as colour-coded maps covering many increments of concentration in the ranges $< 0.05$ to $> 1.0$ mg (chlorophyll + phaeo-pigments) m$^{-3}$ or 0.1 to $> 10.0$ mg m$^{-3}$, depending on whether the 443/550 nm or 520/550 nm radiance ratios are used. Smith & Baker (1982) have also used CZCS data to produce maps of phytoplankton distribution

Fig. 7.10. Phytoplankton distribution in the California Bight on 6th March 1979, derived from Coastal Zone Colour Scanner (Nimbus-7 satellite) data (by permission, from Smith & Baker (1982), *Marine Biology*, **66**, 269–79). The relation between chl *a* conc. (mg m$^{-3}$) and the density of the image is indicated on the scale at the top of the figure.

in the California Bight: one such map, for a day in March 1979, is shown in Fig. 7.10.

The information provided by the CZCS shows aspects of large-scale phytoplankton distribution previously quite unknown. The regions of transition from highly productive coastal waters to less productive oceanic waters are sharply delineated, and within the oceanic regions a complex array of whorls and eddies of varying productivity is revealed.

## 7.7    Further outlook

The development of remote sensing scanners such as the CZCS for mapping phytoplankton distribution is one of the most important developments in the aquatic sciences in recent years. It promises to provide detailed synoptic information on the productivity of the oceans on a scale that could never be envisaged previously. The waters on which the procedures developed so far work best are those oceanic waters in which the phytoplankton are, apart from the water itself, the major influence on the upwelling light flux. Extending the method to marine ecosystems, such as those bordering northern Europe, with much higher levels of dissolved yellow colour from river discharge, will be more difficult. The phytoplankton influence on upwelling flux in the blue region will be hard to separate from the background absorption due to gilvin. Other wavebands will have to be sought in which there is greater selectivity.

Background absorption by yellow substances will be an even greater problem in the remote sensing of phytoplankton in inland waters. Turbidity due to suspended sediments will also interfere.

Inland waters are typically more productive than marine waters and so there are higher concentrations of phytoplankton available for detection. A finer spatial resolution than the 825 × 825 m provided by the CZCS will, however, be required for inland waters.

Even when the phytoplankton signal cannot be separated from the background, remote sensing data can provide useful information about natural waters. Maps of turbidity distribution such as those generated already from Landsat data are of great interest, and it seems possible that with the new generation of scanners which can measure radiance in the blue region, the distribution of total yellow colour could also be mapped.

# PART II

PHOTOSYNTHESIS IN THE AQUATIC
ENVIRONMENT

# 8

.........................................................................................

# *The photosynthetic apparatus of aquatic plants*

⟋

In Part I, we considered the underwater light climate: the particular characteristics that it has in different types of natural water bodies, the scattering and absorption processes that take place in the aquatic medium, and the ways that these operate upon the incident light stream to produce the kinds of underwater light field that we observe. Now, in Part II, we turn our attention to the utilization of this underwater light for photosynthesis by aquatic plants. We begin, in this chapter, by asking: with what intracellular structures, from the level of organelles down to that of molecules, do aquatic plants harvest radiant energy from the underwater light field and convert it to chemical energy?

## 8.1   Chloroplasts

In eukaryotic plants, photosynthesis is carried out by the organelles known as *chloroplasts*, the best-known members of the great class of related and interconvertible organelles known as *plastids*. A detailed account of these organelles may be found in Kirk & Tilney-Bassett (1978): we shall here content ourselves with a rather brief treatment.

The chloroplasts contain the pigments which capture the light, the electron carriers which use the absorbed energy to generate reducing power in the form of $NADPH_2$ and biochemical energy in the form of ATP, and the enzymes which use the $NADPH_2$ and the ATP to convert $CO_2$ and water to carbohydrate. The pigments and electron carriers are contained in a specialized type of membrane known as the *thylakoid*. The enzymes of $CO_2$ fixation are distributed throughout the rest of the inner volume – the *stroma* – of the chloroplast, although in certain algal chloroplasts some of the enzymes may be located in a specific structure within the chloroplast known as the *pyrenoid*.

Chloroplasts occupy a substantial proportion of the cytoplasmic volume outside the vacuole. There is usually enough chloroplast material per cell to ensure that most of the light incident on the cell passes through one or more chloroplasts. As we shall see in a later chapter, however, the disposition of the chloroplast(s) within the cell, and hence the proportion of the cellular cross-section occupied by chloroplast substance, can vary with light intensity.

In higher plants, including aquatic species, chloroplasts are typically lens-shaped bodies, 4–8 $\mu$m in diameter and about 2 $\mu$m thick in the centre. There are commonly 20–400 chloroplasts per mature leaf cell, depending partly on the size of the cell. In terrestrial plants, the chloroplasts occur mainly in the palisade and mesophyll cells of the leaves, whereas the plastids in the epidermal cells usually do not develop into chloroplasts. In submerged aquatic plants, however (the only kind with which we are concerned in this book), the epidermal cells often have well-developed chloroplasts. The palisade layer within the leaf in some cases does not develop at all, being replaced by spongy parenchyma.[196] Stomata are scarce or absent in submerged leaves, the $CO_2$ presumably being taken up directly through the epidermal cell walls.

Amongst the eukaryotic algae, chloroplasts vary greatly in size, number and morphology. Comprehensive accounts may be found in Fritsch (1948), Bold & Wynne (1978) and Raymont (1980). In both unicellular and multicellular algae, chloroplasts can occur singly, in pairs, or in numbers of up to 100 or more per cell. In many, algal species, the chloroplasts contain dense bodies, 1–5 $\mu$m in diameter – the pyrenoids. These are commonly the site of polysaccharide deposition.

Fig. 8.1. Some algal chloroplasts. c = chloroplast, p = pyrenoid. (*b*)–(*e*) after West, G. S. (1904), *The British freshwater algae*, Cambridge University Press.

The chloroplasts may be lens-shaped (*Euglena*, *Coscinodiscus*), helical (*Spirogyra*), star-shaped (*Zygnema*), plate-like (*Pinnularia*), in the form of an irregular network (*Oedogonium*), or various other shapes (Fig. 8.1). Where they occur singly they are usually large as, for example, the flat, approximately rectangular chloroplast of *Mougeotia* which is commonly 100–150 $\mu$m long. Where they occur in large numbers they are usually lens-shaped and of similar size (a few $\mu$m in diameter) to higher plant chloroplasts.

If we direct our attention to individual groups of algae, some useful generalizations are possible. Most of the green algae (Chlorophyta), major components of both the attached and the planktonic flora, contain one chloroplast per cell. This may be a cup-shaped structure as in *Chlamydomonas*, *Chlorella* and other unicellular planktonic species, or it may have a spiral, flat or other shape in the large-celled filamentous forms. Some desmids contain two chloroplasts per cell. Three of the 15 or so orders of green algae, the Caulerpales, Siphonocladales and Dasycladales, have a *coenocytic* construction. As well as having some cells of normal construction they have tubular filaments lacking cross-walls. These tubular structures contain large numbers of, usually lens-shaped, chloroplasts.

The brown algae (Phaeophyta) may contain one, a few, or many chloroplasts per cell. The Laminariales and Fucales, two orders of particular importance for coastal primary production, both contain numerous chloroplasts per cell in their large thalli. A characteristic of these brown-algal chloroplasts is that when they contain a pyrenoid, this forms a bulbous structure protruding out from the otherwise lens-shaped chloroplast.

In that very diverse division the Chrysophyta (yellow, yellow-green or yellow-brown single cells), we shall consider, on the grounds of their major contribution to planktonic primary production, two of the six classes. The class Prymnesiophyceae (formerly Haptophyceae) includes the coccolithophorids, major constituents of the marine phytoplankton, particularly in the warmer regions. Members of this class typically have two, but sometimes four or more, lens-shaped chloroplasts. The class Bacillariophyceae, or diatoms, often forms the major component of the phytoplankton in both marine and fresh waters. In the centric (circular) diatoms the chloroplasts are commonly small, lens-shaped and numerous. In the pennate (elongated) diatoms, fewer (often only two) and larger (lobed or plate-like) chloroplasts are present.

The division Pyrrophyta, or dinoflagellates, is a major contributor to aquatic, especially marine, primary production; being comparable in importance to the diatoms. The Pyrrophyta are made up of two classes, the Dinophyceae with numerous small, lens-shaped chloroplasts and the Desmophyceae with few, plate-like chloroplasts.

Members of the Euglenophyta, unicellular flagellates of common occurrence in ponds, have a variety of types of chloroplast structure, ranging from large numbers of lens-shaped plastids (Fig. 8.1) to small numbers of large plate-like or ribbon-like bodies. Another group of unicellular flagellates, the Cryptophyta, occasionally significant components of the phytoplankton in coastal waters, contain usually two, but sometimes one, chloroplast(s) per cell.

The Rhodophyta, or red algae, are predominantly thalloid in form and occur mainly in marine habitats. They are major components of the benthic (growing, attached, on the bottom) flora of coastal waters. They can be divided into two groups: the Bangiophycidae which usually have a single, stellate, axial chloroplast; and the Florideophycidae which normally contain numerous small lens-shaped chloroplasts around the periphery of the cell.

## 8.2    Membranes and particles

In electron micrographs of sections, chloroplasts are seen to be bounded by a double- or, in some algae, triple-membraned envelope and to be filled with a granular matrix. This matrix is referred to as the *stroma* and is composed of a concentrated solution or gel of proteins, consisting mainly of the enzymes used in carbon dioxide fixation. Embedded within the stroma there is an array of flattened, membrane-bounded sacs known as *thylakoids*. These lie approximately parallel to the main plane of the chloroplast and when seen in electron microscope sections at right angles to that plane, they present the appearance of membranes in pairs. There are large numbers of thylakoids in each chloroplast. In the algae most of the thylakoids extend from one end of the chloroplast to the other.

The different classes of algae differ from one another in the way in which the thylakoids are grouped together. The simplest arrangement is that found in the red algae (Fig. 8.2b). These have single thylakoids, about 20 nm thick, lying separately in the stroma. Attached to the outer surface of each thylakoid is an array of particles 30–40 nm in diameter, known as *phycobilisomes* and consisting mainly of biliprotein molecules.

The next simplest type of thylakoid arrangement is that found in the Cryptophyta, which have thylakoids loosely associated in pairs (Fig. 8.3a). They are thicker than the thylakoids of other algae, being 19–36 nm across, and their interior space is filled with a finely granular, electron-dense material which is thought to consist of the biliprotein pigments.

In all other classes of algae, with the exception of some of the

Fig. 8.2. Parallel single thylakoids with adherent phycobilisomes in blue-green and red algae (by courtesy of Dr M. Vesk). pb – phycobilisomes. Bar = 0.5 μm. (*a*) Thylakoids lying free within the cytoplasm of the blue-green alga *Oscillatoria brevis*. (*b*) Part of a chloroplast in the red alga *Haliptilon cuvieri*, showing the regularly spaced single thylakoids.

(*a*)

(*b*)

Fig. 8.3. Algal chloroplasts with thylakoids grouped in pairs and triplets (by courtesy of Dr M. Vesk). Bar = 0.5 $\mu$m. (*a*) Part of a chloroplast of the cryptophyte alga *Chroomonas* sp., showing the paired thylakoids (arrows) filled with a phycobiliprotein matrix. (*b*) Part of a chloroplast of the dinoflagellate *Gymnodinium splendens*, showing the compound lamellae, each consisting of three closely appressed thylakoids.

(*a*)

(*b*)

Fig. 8.4. Chloroplast structure in green aquatic plants. (*a*) Part of a chloroplast of the unicellular green flagellate *Dunaliella tertiolecta* (Chlorophyta), showing the somewhat irregular stacking of thylakoids to form rudimentary grana (by courtesy of Dr M. Vesk).
Bar = 0.5 $\mu$m. (*b*) Chloroplasts in the leaf of the submerged aquatic plant *Vallisneria spiralis* (by courtesy of Mr D. Price and Mr S. Craig), showing the stacking of thylakoids to form grana (g), and the stroma thylakoids connecting the grana. A large starch grain is evident in the left-hand chloroplast.

(*a*)

(*b*)

Chlorophyta, the thylakoids are grouped, or stacked, in threes (Fig. 8.3*b*). These stacks of thylakoids are referred to as compound lamellae. In aquatic higher plants and in some green algae a more complex arrangement of the thylakoids is found. They occur in much bigger stacks consisting typically of 5–20 thylakoids, but the individual thylakoids in the stacks are of much smaller diameter than a typical algal thylakoid. Each such stack is referred to as a *granum*. There may be 40–60 *grana* in a typical chloroplast. There are numerous interconnections between the different grana and between the different thylakoids within a stack (Fig. 8.4). In this type of chloroplast the thylakoids can best be regarded as individual compartments in a complex, ramifying system of interconnected compartments that constitutes the photosynthetic membrane system of the chloroplast. The thylakoid system in the chloroplast of a submerged aquatic higher plant may be seen in Fig. 8.4*b*.

Blue-green algae (Cyanophyta) are prokaryotes and so the various cellular functions are not compartmented within separate membrane-bounded organelles. These algae also have thylakoids but these lie free within the cytoplasm (Fig. 8.2*a*). They are similar in type to those of the red-algal chloroplasts, occurring as separate, single thylakoids with phycobilisomes on the surface. A type of prokaryotic alga, *Prochloron*, which occurs in symbiotic association with certain marine ascidians, does not contain biliproteins and so lacks phycobilisomes. It has been proposed by Lewin (1976) that these algae should be placed in a new algal division, the Prochlorophyta, but the counterproposal has been put by Antia (1977) that the criteria defining the Cyanophyta be widened so that the group could include algae such as *Prochloron*. Different isolates of *Prochloron* have thylakoids lying within the cytoplasm singly, in pairs, or in stacks of up to 12 thylakoids.[501, 541]

The thylakoid membrane in all plants consists of a lipid bilayer with protein particles embedded within it. The thylakoids of higher plant

Fig. 8.5. Possible mode of organization of thylakoid membrane in algae and aquatic higher plants.

chloroplasts are made up of about 32% colourless lipid, 9% chlorophyll, 2% carotenoid and 57% protein, and it is likely that algal thylakoids have a broadly similar average chemical composition. The nature of the polar lipids constituting the bilayer varies from one class of alga to another.

The light absorption/electron transport system of photosynthesis consists of two subsystems, referred to as photosystem I and photosystem II (see §8.5). It is thought that these exist within the thylakoid membrane as two distinct, but interacting, types of particle, each containing a reaction centre and light-harvesting pigment–proteins, and having specific types of electron transfer components associated with it. The molecular structure of the photosynthetic membrane is likely to vary markedly from one algal class to another, and is not known in any detail for any type of plant. There are, however, grounds for supposing that the mode of organization is something along the lines indicated in Fig. 8.5.

## 8.3    Photosynthetic pigment composition

The task of collecting light energy from the underwater light field is carried out by the photosynthetic pigments – molecules whose structures are such that they efficiently absorb light somewhere in the 400–700 nm range. There are three chemically distinct types of photosynthetic pigment: the chlorophylls, the carotenoids and the biliproteins. All photosynthetic plants contain chlorophyll and carotenoids; the red algae, the blue-green algae and the cryptophytes contain biliproteins as well.

### The chlorophylls

The chlorophylls are cyclic tetrapyrrole compounds with a magnesium atom chelated at the centre of the ring system. Chlorophylls *a* and *b* are derivatives of dihydroporphyrin: their structures are shown in Fig. 8.6*a*. Chlorophyll *d*, which is found only in small amounts in certain red algae, has a structure similar to that of chlorophyll *a*, except that the vinyl group on ring I is replaced by a formyl (—CHO) group. Chlorophylls *a*, *b* and *d* are rendered hydrophobic by the presence of a $C_{20}$ isoprenoid alcohol, phytol, esterified to the propionic acid residue on ring IV. Chlorophylls $c_1$ and $c_2$ are porphyrins rather than dihydroporphyrins and lack the phytol group: their structures are shown in Fig. 8.6*b*.

All photosynthetic plants contain chlorophyll *a*, and most classes of plant contain in addition, either chlorophyll *b*, or one or both of the chlorophyll *c*'s, or (rarely) chlorophyll *d*. The distribution of the chlorophylls (other than chlorophyll *d*, which may have no photosynthetic function) amongst the different plant groups is summarized in Table 8.1. Chlorophyll *a* normally constitutes most of the chlorophyll present.

Table 8.1. *Distribution of chlorophylls amongst different groups of plants.* (*After Kirk & Tilney-Bassett, 1978*)

| Plant group | Chlorophyll | | | |
| --- | --- | --- | --- | --- |
| | $a$ | $b$ | $c_1$ | $c_2$ |
| Angiosperms, Gymnosperms Pteridophytes, Bryophytes | + | + | − | − |
| Algae | | | | |
|   Chlorophyta | + | + | − | − |
|   Euglenophyta | + | + | − | − |
|   Chrysophyta | | | | |
|     Chrysophyceae/Haptophyceae | + | − | + | + |
|     Xanthophyceae | + | − | − | − |
|     Eustigmatophyceae | + | − | − | − |
|     Bacillariophyceae | + | − | + | + |
|   Pyrrophyta | + | − | −(+) | + |
|   Phaeophyta | + | − | + | + |
|   Cryptophyta | + | − | − | + |
|   Rhodophyta | + | − | − | − |
|   Cyanophyta | + | − | − | − |
|   Prochlorophyta | + | + | − | − |

The majority of dinoflagellates (Pyrrophyta), those containing peridinin as the major carotenoid, with one known exception contain only chlorophyll $c_2$: those dinoflagellates in which peridinin is replaced by fucoxanthin contain both $c_1$ and $c_2$[209]

Fig. 8.6. The chlorophylls. (*a*) *a/b*. In chlorophyll *b* the —$CH_3$ on ring II is replaced by a —CHO. (*b*) $c_1/c_2$. In $c_1$ R is —$C_2H_5$; in $c_2$ R is —CH=$CH_2$.

Amongst the algae the chlorophyll *a* content varies widely. The ranges of concentrations found among the three major pigment classes of the littoral multicellular marine algae off Helgoland were, as a percentage of dry mass: red, 0.09–0.44; brown, predominantly 0.17–0.55; green, 0.28–1.53.[108] The amounts per unit area of thallus, in mg dm$^{-2}$, were: red, 0.5–2.8; brown, predominantly 4.3–7.6; green, 0.5–1.4. So far as the phytoplankton in natural waters are concerned, the chlorophyll *a* content is at its highest in nutrient-rich waters favouring rapid growth. Steele & Baird (1965) found that in the northern North Sea the ratio of carbon to chlorophyll *a* in the mixed phytoplankton population was at its lowest value of 20:1 in the spring, after which it decreased to a value of about 100:1 in late summer. Assuming the phytoplankton to contain about 37% carbon, and disregarding the silica of the frustules of the diatoms present, these ratios correspond to chlorophyll *a* contents of about 1.8% and 0.37% of the dry mass, respectively. In the upwelling area off Peru, the average carbon/chlorophyll *a* ratio of the phytoplankton throughout the euphotic zone was 40 ($\equiv$ chlorophyll $\sim 0.9\%$ non-$SiO_2$ dry mass): this population consisted almost entirely of diatoms.[293] In the eastern Pacific ocean, off La Jolla, California, USA, the carbon/chlorophyll *a* ratio of the phytoplankton average about 90 ($\equiv$ chlorophyll $\sim 0.4\%$ non-$SiO_2$ dry mass) in nutrient-depleted surface waters, and about 30 ($\sim 1.2\%$ non-$SiO_2$ dry mass) in deeper, nutrient-rich waters.[115]

When grown in culture under conditions favouring high pigment content – low light intensity, high nitrogen concentration in the medium – algae usually have chlorophyll levels higher than those observed under natural conditions. Unicellular green algae such as *Chlorella* commonly have chlorophyll contents in the range 2 to 5% of the dry mass and *Euglena gracilis* has been observed to contain 3.5% chlorophyll, when grown under these conditions.

The molar ratio of chlorophyll *a* to chlorophyll *b* is about 3 in higher plants and in fresh-water green algae. Marine species of green algae, both multicellular and unicellular, are characterized by low *a/b* ratios, in the range 1.0 to 2.3.[206, 340, 560] Outside the Chlorophyta, chlorophyll *b* occurs only in the Euglenophyta and in the prokaryotic *Prochloron*. In *Euglena gracilis*, *a/b* is commonly about 6, and in *Prochloron* values of 2.6 to 12.0 have been reported.[501]

In those algae which contain chlorophyll *c*, this pigment constitutes (on a molar basis) about the same proportion of the total chlorophyll as chlorophyll *b* does in the green algae and higher plants. In surveys by Jeffrey (1972, 1976) and Jeffrey, Sielicki & Haxo (1975), the range of values of the molar ratio of *a* to *c* (where $c = c_1 + c_2$) found for different classes

Fig. 8.7. Absorption spectra of chlorophylls *a* and *b* in diethyl ether at a concentration of 10 $\mu$g ml$^{-1}$ and 1 cm pathlength. Calculated from data given by French (1960). (Chlorophyll *a* ——. Chlorophyll *b* ------.)

Fig. 8.8. Absorption spectra of chlorophyll $c_1$ and $c_2$ in acetone containing 2% pyridine with a pathlength of 1 cm (Jeffrey, S. W., unpublished data). (Chlorophyll $c_1$ (2.68 $\mu$g ml$^{-1}$) ------; chlorophyll $c_2$ (2.74 $\mu$g ml$^{-1}$) ——.)

of marine algae were: diatoms, 1.5–4.0 (mean, 3.0); peridinin-containing dinoflagellates, 1.6–4.4 (mean, 2.3); fucoxanthin-containing dinoflagellates, 2.6–5.7 (mean, 4.2); chrysomonads, 1.7–3.6 (mean, 2.7); cryptophytes, 2.5; and brown algae, 2.0–5.5 (mean, 3.6). In the majority of those algae which contained both $c_1$ and $c_2$ they were present in approximately equal amounts: however, ratios $(c_1:c_2)$ from 2:1 to 1:5 were found. Most of the dinoflagellates and cryptomonads were found to lack chlorophyll $c_1$: no alga has yet been found which contains $c_1$ and not $c_2$.

The absorption spectra of chlorophylls $a$ and $b$ in organic solvent are shown in Fig. 8.7. They each have a strong absorption band ($\alpha$) in the red and another stronger band (the Soret band) in the blue region, together with a number of satellite bands. Absorption is very low, but not zero, in the middle, green, region of the spectrum, hence the green colour of these

Table 8.2. *Major chloroplast carotenoids in various algal classes.* (*After Kirk & Tilney-Bassett, 1978*)

| Carotenoid | Chlorophyta[a] | Chrysophyta | | | | Chloromonadophyta | Euglenophyta | Phaeophyta | Pyrrophyta | Cryptophyta | Rhodophyta | Cyanophyta |
| | | Xantho-phyceae | Eustigmato-phyceae | Bacillario-phyceae | Chryso-phyceae[b] | | | | | | | |
| --- | --- | --- | --- | --- | --- | --- | --- | --- | --- | --- | --- | --- |
| $\alpha$-Carotene | + | – | – | – | + | – | – | – | – | + | + | – |
| $\beta$-Carotene | + | + | + | + | + | + | + | + | + | – | + | +[c] |
| Echinenone | – | – | – | – | – | – | – | – | – | – | – | + |
| Lutein | +[c] | – | – | – | – | – | – | – | – | – | +[c] | – |
| Zeaxanthin | + | – | – | – | – | – | – | – | – | – | + | + |
| Neoxanthin | + | – | – | – | – | – | + | – | – | – | – | – |
| Vaucheriaxanthin | – | + | +[c,d] | – | – | – | – | – | – | – | – | – |
| Violaxanthin | + | – | +[c] | – | – | – | – | + | – | – | – | – |
| Heteroxanthin | – | + | – | – | – | – | – | – | – | – | – | – |
| Fucoxanthin | – | – | – | +[c] | +[c] | – | – | +[c] | (+)[e] | – | – | – |
| Diatoxanthin | – | + | – | + | + | – | – | – | – | – | – | – |
| Diadinoxanthin | – | +[c] | – | + | + | +[c,f] | +[c] | – | + | – | – | – |
| Peridinin | – | – | – | – | – | – | – | – | +[c] | – | – | – |
| Alloxanthin | – | – | – | – | – | – | – | – | – | +[c] | – | – |
| Myxoxanthophyll | – | – | – | – | – | – | – | – | – | – | – | + |

[a] These data for the Chlorophyta apply to higher plants also. Sublittoral species of green algae usually possess siphonaxanthin or siphonein, instead of, or as well as, lutein[351, 559]
[b] $\delta$-Carotene frequently present; Haptophyceae have similar carotenoid composition to Chrysophyceae
[c] Predominant carotenoid(s) present
[d] Predominantly esterified
[e] In some dinoflagellates fucoxanthin replaces peridinin as the major carotenoid[209]
[f] Identity of xanthophylls still in some doubt

pigments. As chlorophylls $c_1$ and $c_2$ are metalloporphyrins rather than metallodihydroporphyrins, they have spectra (Fig. 8.8) in which the Soret band is more intense, and the $\alpha$ band less intense, then the corresponding bands in chlorophylls $a$ and $b$. Also, the $\beta$ band ($\sim 580$ nm) in the $c$ chlorophylls is comparable in intensity to the $\alpha$ band ($\sim 630$ nm).

Fig. 8.9. Structures of chloroplast carotenoids. (*a*) β-carotene. (*b*) Lutein. (*c*) Siphonaxanthin. (*d*) Vaucheriaxanthin. (*e*) Fucoxanthin. (*f*) Diadinoxanthin. (*g*) Peridinin. (*h*) Alloxanthin.
Each of carotenoids (*b*)–(*h*) is the major carotenoid in at least one algal group (see Table 8.2).

It can be seen from Figs 8.7 and 8.8 that chlorophyll *a* absorbs only weakly between 450 and 650 nm, and that chlorophylls *b* or *c*, when present, have the effect of increasing absorption within this window, at both the long- and the short-wavelength ends.

### The carotenoids

The carotenoids are another class of photosynthetic pigment, which extend absorption still farther into the 'window', at the short-wavelength end. Chemically they are quite distinct from the chlorophylls, being $C_{40}$ isoprenoid compounds. There are far more known carotenoids than there are chlorophylls. The distribution of the various chloroplast carotenoids amongst different plant groups is shown in Table 8.2. $\beta$-carotene is present in all except the Cryptophyta. In higher plants and green algae, which rely mainly on chlorophylls for light harvesting, the molar ratio of carotenoid to chlorophyll $(a+b)$ is about $1:3$. Other classes of algae, apart from those which contain biliproteins, depend to a greater extent on their carotenoids to capture light, and this is evident in the pigment composition. The molar ratio of carotenoid to chlorophyll $(a+c)$ is about $1:0.5$ in the diatom *Phaeodactylum*,[164] $1:1.4$ in the dinoflagellate *Gyrodinium*,[209] and amongst the brown algae is about $1:2$ in *Hormosira*[256] and $1:0.5$ in

Fig. 8.10. Absorption spectrum of $\beta$-carotene in petrol at a concentration of 10 $\mu$g ml$^{-1}$ and pathlength 1 cm.

*Laminaria*.[4] Even amongst the red algae, which have biliproteins, the carotenoid/chlorophyll molar ratio varies from 1:2.6 to 1:1.[44]

The structures of some of the most important carotenoids in higher plants and algae are shown in Fig. 8.9. Their ability to absorb visible quanta is due to their possessing systems of up to 11 conjugated double bonds. They absorb at the short-wavelength end of the visible range, which is why their characteristic colours are yellow, orange or red. The absorption spectrum of $\beta$-carotene is shown in Fig. 8.10.

### Chlorophyll/carotenoid–protein complexes

Essentially all the chlorophyll and most of the carotenoid in chloroplasts occur complexed to protein. There is good reason to suppose that these pigments can function in photosynthesis only as components of specific complexes with protein. All such photosynthetic complexes that have been examined so far, contain chlorophyll *a*: some also contain chlorophyll *b* or *c*. The great majority, but possibly not all, of these complexes, contain one or more carotenoids as well as chlorophyll. There are usually several chlorophyll molecules per polypeptide together with one, or several, carotenoid molecules. Some of the complexes occur in the photosynthetic reaction centres and may participate in the light reactions of photosystem I or photosystem II. However, the bulk of the chlorophyll/carotenoid–protein consists of various light-harvesting pigment–proteins whose role is to collect light from the prevailing field and transfer the absorbed energy to the reaction centres.

Although the molecular structures of these pigment–protein complexes are not well understood, it seems likely that the chlorophylls and probably also the carotenoids are enclosed within the protein, rather than being attached at its surface. The bonding between the pigments and the protein, although no doubt highly specific, is not covalent. Some, perhaps all, of the chlorophyll molecules, are likely to have a ligand from the polypeptide chain to the central magnesium atom. Amino acid side chains such as histidine imidazole are believed to be the source of the ligands. In addition there must be hydrogen bonding between oxygen-containing ring substituents in the chlorophyll molecules, and appropriate groups in the polypeptide chains, as well as hydrophobic associations between the phytyl chains and the non-polar regions of the protein.

In the case of the carotenoids there must be hydrophobic associations between the central hydrocarbon chain region of the carotenoid and non-polar amino acid side chains of the polypeptide. The polar groups at each end of the carotenoid molecule presumably participate in hydrogen bonding with polar groups in the polypeptide. In the case of $\beta$-carotene, which has no polar groups, all interactions with the protein are presumably hydrophobic.

A very important consequence of the interaction between the pigments and the protein is that the absorption spectra of the former are modified. The absorption peaks are, to varying degrees, shifted to longer wavelength, with, in the case of the chlorophylls, some increase in complexity as well. Chlorophyll *a* in living cells or chloroplasts has a rather broad absorption peak in the red at about 676 nm, a shift of 9–15 nm relative to its position in organic solvents (usually 661–667 nm). Computer deconvolution of the absorption curve indicates that the peak is made up of four major forms of chlorophyll *a* with peaks at 662, 670, 677 and 684 nm, together with two minor forms at 692 and 703 nm.[128] Isolated chlorophyll–protein complexes show similar complexity: it may possibly be due to exciton interaction between the chlorophyll molecules within each pigment–protein complex, leading to splitting of the absorption bands and/or to the presence of different types of chlorophyll–protein binding.

In the case of some chloroplast carotenoids, such as *β*-carotene, the absorption peaks *in vivo* are shifted, relative to their positions in organic solvents, to longer wavelengths to about the same extent as the chlorophyll peak. However, certain major light-harvesting carotenoids in the algae – fucoxanthin, peridinin, siphonaxanthin – have *in vivo* spectra shifted by about 40 nm to longer wavelength relative to their spectra in organic solvents. This has the effect of substantially increasing absorption in the green window between the chlorophyll blue and red peaks. The diatoms and the Phaeophyta, for example, are brown in colour rather than green precisely because of their higher absorption in the green waveband. The shift in absorption to longer wavelengths is a consequence of the specific association between the carotenoid and the protein: if this association is disrupted by, say, heat denaturation, then the carotenoid spectrum reverts to the *in vitro* type. When a piece of brown-algal thallus is dipped in hot water it rapidly turns green. One possibility is that the binding between protein and carotenoid is such that the polyene chain of the latter is twisted: apparently this could account for the shift of the absorption band to longer wavelength.[65]

Fucoxanthin and peridinin are the dominant carotenoids in those algae which contain them and show up in the *in vivo* spectra as a major shoulder in the 500–560 nm region (see Figs 9.3, 9.4*b*). Siphonaxanthin does not to the same extent dominate the carotenoid make-up of those algae which possess it, and shows up as a subsidiary peak at about 540 nm in the *in vivo* spectrum.[559]

Many different chlorophyll/carotenoid–proteins from higher plants and algae have been described in recent years. They have in most cases been liberated from the thylakoid membranes, in which they are normally embedded, with the help of detergents. We shall here briefly consider only some of the better-characterized examples. Reviews[263, 500] should be consulted for further details. All algae and higher plants contain a chlorophyll $a/\beta$-carotene–protein (Complex I) which functions in photosystem I, its absorbed light energy being transferred to the photosystem I reaction centre known as $P_{700}$ (see §8.4). It constitutes 25–30% of the total chlorophyll in higher plants and contains about nine chlorophyll molecules and one $\beta$-carotene molecule per 67 kilodalton (kd) polypeptide. Another chlorophyll $a$/carotenoid–protein with a polypeptide moiety of 47 kd, accounts for 5–10% of the total chlorophyll in higher plants, and may be a component of the photosystem II reaction centre. The major pigment–protein (Complex II), accounting for 40–60% of the chlorophyll and most (possibly all) of the chlorophyll $b$ in higher plants and green algae, contains five to six molecules of chlorophyll ($3a + 3b$ or $3a + 2b$) and one or two carotenoid molecules (lutein and/or other xanthophylls) per 25 kd polypeptide. This light-harvesting protein feeds its energy primarily into photosystem II, but spillover of surplus energy to photosystem I can occur.

A brown chlorophyll $a$/chlorophyll $c$/fucoxanthin complex, containing most of the fucoxanthin and about 40% of the total chlorophyll, has been isolated from brown-algal chloroplasts.[28, 29, 256] This may well be the major light-harvesting complex of the brown algae, analogous to Complex II of the green plants. Use of a more vigorous procedure has led to the isolation of two complexes: one – apparent mol wt 22–27 kd – containing fucoxanthin and chlorophyll $a$ (molar ratio 4–5:1) but no chlorophyll $c$; the other – apparent mol wt 35–40 kd – containing chlorophyll $a$ and $c$ (molar ratio 1:2) but no fucoxanthin.[4] The first of these contains all the fucoxanthin and about half the chlorophyll $a$ of the cell; the other contains all the chlorophyll $c$ and about 9% of the chlorophyll $a$ of the cell. It is possible that these two pigment–proteins are subunits of the chlorophyll $a/c$/fucoxanthin complex described above: they are present in diatoms as well as brown algae.

In some dinoflagellate species, a substantial proportion (20–50%) of the peridinin can be isolated from disrupted cells without the use of detergents as a brick-red water-soluble peridinin/chlorophyll $a$–protein complex.[182, 380] In *Amphidinium* there are nine molecules of peridinin and two molecules of chlorophyll $a$ in a 32 kd polypeptide, whereas in certain other species it appears that there are only four molecules of peridinin and one of chlorophyll $a$ per 32 kd of polypeptide. It is possible that the peridinin which is not released from the thylakoid membranes is present in molecules of the same peridinin/chlorophyll $a$–protein which are more firmly bound in the membrane. Making the assumption that all the peridinin is present in the form of this complex, it may be estimated that it accounts for 30% of the total chlorophyll *a* in *Gonyaulax* and 50% in *Glenodinium*.[379]

Another dinoflagellate pigment–protein is a chlorophyll *a*/chlorophyll *c*/carotenoid complex from *Glenodinium* with a *c*/*a* ratio of about 5.[49] This contains ~ 30% of the chlorophyll *a* and most of the chlorophyll *c* of the cells, together with (probably) the carotenoids dinoxanthin and diadinoxanthin, in a complex of apparent mol wt 50 kd. This may be analogous to the chlorophyll *a*/*c*–protein found in brown algae and diatoms.

Table 8.3. *Spectroscopic properties of the different biliproteins and the distribution of phycobilin chromophores amongst their sub-units*[147, 148, 346, 350]

| Biliprotein | Algal type[a] | Absorption maxima in photosynthetic range[b] (nm) | α Subunit[c] | β Subunit[c] | γ Subunit[c] |
|---|---|---|---|---|---|
| *Red and blue-green algae* | | | | | |
| Allophycocyanin | R, B | 650, (620) | 1 PCB | 1 PCB | – |
| Allophycocyanin B | R, B | 671, 618 | 1 PCB | 1 PCB | – |
| C-Phycocyanin | R, B | 620 | 1 PCB | 2 PCB | – |
| R-Phycocyanin | R | 618, 553 | 1 PCB | { 1 PCB<br>{ 1 PEB | – |
| Phycoerythrocyanin | B | (590), 568 | 1 PBV | 2 PCB | – |
| C-Phycoerythrin | B | 562–565 | 2 PEB | 4 PEB | – |
| b-Phycoerythrin | R | (563), 545 | 2 PEB | 4 PEB | – |
| B-Phycoerythrin | R | (565), 546, (498) | 2 PEB | 4 PEB | { 2 PEB<br>{ 2 PUB |
| R-Phycoerythrin | R | 568, 540, 498 | n.d. | n.d. | n.d. |
| *Cryptophytes* | | | | | |
| Phycocyanin-615 | | 615, 585 | { PCB<br>{ PBV | { PCB<br>{ PBV | – |
| Phycocyanin-630 | | 630, 585 | n.d. | n.d. | – |
| Phycocyanin-645 | . | 645, (620), 585 | U | { PCB<br>{ PBV | – |
| Phycoerythrin-544 | | (565), 544 | PEB | PEB | – |
| Phycoerythrin-555 | | 555 | PEB | PEB | – |
| Phycoerythrin-568 | | 568 | PEB | PEB | |

[a]R: red algae; B: blue-green algae
[b]Minor maxima or shoulders are in brackets
[c]PCB: phycocyanobilin; PEB: phycoerythrobilin; PUB: phycourobilin; PBV: phycobiliviolin; U: unknown phycobilin chromophore, distinct from the previous four. The absence of a number means that the number of chromophores is unknown. n.d.: no data

*Biliproteins*

The biliprotein chloroplast pigments are found only in certain algae: the Rhodophyta, Cryptophyta and Cyanophyta. They are either red (phycoerythrins, phycoerythrocyanin) or blue (phycocyanins, allophycocyanins) in colour. They have been reviewed by Bogorad (1975), Gantt (1975, 1977), O'Carra & O'h Eocha (1976) and by Glazer (1981). These are much better characterized than the chlorophyll/carotenoid–proteins; in some cases the complete amino acid sequence is known.

The biliproteins of the red and blue-green algae are closely related and we shall consider them together. They are divided into three classes on the basis of the position of their absorption bands: these are, in order of increasing wavelength, the phycoerythrins (and phycoerythrocyanin), the phycocyanins and the allophycocyanins. The position of the main absorption peaks are listed in Table 8.3, and absorption spectra of a red-algal phycoerythrin and phycocyanin are shown in Fig. 8.11.

Fig. 8.11. Absorption spectra of biliproteins (after O'h Eocha, 1965). (*a*) R-phycoerythrin from *Ceramium rubrum*. (*b*) R-phycocyanin from *Porphyra laciniata*.

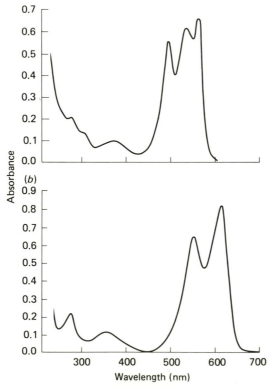

The general principle governing the distribution of these pigments appears to be that allophycocyanin and one or other of the phycocyanins occur in all red- and blue-green-algal species: a phycoerythrin (or phyco-erythrocyanin in some blue-green algae) may or may not, in addition, be present. Phycoerythrin is rarely absent in the red algae but is frequently absent in the blue-green algae. The typical situation is that in the red algae, phycoerythrin constitutes most of the biliprotein present, whereas in the blue-green algae, phycocyanin or (less commonly) phycoerythrin (or phycoerythrocyanin) is the major component: allophycocyanin is nearly always a minor component.

The chromophores to which biliproteins owe their colour are open-chain tetrapyrrole compounds known as phycobilins. There are three main chromophores – phycocyanobilin, phycoerythrobilin and phycourobilin. Unlike chlorophylls and carotenoids, they are covalently bound to their proteins. Their structures and modes of linkage to the apoproteins are shown in Fig. 8.12. Another phycobilin chromophore (provisionally phycobiliviolin) occurs in phycoerythrocyanin.

All the algal biliproteins contain equimolar amounts of two kinds of subunit, referred to as $\alpha$ and $\beta$: some of the phycoerythrins contain in addition much lesser amounts (one molecule for every six of $\alpha$ or $\beta$) of a third class of subunit, $\gamma$. The $\alpha$ and $\beta$ subunits have mol wts of about 17–22 kd but the $\gamma$ subunit may be larger. Each $\alpha$ or $\beta$ subunit in a biliprotein has at least one phycobilin chromophore attached. The $\alpha$

Fig. 8.12. Structures, and probable mode of binding to protein, of the phycobilin chromophores of algal biliproteins (after Killilea *et al.*, 1980). The systems of conjugated double bonds, upon which the spectral properties of the chromophores depend, are emphasized by heavy lines. Phycourobilin is likely to have $CH_2$ rather than CH groups joining rings A to B and C to D.

(*a*) Phycoerythrobilin

(*b*) Phycocyanobilin

subunit can have one or two chromophores; the $\beta$ subunit has one, two or four, and the $\gamma$ subunit has four. The distribution of chromophores amongst the subunits in the different proteins is shown in Table 8.3. This distribution may not be immutable. In phycoerythrin from the red alga *Callithamnion roseum* it seems that the proportion of phycoerythrobilin to phycourobilin varies with light intensity during growth:[561] it is possible that at some sites within the protein, either of these two chromophores may be attached.

Biliproteins occur as aggregates of the basic $\alpha\beta$ structure. In the isolated, soluble state, various sizes are found of which the most common appear to be the hexamer $\alpha_6\beta_6$ and the trimer $\alpha_3\beta_3$. Those phycoerythrins which have a $\gamma$ subunit are likely to occur as $\alpha_6\beta_6\gamma$ aggregates in the isolated state. *In vivo*, biliproteins occur as much larger aggregates, the phycobilisomes, particles of diameter 30–40 nm attached to the outer surface of the thylakoids (Fig. 8.2). These are ordered, specific structures rather than random aggregates, and it seems likely that the three or so different biliproteins present in the alga are all present in each phycobilisome, together with small amounts of some colourless proteins. Allophycocyanin, which as we shall see later transfers the energy harvested by the phycobilisome to chlorophyll, is believed to be located at the base of the particle, adjoining the membrane: the other, more plentiful proteins form a shell around the allophycocyanin, with the phycocyanin being next to the allophycocyanin and the phycoerythrin or phycoerythrocyanin forming the periphery of the particle.

Turning to the Cryptophyta we find some similarities to, as well as marked differences from, the situation in the Rhodophyta and Cyanophyta. The cryptophytan pigments are either red or blue and accordingly are classified either as phycoerythrins or phycocyanins. Normally only one biliprotein is detected in any of these algal species. No allophycocyanin-like pigments have yet been found in cryptophytes. The main absorption peaks are in the 612–645 nm range in the case of the phycocyanins, and the 544–568 nm range in the case of the phycoerythrins.

The cryptophyte biliproteins appear to be organized on the same $\alpha\beta$ subunit principle as the red- and blue-green-algal proteins: in the isolated state they exist predominantly as $\alpha_2\beta_2$ dimers. Where they differ most strikingly from the corresponding proteins in the other algae is their location in the photosynthetic apparatus: instead of being present as phycobilisomes on the surface of the thylakoids, the cryptophyte biliproteins occur as a dense granular matrix filling the interior of the thylakoid (Fig. 8.3).

## 8.4 Reaction centres and energy transfer

The crucial step within photosystems I and II is the use of the absorbed light energy to transfer an electron from a donor molecule to an acceptor molecule. The particular donor and acceptor molecules are different in the two photosystems. The site in a photosystem at which this event occurs is known as the *reaction centre*. The central role of using the excitation energy to extract an electron from one molecule and transfer it to another is, in each photosystem, carried out by a special form of chlorophyll *a* complexed with a specific protein. When reaction centre chlorophyll receives excitation energy, it is raised to an excited electronic state. In this excited state it can reduce (transfer an electron to) the acceptor molecule. The oxidized chlorophyll then withdraws an electron from the donor molecule, and so returns to its original state. The loss of an electron is accompanied by a fall in the absorption spectrum of the reaction centre chlorophyll, in the region of its red peak. The spectral change is maximal at about 700 nm in photosystem I and at about 680 nm in photosystem II: accordingly, the specialized forms of chlorophyll are referred to as $P_{700}$ and $P_{680}$, respectively. Since the reaction centre chlorophylls constitute only a very small proportion of the total chlorophyll *a* – one $P_{700}$ and one $P_{680}$ per 500 total chlorophyll molecules in green plants – their spectral changes are insignificant with respect to the total absorption spectrum of the system.

Excitation of reaction centre chlorophyll can be brought about by direct absorption of a photon. Since $P_{700}$ and $P_{680}$ constitute a tiny proportion of the total pigment, however, this does not take place very often. In fact, virtually all the excitation energy received by the reaction centre is energy initially captured by the vastly more numerous light-harvesting or antenna pigment molecules of that photosystem and transferred to the reaction centre. The mechanism by which this takes place is known as inductive resonance transfer, first postulated by the theoretical chemist T. Förster in 1947.

Absorption of a photon by one molecule is followed by vibrational energy dissipation, bringing the excited electron to its lowest excited state. This can be in a state of resonance with one of the upper vibrational levels of the excited state of another molecule (not, initially, in an excited state). The energy is transferred from the first molecule to the second, i.e. the first molecule reverts to the ground state and the second molecule is raised to an excited electronic state. For efficient transfer the fluorescence emission peak of the donating molecule must overlap the absorption spectrum of the receiving molecule. Since the fluorescence emission spectrum of any molecule is a mirror image of the absorption spectrum (on the long-wavelength side), with the peak shifted to longer wavelength, it follows that for efficient energy transfer the absorption peak of the donating molecule should be at a shorter wavelength than that of the receiving molecule. In addition the molecules must not be too far apart: efficient transfer can take place at distances up to about 5 nm.

The light-harvesting, as opposed to reaction centre, pigment molecules are made up of the great majority of the chlorophyll *a* molecules, chlorophylls *b*, $c_1$ and $c_2$, the various carotenoids, and the biliproteins. All these (except carotenoids) fluoresce actively *in vitro* and so the assumption that energy transfer amongst them takes place by inductive resonance presents no problem. Carotenoids, however, do not fluoresce measurably *in vitro*, and so the view is sometimes expressed that energy transfer from carotenoids to other pigments must involve some other mechanism. This need not be so. The lack of detectable fluorescence by the carotenoids simply means that their excitation energy decays quickly: it is, however, entirely possible that energy transfer by inductive resonance is even faster, so that transfer can still take place. The carotenoid molecules might, however, need to be rather close to the other pigment molecules to transfer energy efficiently to them.

In higher plants and green algae, energy absorbed by chlorophyll *b* is transferred to chlorophyll *a* with about 100% efficiency: energy absorbed by carotenoids is transferred (with lower efficiency), probably first to chlorophyll *b* and then to chlorophyll *a*. In all those algae containing chlorophyll *c*, the energy absorbed by this pigment is transferred efficiently to chlorophyll *a*. In those algae which contain major light-harvesting carotenoids – fucoxanthin, peridinin or siphonaxanthin – with substantial absorption in the 500–560 nm region, there is efficient energy transfer directly from the carotenoid to chlorophyll *a*. In red and blue-green algae, the sequence of transfer, 80–90% efficient overall, is phycoerythrin (or phycoerythrocyanin) → phycocyanin → allophycocyanin → chlorophyll *a*. In cryptophytes, which usually have only one biliprotein, this sequence is not possible: direct energy transfer from the biliprotein to chlorophyll *a* may occur, with quite high efficiency from phycocyanin, but lower efficiency from phycoerythrin.

In every case, no matter which pigment first captures the light, the absorbed energy always ends up in chlorophyll *a*. This is to be expected since chlorophyll *a* has its absorption peak at a longer wavelength than any of the other pigments and, as noted earlier, energy migration by this mechanism is in the direction of the molecules absorbing at the greatest wavelength. Amongst the bulk chlorophyll *a*–protein complexes, the excitation energy moves at random until it reaches a reaction centre, where it is immediately trapped and used for electron transfer.

## 8.5    The overall photosynthetic process

The photosynthetic process can be divided into two parts, the light reactions and the dark reactions. In the light reactions, which take place in the thylakoid membrane system, hydrogen is withdrawn from water and passed along a series of hydrogen carriers to NADP, so that $NADPH_2$ is formed and oxygen is liberated. Associated with this hydrogen (or electron) transport there is a conversion of ADP and inorganic phosphate to ATP, probably two (or, on average, some fractional number between one and two) ATP molecules being formed for every two electrons transferred or molecule of NADP reduced. These chemical changes are associated with a considerable increase in free energy: this is made possible by the light energy absorbed by the chloroplast pigments. Thus we may summarize the light reactions by the equation

$$H_2O + NADP \xrightarrow{\quad \sim 4h\nu \quad} \tfrac{1}{2}O_2 + NADPH_2$$
$$2ADP + 2P_i \qquad\qquad 2ATP$$

In the dark reactions, which take place in the stroma of the chloroplast, the $NADPH_2$ produced in the light reactions is used to reduce $CO_2$ to the level of carbohydrate. This too is associated with an increase in free energy, the energy being supplied by the concomitant breakdown of ATP produced in the light reactions. The dark reactions can be summarized by the equation

$$CO_2 + 2NADPH_2 \xrightarrow{\hspace{4cm}} (CH_2O) + H_2O + 2NADP$$
$$3ATP \qquad\qquad 3ADP + 3P_i$$

Thus, the overall photosynthetic process can be represented by

$$CO_2 + 2H_2O \xrightarrow{\ \sim 8h\nu\ } (CH_2O) + H_2O + O_2$$

### The light reactions

It is now generally believed that there are two photochemical reactions, occurring in series in photosynthesis: a recent review is that of Mathis & Paillotin (1981). Light reaction 1 is associated with the reduction of NADP; light reaction 2 brings about the liberation of oxygen from water. Each of the photoreactions takes place in a reaction centre in association with specialized light-harvesting pigment–proteins and electron transfer agents. The set of specific functional components associated with light reaction 1 is referred to as *photosystem I*; the set associated with light reaction 2 is referred to as *photosystem II*. The functional unit consisting of a single photosystem I and a single photosystem II working together, plus a set of light-harvesting pigment–proteins, is commonly referred to as a *photosynthetic unit*.

Each photosystem has a chlorophyll *a*/carotenoid–protein which both harvests light and is intimately associated with the reaction centre. This pigment–protein represents a funnel through which all the excitation energy collected by other pigment–proteins must pass before it is delivered to the reaction centre. Complex I, for example, the chlorophyll *a*/β-carotene–protein present in all plants (see §8.3) performs this role for photosystem I. The light-harvesting chlorophyll *a*–protein which is specifically associated with the photosystem II reaction centre is not yet so well characterized.

The extent to which each of the other light-harvesting pigment–proteins of algae and higher plants, described in §8.3, transfers energy specifically to one photosystem or the other, or to both, is uncertain. There is some

Fig. 8.13. The light phase of photosynthesis. The details are discussed in the text. The broken line indicates the possible pathway of the electron transfer loop involved in a form of photosphorylation which depends on a cyclic flow of electrons rather than on the net transfer of electrons from water to NADP.

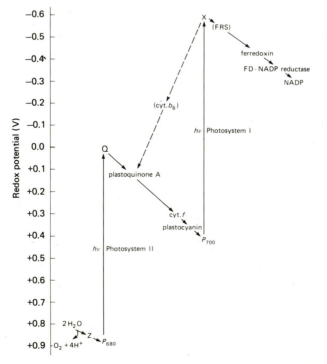

evidence that the major light-harvesting pigment–proteins, such as the chlorophyll $a/b$-containing Complex II of higher plants and green algae, and the biliproteins of red algae, transfer most, but not all, of their energy initially to photosystem II. Transfer of some of this energy from photosystem II to photosystem I (referred to as 'spillover') then takes place, thus making it possible for the two photoreactions to continue at the same rate. The review by Butler (1978) may be consulted for a detailed discussion of this topic.

Turning now to what actually happens in the light phase of photosynthesis, the current view of the sequence of events is summarized in Fig. 8.13. In photosystem I when the reaction centre chlorophyll, $P_{700}$, acquires excitation energy it loses an electron to a primary acceptor, X, not yet identified. From the primary acceptor, electrons are transferred, probably via one or more other electron carriers, to ferredoxin and from there, via ferredoxin–NADP reductase, to NADP. The oxidized reaction centre chlorophyll, $P_{700}^+$, is restored to its original state by an electron from the copper–protein, plastocyanin, which is then in turn reduced by cytochrome $f$.

When the reaction centre chlorophyll in photosystem II, $P_{680}$, acquires excitation energy it loses an electron to a primary acceptor, $Q$, which may be a bound form of plastoquinone. The electron is then transferred, possibly via another specialized plastoquinone molecule, to the major pool of plastoquinone, from which in turn electrons are transferred to reduce the cytochrome $f$ molecules which were oxidized by photosystem I. The oxidized reaction centre chlorophyll, $P_{680}^+$, is reduced by transfer of an electron from a donor, Z, of unknown nature, and the oxidized form of Z, probably in association with other catalytic components, removes electrons from water, liberating oxygen.

Putting all these processes together, the transfer of hydrogen from water to NADP, giving rise to oxygen and $NADPH_2$, is now complete. As a consequence of the particular location of electron carriers within the membrane, and the particular reactions that occur, there is a movement of eight protons from outside the thylakoid to the intrathylakoid space for every oxygen molecule liberated. It is believed that the pH gradient and electric potential set up in this way operate, by means of a reversible ATPase in the membrane, working backwards, to bring about ATP synthesis, i.e. photophosphorylation.

### The dark reactions

The pathway by which, in the stroma, or possibly in some algae in the pyrenoid region, of the chloroplast, the $NADPH_2$ and ATP produced in the light reactions are used to convert $CO_2$ to carbohydrate was elucidated largely by the work of Calvin & Benson, and is outlined in Fig. 8.14. The cycle is somewhat involved and needs several turns to produce one molecule of hexose. To understand the overall effect of the dark reactions let us consider the fate of six molecules of $CO_2$. These react

Fig. 8.14. The photosynthetic $CO_2$ fixation cycle. P = phosphate group; Pi = inorganic phosphate. The enzymes involved in each step are: (1) RuDP carboxylase; (2) 3-phosphoglyceric acid kinase; (3) glyceraldehyde-3-phosphate dehydrogenase; (4) triose-phosphate isomerase; (5) fructose-diphosphate aldolase; (6) fructose 1,6-diphosphatase; (7) transketolase; (8) aldolase; (19) sedoheptulose 1,7-diphosphatase; (10) transketolase; (11) xylulose-5-phosphate epimerase; (12) ribose-5-phosphate isomerase; (13) ribulose-5-phosphate kinase.

with six molecules of ribulose diphosphate ($C_5$) to give 12 molecules of phosphoglyceric acid ($C_3$). The phosphoglyceric acid is, with the help of ATP and $NADPH_2$, reduced to triose phosphate. For simplicity we can regard these as being converted to six molecules of hexose phosphate ($C_6$). One of these molecules of hexose can be removed to form starch or other reserve carbohydrate, leaving five hexose phosphate molecules. These are now rearranged to form six pentose phosphate ($C_5$) molecules which are, with the help of ATP converted to six molecules of ribulose diphosphate. Thus the cycle is completed with six molecules of $CO_2$ being converted to one of hexose. The overall process in terms of carbon atoms may be summarized

Some higher plants, known as $C_4$ plants, possess a variant of this cycle in which $CO_2$ is first fixed in the form of a $C_4$ acid, such as malic acid. The $C_4$ acid is then translocated to another tissue where $CO_2$ is liberated from it again and converted to carbohydrate by the normal cycle. It seems very unlikely that this $C_4$ pathway occurs in any alga, and there is no good evidence for its existence in any aquatic higher plant, so we may disregard it as a contributor to aquatic photosynthesis.

In bright sunlight, photosynthesis produces carbohydrates faster than they can be used in respiration or growth. Accordingly the plant must store the fixed carbon in some form that it can utilize later, most commonly as grains of polysaccharide. The chemistry of algal storage products has been reviewed by Craigie (1974). In higher plants and green algae, fixed carbon is accumulated inside the chloroplast as starch grains in the stroma, or forming a shell around the pyrenoid in those algae which have them. In algal phyla other than the Chlorophyta, the photosynthetic end products accumulate outside the chloroplast. Where a pyrenoid is present the polysaccharide grains are usually found in the cytoplasm but in close contact with the pyrenoid region of the chloroplast.

The Floridean starch of the Rhodophyta is an $\alpha$-D-($1 \rightarrow 4$)-linked glucan with $\alpha$-D-($1 \rightarrow 6$) branch points, and can thus be regarded as an amylopectin, like the major component of higher plant starch. Cryptophytes accumulate a starch of the higher plant type, containing both amylose and amylopectin. Members of the Pyrrophyta accumulate grains of a storage polysaccharide with staining properties similar to starch. Brown algae (Phaeophyta) accumulate the polysaccharide laminaran, a $\beta$-D-($1 \rightarrow 3$)-linked glucan with 16–31 residues per molecule and a mannitol residue at the reducing end of each chain: some of the laminaran molecules are unbranched, some have two to three $\beta$-D-(1 → 6) branch points per molecule. In addition, these algae accumulate large amounts of free mannitol. In the Euglenophyta, the photosynthetic storage product accumulates as granules of paramylon, a $\beta$-D-(1 → 3) glucan, with 50–150 residues per molecule. The photosynthetic storage product of the Chrysophyceae is chrysolaminaran (also known as leucosin), a $\beta$-D-(1 → 3)-linked glucan with possibly two 1 → 6 branch points per 34-residue molecule. Diatoms also contain polysaccharides of the chrysolaminaran type.

# 9

................................................................................................

# Light capture by aquatic plants

The collection of light energy for photosynthesis by aquatic plants is, as we have seen, carried out by the photosynthetic pigments. We have already examined the spectral absorption capabilities of each of the different classes of pigment. We shall now consider the light-harvesting properties of the complete photosynthetic system, with particular reference to the dependence of these on the particular combination of pigments present and, in the case of phytoplankton, on the size and shape of the cells or colonies.

## 9.1    Absorption spectra of photosynthetic systems

We might measure the absorption spectrum of, say, phytoplankton or a multicellular algal thallus, for a number of different reasons. We might seek information on what pigments are present. We might wish to compare the spectral position and shape of an *in vivo* absorption peak with those of the same peak in the isolated pigment with a view to assessing the extent to which the absorption properties are modified by binding to protein. We might want to know to what extent an alga is equipped to efficiently harvest light from the underwater radiation field in which it lives.

An absorption spectrum is the variation of some measure of light absorption by a system with wavelength. Light absorption might be expressed in terms of the absorptance, $A$, the percent absorption ($100\,A$), the absorbance, $D$ (where $D = -\log_{10}(1-A)$), or some other function such as the first derivative of the absorbance. The particular light absorption parameter chosen will depend on the purpose of the absorption spectrum. If, for example, we wish to be able to estimate the contribution of a planktonic alga to the total absorption coefficient of the aquatic medium at any wavelength, then we would measure the absorbance of a suspension of known concentration of the algae, as a function of wavelength, and (since absorbance is proportional to the absorption coefficient) calculate the specific absorption coefficient per unit algal biomass or pigment at the wavelength of interest. If, on the other hand, we require information on the rate at which an algal thallus is absorbing quanta from a particular incident light field, then the absorptance, or percent absorption, spectrum of that thallus is what is needed.

We shall consider absorbance spectra of algae in some detail before returning to the absorptance spectra. At this point it should be noted that since living cells scatter, as well as absorb, light then all measurements, of whatever absorption parameter, must be carried out with procedures which

Fig. 9.1. Absorbance spectrum of whole cells of *Euglena gracilis* compared with that of disrupted cells in which absorption is essentially due to thylakoid fragments (Kirk, unpublished). The spectra of intact cells (——) and of cells fragmented by ultrasonication (-----) have been corrected for scattering and in both cases correspond to 12 $\mu$g chlorophyll *a* ml$^{-1}$ and 1 cm pathlength.

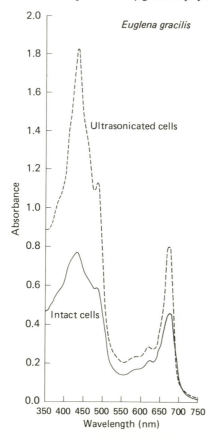

eliminate the effect of scattering on the absorption spectrum, as described in §3.2. Except where otherwise specified, it should be assumed in the remainder of this chapter that all absorption spectra referred to have been corrected for scattering.

We saw in the previous chapter that what actually carries out the light absorption in aquatic plants, the fundamental light-absorbing system, is the thylakoid. The absorbance spectrum of the thylakoid is determined by the particular kind and quantity of chlorophyll/carotenoid–protein and, in some cases, biliprotein complexes present within, or attached to, the membrane. It might therefore be thought that two species of planktonic algae which have the same array of pigment–proteins, and consequently the same absorbance spectrum at the thylakoid level, must also have the same *in vivo* absorbance spectrum in suspensions of cells or colonies at the same total pigment concentration. This is not the case. While two such algal species would certainly have rather similar spectra with the peaks and the troughs in the same positions, the extent to which the peaks rise above the troughs, and the specific absorption coefficient per unit pigment at any wavelength, can differ markedly between the two species. This is because the *in vivo* absorbance spectrum is, as we shall see, determined by the size and shape of the chloroplasts, cells or colonies, as well as by the pigment composition.

It is not feasible to determine the absorbance spectrum of a single thylakoid directly. However, in many cases it is possible to disperse the chloroplasts, by physical disruption and/or detergents, into particles so small that the effects of size and shape on the spectrum are eliminated, but with the pigment–protein interaction, so important for spectral characteristics, being unaffected. Spectra of such preparations may be regarded as being reasonable approximations to the true *in situ* absorbance spectra of the thylakoids. Fig. 9.1*a* shows a spectrum of this type for thylakoid fragments of the green planktonic alga *Euglena gracilis*.

## 9.2    The package effect

The absorbance spectrum of a cell or colony suspension (in the case of unicellular algae) or of a segment of thallus or leaf (in the case of multicellular aquatic plants) will be found to differ noticeably from that of dispersed thylakoid fragments. The *in vivo* spectra (e.g. *Euglena*, Fig. 9.1) will be found to have peaks which are less pronounced with respect to the valleys, and to have, at all wavelengths, a lower specific absorption per unit pigment. These changes in the spectra are due to what we shall refer to as the *package* effect – sometimes, inappropriately, called the *sieve* effect. It is a consequence of the fact that the pigment molecules, instead of being uniformly distributed, are contained within discrete packages: within chloroplasts, within cells, within cell colonies. This, as one might intuitively suspect, lessens the effectiveness with which they collect light from the prevailing field – hence the lowered specific absorption. It is in the nature of the package effect to be proportionately greatest when absorption is strongest[107] – hence the flattening of the peaks. The influence of the package effect on the absorption spectra of algal suspensions was first studied, experimentally and theoretically, by Duysens (1956), and its implications for the absorption of light by phytoplankton populations have been analysed by Kirk (1975a, b, 1976a) and Morel & Bricaud (1981).

To better understand the package effect, let us compare the absorption properties of a suspension of pigmented particles with those of the same amount of pigment uniformly dispersed – in effect, in solution. We shall for simplicity ignore the effects of scattering, and absorption by the medium. Consider a suspension containing $n$ particles per ml, illuminated by a parallel beam of monochromatic light. The $j$th particle in the light beam would, in the absence of the other particles, absorb a proportion $\alpha_j A_j$ of the light in unit area of beam, where $\alpha_j$ is the projected area of the particle in the direction of the beam, and $A_j$ is the particle absorptance – the fraction of the light incident on it which it absorbs. $\alpha_j A_j$ has the dimensions of area and is the absorption cross-section of the particle (see §4.1). The natural logarithm absorbance $(\ln 1/(1-A))$ of the suspension due to the $j$th particle is $\ln 1/(1-\alpha_j A_j)$, which, since $\alpha_j A_j$ is small, is approximately equal to $\alpha_j A_j$. Assuming that Beer's Law applies to a suspension of particles, i.e. that the absorbance of the suspension is equal to the sum of the absorbances due to all the individual particles,[107, 472] then the natural logarithm absorbance of the suspension, for a 1 cm pathlength, is

$$\sum_{j=1}^{n} \alpha_j A_j = n\overline{\alpha A}$$

where $\overline{\alpha A}$ is the mean value, for all the particles in the suspension, of the product of the projected area and the particle absorptance, i.e. the average absorption cross-section. The base-10 logarithm absorbance $(\log_{10} 1/(1-A))$ of the suspension is given by

$$D_{\text{sus}} = 0.434 \, n\overline{\alpha A} \tag{9.1}$$

If the particles have an average volume of $\bar{v}$ cm$^3$, and the pigment concentration within the particle is $C$ mg/ml, then if the pigment was to be uniformly dispersed throughout the medium, it would be present at a concentration of $nC\bar{v}$ mg/ml. If the specific absorption coefficient of the pigment (the absorption coefficient due to pigment at a concentration of

1 mg/ml) at the wavelength in question is $\gamma$ cm$^2$ mg$^{-1}$, then the absorbance due to the dispersed pigment is given by

$$D_{sol} = 0.434 \, nC\bar{v}\gamma \tag{9.2}$$

The extent of the package effect can be characterized by the ratio of the absorbance of the suspension to that of the solution. From eqns 9.1 and 9.2, it follows that this ratio is given by

$$\frac{D_{sus}}{D_{sol}} = \frac{\overline{A\alpha}}{C\bar{v}\gamma} \tag{9.3}$$

It can be shown[251] that for particles of any shape or orientation, $D_{sus}/D_{sol}$ is always less than 1.0. However, it can also be shown[253] that if the individual particles absorb only weakly, e.g. because they are small or because the pigment concentration within the particles is low, then

$$\overline{A\alpha} \simeq C\bar{v}\gamma$$

and so, for such particles $D_{sus} \simeq D_{sol}$.

We shall now consider what happens if we make the particles absorb progressively more strongly. For simplicity we shall keep the size and shape of the particles constant, i.e. we shall keep $\alpha$ and $\bar{v}$ constant. Both $C$ and $\gamma$ can be increased in value almost indefinitely by increasing the pigment concentration within the particles, and by changing to a more intensely absorbed wavelength, respectively: thus the denominator in eqn 9.3 can be increased in value many-fold. As $C$ or $\gamma$ increases, so $A$ in the numerator also increases, but since this is the *fraction* of the light incident on the particle which is absorbed, it can never exceed 1.0, and cannot increase in proportion to the increase in $C$ or $\gamma$. When $A$ is very low (weakly absorbing particles) a doubling in $C$ or $\gamma$ can bring about an almost commensurate increase in $A$, but the closer $A$ gets to 1.0 the less leeway it has to increase in response to a given increase in $C$ or $\gamma$. This is why the package effect, the discrepancy between the spectrum of the particle suspension, and the corresponding solution, becomes more marked as absorption by the individual particles increases: this in turn explains why the peaks of the spectrum are affected more than the troughs. In short, it is the dependence of the absorption spectrum of a suspension on the *fractional light absorption per particle* (particle absorptance) which accounts for the flattening of the spectrum and the lowered specific absorption per unit pigment concentration. A more extensive treatment of this phenomenon can be found elsewhere.[251, 252, 253] It should be noted that the package effect still affects the absorption spectrum even in suspensions so dense that no photon can avoid traversing a chloroplast while passing through: this is one reason why the term 'sieve effect' is inappropriate.

In the special case of spherical cells or colonies, which present the same cross-section to the light, whatever their orientation, an explicit expression for the particle absorptance was derived by Duysens (1956)

$$A = 1 - \frac{2[1 - (1 + \gamma Cd)e^{-\gamma Cd}]}{(\gamma Cd)^2}$$

where $d$ is the diameter of the particle in cm. Morel & Bricaud (1981) and Kirk (1975$b$) have used this relation to carry out an analysis of the package effect and its implications for light absorption for spherical phytoplankton cells.

It is apparent from eqn 9.3 that the extent to which the package effect influences absorbance depends only on the absorption properties of the individual particles. It is therefore just as true for a suspension of particles as it is for a solution, that the *shape* of the absorbance spectrum, and the specific absorbance per unit pigment, are independent of concentration. Measured spectra of suspensions of algal cells might sometimes seem not to conform to this rule: concentrated suspensions may have higher absorbances than anticipated, even when spurious absorbance due to scattering of light away from the detector is eliminated. This is because there is a residual scattering artifact in dense suspensions which cannot be overcome by instrumental means; namely, that as a result of the multiple scattering that takes place within such suspensions, the pathlength of the photons and hence the number absorbed within the suspension is increased.

We noted at the beginning of this section that the package effect can be observed in the absorbance spectra of multicellular photosynthetic tissues as well as in those of suspensions. This is because within the tissue the pigments are also segregated into packages – the chloroplasts. However, since the chloroplasts are not randomly distributed in space and since, moreover, in many cases they change their position within the cells in response to changes in light intensity, no simple mathematical treatment of the light-absorption properties of multicellular systems is possible.

## 9.3    Effects of variation in cell/colony size and shape

We have seen in the previous section that the absorbance and hence the light-harvesting efficiency of the photosynthetic pigments are lower when they are segregated into packages than when they are uniformly dispersed. The kind of packages within which the photosynthetic pigments occur in the aquatic biosphere vary greatly in size, shape and internal pigment concentration, however, and so we need some more general rules to assist our understanding of the light-intercepting capabilities of these different forms.

We already have two rules:

(i) $D_{sus} < D_{sol}$

(ii) At constant cell/colony size and shape (constant $\alpha_p$), when $A_p$ is increased (by raising the intracellular pigment concentration or altering the wavelength), $D_{sus}/D_{sol}$ decreases.

It is possible also to deduce the following:[251, 253]

(iii) At constant total pigment and biomass in the system, when both $\alpha_p$ and $A_p$ are increased (by decreasing the number and thus increasing the size of the cells or colonies, without changing their shape), $D_{sus}/D_{sol}$ decreases.

(iv) At constant total pigment in the system, when biomass is increased (by increasing cell/colony number or volume, without changing the shape), $D_{sus}/D_{sol}$ increases.

(v) At constant total pigment and biomass in the system, and constant cell/colony volume, as the shape becomes more extended, e.g. more elongated, so $D_{sus}/D_{sol}$ increases.

In both (iv) and (v) the diminution in absorptance associated with the dilution of the pigment into more biomass, or the stretching-out of the cells/colonies, is proportionately less than the increase in projected area and so, overall, the absorption cross section $\alpha_p A_p$ increases.

Having examined the underlying mechanism and the general rules governing the expression of the package effect, we shall now consider, in quantitative terms, just how much it can influence the light-harvesting capacity of planktonic algae. A convenient parameter to look at is the absorption cross-section of a given amount of algal biomass organized into packages of different sizes and shapes.[253] Fig. 9.2a shows the absorption cross-section at wavelengths from 350 to 700 nm calculated for model randomly oriented blue-green algal colonies of various geometrical forms: the volume of biomass is $100\,000\ \mu m^3$ in each case, and the pigment concentration is constant at 2% (of the dry mass) chlorophyll $a$. It can be seen that the most inefficient arrangement for light collection is the large, spherical colony, 58 $\mu m$ in diameter (lowest curve). Matters are improved somewhat if the spheres are elongated into prolate spheroids, $230 \times 29\ \mu m$. A much greater increase in efficiency is achieved if the spheres are transformed into long, thin, cylindrical filaments, 3500 $\mu m$ long and 6 $\mu m$ in diameter. Only a marginal further increase in light-harvesting capacity is brought about if the filaments are chopped up into about 900 pieces, each piece being rounded up into a sphere of 6 $\mu m$ diameter (uppermost curve). The advantages of the more extended package are much more evident at strongly absorbed, than weakly absorbed, wavelengths. For example, the ratio of the absorption cross-section of the thin cylinder to that of the large sphere is 3.82 at 435 nm but only 1.16 at 695 nm. A given change in shape cannot increase the absorption cross-section by more (proportionately) than it increases the average projected area. As the light

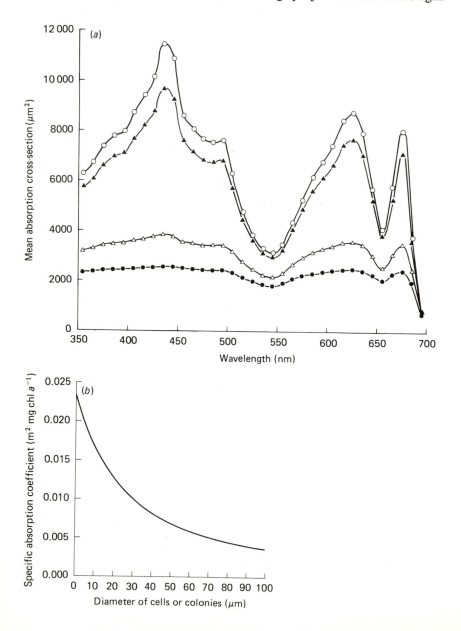

absorption by the cell is intensified (i.e. as $A_p$ approaches 1) so the effect of a given geometrical change on absorption cross-section tends to become identical to the effect on average projected area.

As another quantitative illustration of the significance of the package effect, Fig. 9.2b shows how the specific absorption coefficient per mg chlorophyll *a* present in the form of a suspension of spherical cells or colonies, at its red peak, decreases as the diameter of the cells/colonies increases.

## 9.4    Rate of light absorption by aquatic plants

The rate of photosynthesis by an aquatic plant must ultimately be limited by (although it is not always simply proportional to) the rate at which the higher plant leaf, or multicellular algal thallus, or individual phytoplankton cell or colony, is absorbing quanta from the underwater light field. In the case of the leaf or the algal thallus, the rate of absorption of quanta of a given wavelength incident at a particular angle on a particular element of tissue surface is equal to $E(\lambda, \theta, \phi) . \delta\alpha . A(\lambda, \theta, \phi)$, the product of the irradiance at that angle, the area of the element, and the absorptance for that wavelength and that angle. The total rate of absorption of light of that wavelength by the whole leaf or thallus is the sum of this product over all angles of incident light for each element of surface, and over all the elements which constitute the total area of the leaf or thallus.

It is clear that the crucial optical property of the tissue is the absorptance (the fraction of incident light absorbed) rather than the absorbance. Because of the variation in optical path through the tissue with angle, the

Fig. 9.2. Effect of size and shape on light absorption properties of phytoplankton. (*a*) The absorption cross-section spectra of randomly-oriented blue-green-algal colonies of various shapes and sizes (calculated by Kirk (1976*a*) for idealized colonies containing 2%, dry mass, chlorophyll *a*). In every case the data apply to 100 000 $\mu m^3$ of algal volume. This corresponds to one particle in the cases of the 57.6 $\mu m$ diameter spheres (●), the 230.4 × 28.8 $\mu m$ prolate spheroids (△), and the 3537 × 6 $\mu m$ cylinders (▲); and to 884 particles in the case of the 6 $\mu m$ diameter spheres (○). (*b*) Specific absorption coefficient of phytoplankton chlorophyll at the red maximum (670–680 nm) as a function of cell or colony size (Kirk, unpublished). The values were obtained from the particle absorptance values calculated using the equation of Duysens (1956), assuming that the cells/colonies contained 2%, dry mass, ($\sim 4$ g l$^{-1}$ cell volume) of chlorophyll *a* and that the (natural logarithm) specific absorption coefficient of chlorophyll *a* in solution at its red peak is about 0.0233 m$^2$ mg$^{-1}$, as it is, for example, in diethyl ether.[127] Similar calculations have been carried out by Morel & Bricaud (1981).

absorptance of the leaf or thallus at any point will in fact vary somewhat with the angle of incidence of the light. The effective absorptance of the tissue at a given point for the total light of a given wavelength incident at all angles is a function of the angular distribution of the light field, and so it is not, strictly speaking, possible to attribute a particular absorptance to the tissue for underwater light of a given wavelength regardless of the radiance distribution of that light. The absorptance is in fact usually determined with the measuring beam at right angles to the plane of the photosynthetic tissue. This is considered to provide at least an approximate measure of the absorptance that the tissue will present to the light incident on it within the water. Due to variations in thickness, and/or chloroplast number and pigment composition, there can be variation in absorptance from place to place within a leaf or thallus. This is particularly likely to be the case with long algal thalli, different parts of which are normally exposed to different light environments, e.g. because the lower parts are shaded. The increase in photon pathlength resulting from multiple scattering within the tissue can accentuate absorption by multicellular plants.

Fig. 9.3 shows the 90° absorptance spectrum for an aquatic higher plant; corresponding spectra for various kinds of multicellular algae are shown in Fig. 10.6*a*, *b* and *c*. The differences between the spectra are in part due to differences in concentration of chloroplast pigments per unit area, and this can of course vary markedly within any algal class as well as from one

Fig. 9.3. Absorption spectrum of the leaf of a fresh-water macrophyte (Kirk, unpublished). The spectrum was measured on a piece of leaf of *Vallisneria spiralis* (Hydrocharitaceae), free of epiphytic growth, from Lake Ginninderra, ACT, Australia, with the sample cell close to the photomultiplier: the spectrum has been corrected for scattering.

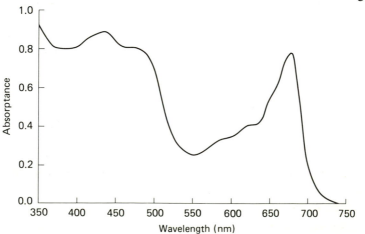

class to another. Some of the differences in the shape of the spectra are, however, attributable to the different types of pigment present. The relatively greater absorption in the 500–560 nm region in the brown alga compared to the green alga and higher plant, is due to the presence of fucoxanthin in the former. The broad peak at 520–570 nm in the spectrum of the red alga is due to the presence of the biliprotein, phycoerythrin.

In the case of phytoplankton, the rate of absorption by an individual cell or colony of light quanta of a given wavelength coming from a given direction, is equal to $E(\lambda, \theta, \phi) . \alpha_p(\theta, \phi) . A_p(\lambda, \theta, \phi)$, the product of the irradiance in that direction, the cross-sectional area (in the specified direction) of the cell or colony and the absorptance of the particle in its particular orientation with respect to the light stream. Since the cells or colonies are randomly oriented, each will present a somewhat different projected area, and absorptance, to light flowing in the specified direction. The average rate of absorption of this light per particle is $E(\lambda, \theta, \phi) . \overline{\alpha_p A_p}$. It will be recalled that $\overline{\alpha_p A_p}$ is the average absorption cross-section of the particles (§9.2). It follows from the random orientation of the cells or colonies that they have the same average absorption cross-section for light at all directions. We can therefore validly attribute an average absorption cross-section to a phytoplankton population, regardless of the angular distribution of the underwater light field. It is in fact possible to attribute an average absorptance to the individual particles in a plankton population, but this is not a useful thing to do, since it is the product of absorptance and cross-sectional area (which vary together, with orientation) rather than absorptance alone, which determines the rate of collection of quanta from a particular light stream.

What, in the present context, we wish to know about a phytoplankton population is the average absorption cross-section of the cells or colonies composing the population, at all wavelengths in the photosynthetic range. To carry out measurements on individual cells or colonies, although possible, is technically difficult and does not give results of high accuracy. We must therefore rely on spectroscopic measurements carried out on suspensions of phytoplankton. The concentrations at which phytoplankton normally occur are too low for accurate absorption measurements. It is therefore necessary to prepare more concentrated suspensions by filtration or centrifugation, followed by resuspension in a smaller volume.

Given a reasonably concentrated suspension of phytoplankton, by what sort of measurement can we determine the average absorption cross-section? Despite the importance of absorptance of the individual cell or colony in determining the absorption cross-section, measurements of the absorptance of the whole suspension tell us relatively little about the absorption properties of the individual particles in the suspension. The absorptance spectrum of the suspension changes shape as the phytoplankton concentration changes, and at very high concentrations tends to become a straight line, with $A_{sus} \simeq 1.0$ throughout the spectrum (total absorption of light at all wavelengths). The absorptance spectrum of the suspension, however, is of relevance if, for some reason, we need to know the rate of light absorption by the whole suspension – for example, in laboratory studies of photosynthetic efficiency.

To determine the average cross-section per individual free-floating particle, whether cell or colony, in the suspension, what we in fact measure is the absorbance of the suspension, $D_{sus}$. Fig. 9.4 shows the absorbance spectra of suspensions of three planktonic algae: *Chlorella* (green),

Fig. 9.4. Absorbance spectra of cultured cells of three species of planktonic algae measured using an integrating sphere (after Latimer & Rabinowitch, 1959). (*a*) *Chlorella pyrenoidosa* (green). (*b*) *Navicula minima* (diatom). (*c*) *Synechocystis* sp. (blue-green).

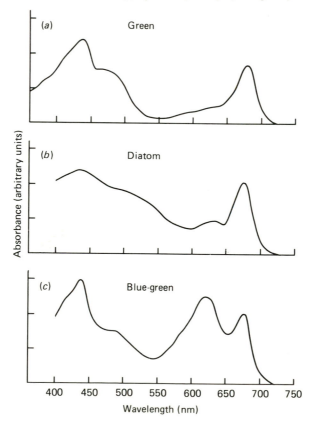

*Navicula* (a diatom) and *Synechocystis* (a blue-green). The absorbance, in a 1-cm pathlength, of a suspension of particles, as we saw earlier (eqn 9.1, §9.2), is equal to $0.434 \, n\overline{\alpha_p A_p}$, where $n$ is the number of particles per ml, and $\overline{\alpha_p A_p}$ is the average absorption cross-section per particle. Thus we may obtain the values of $\overline{\alpha_p A_p}$ throughout the photosynthetic range for any phytoplankton population by preparing a suitably concentrated suspension, measuring the absorbance spectrum and $n$, and applying the relation

$$\overline{\alpha_p A_p} = \frac{D_{sus}}{0.434 \, n} \tag{9.4}$$

at a series of wavelengths from 400 to 700 nm. Since the pathlength is 1 cm then $n$, as well as being the number of cells per ml, is also the number of particles per cm$^2$ in the path of the measuring beam. Thus, in eqns 9.1 and 9.4, $n$ can be taken to mean the number of particles per unit area, provided only that the units in which area is expressed are the same as those in which absorption cross-section $(\overline{\alpha_p A_p})$ is expressed, i.e. with a given number of particles per unit area, the absorbance is the same no matter through what pathlength they are distributed.

The average rate of absorption of light of wavelength $\lambda$, per individual phytoplankton cell or colony at depth $z$ m, is $E_0(\lambda, z) . \overline{\alpha_p A_p(\lambda)}$, where $E_0(\lambda, z)$ is the scalar irradiance of light of that wavelength at that depth and $\overline{\alpha_p A_p(\lambda)}$ is the average absorption cross-section in m$^2$. The rate of light absorption (in W, or quanta s$^{-1}$) per horizontal m$^2$ by all the phytoplankton within a thin layer of thickness $\Delta z$ m, at depth $z$ m is given by

$$E_p = E_0(\lambda, z) . \overline{\alpha_p A_p(\lambda)} . \Delta z N \tag{9.5}$$

where $N$ is the number of phytoplankton cells or colonies per m$^3$.

In the field of aquatic primary production, phytoplankton concentration is more commonly expressed in terms of mg chlorophyll $a$ m$^{-3}$, than cells or colonies m$^{-3}$. The absorbance of a system is equal to $0.434 \, ar$, where $a$ is the absorption coefficient and $r$ is the pathlength of the measuring beam (§3.2). The absorbance of a phytoplankton suspension at wavelength $\lambda$ for a 1-m pathlength (to keep the units in metres) can therefore be expressed as

$$D_{sus} = 0.434 \, a_{sus}(\lambda)$$

where $a_{sus}(\lambda)$ is the absorption coefficient due to phytoplankton, or as

$$D_{sus} = 0.434 \, Chl . a_c(\lambda) \tag{9.6}$$

where $Chl$ is the concentration of phytoplankton chlorophyll $a$ in mg m$^{-3}$, and $a_c(\lambda)$ is the specific absorption coefficient of the phytoplankton per mg chlorophyll $a$ m$^{-3}$. The absorption coefficient, $a_{sus}(\lambda)$, due to phytoplankton is equal to $N\overline{\alpha_p A_p(\lambda)}$, and so it follows that

$$N\overline{\alpha_p A_p(\lambda)} = Chl \cdot a_c(\lambda)$$

and

$$E_p = E_0(\lambda, z) \cdot Chl \cdot a_c(\lambda) \cdot \Delta z \qquad (9.7)$$

The specific absorption coefficient, $a_c(\lambda)$, has the units m$^2$ mg chlorophyll $a^{-1}$. A set of values for the specific absorption coefficient of marine phytoplankton across the photosynthesis spectral region may be found in Fig. 3.8 (Chapter 3).

### 9.5    Effect of aquatic plants on the underwater light field

We have concerned ourselves so far in this chapter with the ability of phytoplankton and macrophytes to make use of the underwater light field. By harvesting light from the field, however, the plants in turn modify the light climate for any other plants below them in the water column. Within any substantial stand of aquatic macrophytes, such as a kelp forest, or a bed of sea grass or fresh-water aquatic higher plants, the intensity of PAR is greatly reduced. Phytoplankton also increase the rapidity of attenuation of light with depth, and in productive waters may do so to such an extent that by self-shading they become a significant factor limiting their own population growth. The contribution of phytoplankton to vertical attenuation of PAR must therefore be taken into account in any consideration of the extent to which light availability limits primary production in the aquatic biosphere.

We saw in §6.7 that the total vertical attenuation coefficient for monochromatic irradiance at a given depth can be regarded as the sum of a set of partial attenuation coefficients, each corresponding to a different component of the medium. This proposition does not, strictly speaking, hold for the average vertical attenuation coefficient throughout the euphotic zone, and even less does it hold for the whole photosynthetic waveband, 400–700 nm. Nevertheless, the assumption that the average $K_d$(PAR) for the euphotic zone can be partitioned in this way is so useful that, despite its only approximate truth, it is commonly made, and may be given approximate expression through the equation

$$K_d(PAR) = K_W + K_G + K_{TR} + K_{PH} \qquad (9.8)$$

where $K_W$, $K_G$, $K_{TR}$ and $K_{PH}$ are the partial attenuation coefficients (for PAR) due to water, gilvin, tripton and phytoplankton, respectively. The contribution of phytoplankton to vertical attenuation of PAR can therefore be expressed in terms of the contribution of $K_{PH}$ to the value of $K_d(PAR)$ in eqn 9.8.

To arrive at a quantitative estimate of the contribution of phytoplankton to $K_d(PAR)$ we must determine $K_{PH}$. To do this we make use of a further commonly made assumption, which as we saw in §6.7 is also only approximately valid, namely that the contribution of any component of the medium to $K_d(PAR)$ is linearly related to the concentration of that component. Applying this to the phytoplankton we assume that

$$K_{PH} = B_c k_c \qquad (9.9)$$

where $B_c$ is the phytoplankton biomass concentration expressed in terms of mg chlorophyll $a$ m$^{-3}$ (mg chl $a$ m$^{-3}$), and $k_c$ is the specific vertical attenuation coefficient (units m$^2$ mg chl $a^{-1}$) per unit phytoplankton concentration. We may now write

$$K_d(PAR) = K_W + K_G + K_{TR} + B_c k_c \qquad (9.10)$$

If, for any given water body, an estimate of the specific attenuation coefficient, $k_c$, is available, then from a measurement of the phytoplankton concentration, $B_c$, we obtain a value for $K_{PH}$.

The value of $k_c$ has been calculated for each of the four kinds of blue-green-algal colonies to which the data in Fig. 9.2 apply: $k_c$ was 0.0063 m$^2$ mg$^{-1}$ for the 58 $\mu$m spheres, 0.0084 m$^2$ mg$^{-1}$ for the 230 $\times$ 29 $\mu$m prolate spheroids, 0.0133 m$^2$ mg$^{-1}$ for the 3500 $\times$ 6 $\mu$m cylinders, and 0.0142 m$^2$ mg$^{-1}$ for the 6 $\mu$m spheres.[253] This marked variation of $k_c$ amongst the different types of algae confirms the importance of the package effect in light capture by phytoplankton. These values were all obtained by calculation for idealized algae: $k_c$ can also be determined experimentally for real phytoplankton populations by measuring $K_d$ for PAR in natural water bodies at different times as the algal population waxes and wanes, and determining the linear regression of $K_d$ with respect to phytoplankton chlorophyll $a$ concentration. Not many measurements of $k_c$ covering the whole photosynthetic waveband have in fact been made. There are, however, other data in the literature, on the effect of varying phytoplankton concentration on the value of $K_d$ in particular spectral bands within the 400–700 nm range, obtained with irradiance meters fitted with broad-band filters. Talling (1957$b$) found that for various natural waters an approximate value for $K_d(PAR)$ could be obtained by multiplying the minimum value of $K_d(\lambda)$ for the water body concerned (usually that in the green waveband in inland waters) by 1.33; for Lough Neagh, N. Ireland, a factor of 1.15 was found to be more suitable.[218] Using this sort of relation, measurements of $K_d$ in wavebands within the photosynthetic range can be used to provide estimates of $k_c$. Table 9.1 lists some values of $k_c$ determined in various natural water bodies, from measurements either of irradiance for total PAR or for a particular waveband. It can be seen that $k_c$ varies widely – by a factor of four between the lowest and the highest value – from one alga, one water body, to another.

There are several possible reasons for this variability. One is the influence of cell size and geometry. We noted above that the package effect – within the range that might occur in nature – can vary $k_c$ by a factor of more than two in blue-green algae of identical pigment composition. We may reasonably attribute the low $k_c$ values for the large dinoflagellates in Table 9.1 to the low efficiency of light collection by large pigmented particles. The package effect also, as outlined earlier (§9.2), increases the more strongly the particles absorb (at constant size and shape). Thus, even for algae of similar size, shape and pigment type, $k_c$

Table 9.1. *Values of specific vertical attenuation coefficient for PAR per mg phytoplankton chlorophyll* a, *obtained from* in situ *measurements of irradiance*

| Water body | Phytoplankton type | $k_c$ (m² mg⁻¹) | Reference |
|---|---|---|---|
| L. Windermere, England | *Asterionella* (diatom) | 0.027[a] | 484 |
| Esthwaite Water, England | *Ceratium* (large dinoflagellate) | ~ 0.01[a] | 486 |
| L. George, Uganda | *Microcystis* (blue-green) | 0.016–0.021[a] | 136 |
| Loch Leven, Scotland | *Synechococcus* (blue-green) | 0.011[a] | 40 |
| L. Vombsjon, Sweden | *Microcystis* (blue-green) | 0.021[a] | 141 |
| Lough Neagh, Ireland | *Melosira* (diatom) | 0.014[b] | 218 |
|  | *Stephanodiscus* (diatom) | 0.008[b] |  |
|  | *Oscillatoria* (blue-green) | 0.012–0.013[b] |  |
| L. Minnetonka, Minn., USA | Mixed blue-green (*Aphanizomenon* etc.) | 0.022[c] | 313 |
| Sea of Galilee, Israel | *Peridinium* (large dinoflagellate) | 0.0067 | 100 |
| Various oceanic and coastal waters | Mixed | 0.016 | 440 |

[a]Obtained by multiplying $K_d(\lambda)_{min}$ by 1.33
[b]Obtained by multiplying $K_d(\lambda)_{min}$ by 1.15
[c]Derived from measurements of scalar, rather than downward, irradiance of PAR

will decrease as total pigment content increases. In addition, since $k_c$ is expressed per unit chlorophyll *a*, there can be marked variation in specific attenuation from one alga to another, due to differences in the type of other photosynthetic pigments present and their ratio to chlorophyll *a*. Calculations for model cells having the same chlorophyll *a* content indicated that $k_c$ for diatoms would be about 70% higher than that for green algae because of the increased absorption in the 500–560 nm region due to fucoxanthin; $k_c$ for blue-green algae with substantial levels of the biliprotein phycocyanin, absorbing in the 550–650 nm region, was calculated to be about twice that for diatoms.[253]

The colour of the aquatic medium in which the cells are suspended can also have a marked influence on the values of $k_c$. Green algal cells, for example, absorb strongly in the blue region (Fig. 9.4*a*). In typical inland waters, however, with high levels of yellow substances, the contribution of the blue spectral region to the underwater light field is greatly diminished, and so in such waters, green cells have a low value of $k_c$.[252] The $k_c$ values we have considered so far have been the average values over some considerable optical depth, i.e. a depth in which the downward irradiance falls to some small fraction of that at the surface. In fact, since the spectral distribution of the light changes progressively with depth, so the value of $k_c$ calculated over a small increment of depth also changes.[13] In the case of blue-green algae which, due to the presence of ample levels of biliprotein, absorb strongly in the green, as well as in the blue and red regions, $k_c$ does not vary markedly with depth in any water type. In the case of green algae, however, in an inland water absorbing strongly in the blue, calculations by Atlas & Bannister (1980) indicate that $k_c$ diminishes from about 0.012 $m^2 mg^{-1}$ in the surface layer to about 0.005 $m^2 mg^{-1}$ at the bottom of the euphotic zone: in a typical blue-green marine water, on the other hand, $k_c$ for green algae remains in the region of 0.014 $m^2 mg^{-1}$ down to irradiance levels 5% of the surface value. Calculations for a diatom indicated that $k_c$ remains at 0.015–0.016 $m^2 mg^{-1}$ with depth in a blue-green marine water, but diminishes from 0.014 to 0.010 $m^2 mg^{-1}$ from the top to the bottom of the euphotic zone in an inland water. To summarize, the variation of $k_c$ with depth is not of much importance for any phytoplankton type in blue-green marine waters: in yellow, inland waters it should be taken into account in the case of green algae, but may to a first approximation be ignored in the case of blue-green algae and diatoms.

One further possible cause of variation in the value of $k_c$ from one kind of phytoplankton to another is variation in their light-scattering properties. Scattering, as we saw earlier (§6.7), contributes in various ways to the vertical attenuation of irradiance. In dense algal blooms the contribution of the algal population to total scattering could significantly increase $k_c$. Little data is available on the scattering properties of algae. It seems reasonable, however, to suppose that the amount of scattering – especially per unit chlorophyll $a$ – can vary markedly from one species to another. We might anticipate, for example, that diatoms with their siliceous frustules would scatter light more intensely than algae enclosed within less refractile integuments.

The effects of macrophytes on the underwater light field vary so much with the growth habit of the plants and the morphology of the leaves or thallus that a general theoretical account is not feasible. According to Westlake (1980c) the specific vertical attenuation coefficient per mg chlorophyll is lower for macrophytes than for phytoplankton. Dense stands of emergent and floating macrophytes can make the whole water column virtually aphotic. Within submerged weed beds the spectral distribution of irradiance is predominantly green[540a].

# 10

..............................................................................................

# *Photosynthesis as a function of the incident light*

The rate of photosynthesis achieved by a phytoplankton cell or aquatic macrophyte depends on the rate of capture of quanta from the light field. This is determined by the light absorption properties of the photosynthetic biomass, which we have considered in some detail, and by the intensity and spectral quality of the field. The rate of photosynthesis is not, however, simply proportional to the rate of capture of photons. The efficiency with which the photosynthetic apparatus can make use of the absorbed energy to fix $CO_2$ varies from one plant cell to another and within a given cell as its physiological state changes. Light quanta may be collected by the pigments faster than the electron carriers and enzymes can make use of them. In particularly high light intensities the excess absorbed energy can inactivate the photosynthetic system. The relation between the rate of photosynthesis and the characteristics of the incident light is thus not a simple one: we shall examine it in this chapter.

In order to study the effects of light intensity and spectral quality on photosynthesis, suitable quantitative procedures for determining the photosynthetic rate per unit biomass must be used. Detailed descriptions of such methods for use in the field or the laboratory may be found elsewhere,[412, 526] and so they will only be briefly mentioned here. Photosynthesis can be measured in terms of either carbon dioxide fixed or oxygen released. Because of the stoichiometry of the overall photosynthetic process (§8.5), approximately one $O_2$ molecule is liberated for every molecule of $CO_2$ fixed. However, since the average composition of plant biomass differs somewhat from $CH_2O$ due to the presence of protein, lipid and nucleic acid as well as carbohydrate, the $O_2/CO_2$ ratio (known as the *photosynthetic quotient*) is usually in the range 1.1 to 1.2, rather than exactly 1.0. In the case of the more active photosynthetic systems, it is convenient to measure $O_2$ liberation – using chemical analysis, an oxygen electrode, or manometry. For field measurements of phytoplankton photosynthesis, however, in all except highly productive waters, the much more sensitive procedure of measuring fixation of $^{14}CO_2$ is used: this method was introduced by Steemann Nielsen in 1952. Bottles containing water samples with the natural phytoplankton population present and with small amounts of [$^{14}$C]-bicarbonate (hydrogen carbonate) added, are suspended at a series of depths throughout the euphotic zone, generally for periods of a few hours in the middle of the day. The amount of radioactivity fixed in cells, collected on a filter and treated with acid, is determined.

Alternatively, the incubations of the phytoplankton samples with [$^{14}$C]-$HCO_3$ can be carried out in the laboratory at the same temperature as that in the water body, and at a series of irradiance values designed to correspond to different depths. For marine phytoplankton, Jitts (1963) introduced the technique of carrying out the laboratory incubations under a series of thicknesses of blue glass, to simulate the variation of spectral distribution, as well as total irradiance, with depth.

The rate may be expressed as either *gross* or *net* photosynthesis. Gross photosynthesis is the total rate of carbon dioxide fixation, making no allowance for the fact that some $CO_2$ is simultaneously lost in respiration. Net photosynthesis is the total rate of photosynthetic $CO_2$ fixation minus the rate of loss of $CO_2$ in respiration. The rate of increase in oxygen concentration within an illuminated bottle containing phytoplankton is a measure of net photosynthesis; a value for gross photosynthesis may be obtained by adding to this the rate of respiratory oxygen consumption measured in a darkened bottle incubated in parallel. Whether the $^{14}CO_2$ fixation method measures net or gross photosynthesis, or something between the two, remains a matter of controversy. In short-term incubations, such as are possible in the more productive waters, there are *a priori* as well as experimental grounds for considering $^{14}CO_2$ fixation as an approximate measure of gross photosynthesis.

Rates of photosynthesis, whether net or gross, can be expressed per unit biomass (specific photosynthetic rate) or per unit area or volume of the water. Typical units of specific photosynthetic rate ($P$) for a phytoplankton population are $\mu$moles $CO_2$ (or $O_2$) or mg C (carbon), per mg chl $a$ per h. It is convenient to indicate it as $P(CO_2)$, $P(O_2)$ or $P(C)$ in accordance with the units being used. If the rate per $m^3$ of water ($P_v$) is summed for every 1-m depth interval from the surface to the bottom of the euphotic zone, then the areal (or integral) photosynthetic rate, $P_A$, i.e. the rate of photosynthesis in the whole water column beneath 1 m² of surface, is obtained: its units are moles (or $\mu$moles) $CO_2$ (or $O_2$), or g (or mg) C, m$^{-2}$

$$P_A = \int_{z_{eu}}^{surface} P_v \, dz$$

It may be useful to indicate areal photosynthetic rate as $P_A(CO_2)$, $P_A(O_2)$ or $P_A(C)$ in accordance with the units used. Rates of photosynthesis of benthic macrophytes may be expressed per g dry mass of tissue, or per m² of the substrate on which they are growing.

## 10.1 Photosynthesis and light intensity

The variation of the photosynthetic rate of a phytoplankton population with incident light intensity can be studied using bottles suspended at a series of depths: in this case the attenuation of light with depth provides the necessary range of irradiance values. Alternatively the measurements can be carried out under artificial light in the laboratory.

Fig. 10.1. Relative specific photosynthetic rate ($P/P_m$) of phytoplankton as a function of irradiance ($E_d$). Marine phytoplankton from the mid-point of the euphotic zone in the Sargasso Sea (redrawn from data of Ryther & Menzel, 1959) —●—. Fresh-water diatoms (mainly *Asterionella formosa* and *Fragilaria crotonensis*) in Lake Windermere, England (redrawn from data of Belay, 1981) -----○-----. Appropriate conversion factors have been used to convert the authors' original $E_d$ values to $\mu$einsteins (PAR) m$^{-2}$ s$^{-1}$.

|   | L.S. | L.C.P. | $\tfrac{1}{2}V_{max}$ |
|---|------|--------|------------------|
| ○ *Hydrilla verticillata* | 600 | 15 | 80 |
| ● *Ceratophyllum demersum* | 700 | 35 | 145 |
| △ *Myriophyllum spicatum* | 600 | 35 | 120 |
| ▲ *Cabomba caroliniana* | 700 | 55 | 160 |

### P versus $E_d$ curves

In the dark there is of course no photosynthesis and aquatic plants exhibit a net consumption of $O_2$ and liberation of $CO_2$, due to cellular respiration. As the light intensity is gradually increased from zero, some photosynthetic $O_2$ production and $CO_2$ consumption takes place, but at very low intensities shows up as a diminution in the rate of $O_2$ consumption rather than a net liberation of $O_2$, i.e. there is significant gross photosynthesis but no net photosynthesis. An irradiance value ($E_c$) is eventually reached at which photosynthetic oxygen liberation just equals respiratory oxygen consumption: this is the *light compensation point*. Beyond this point liberation exceeds consumption and net photosynthesis is achieved. The typical pattern of behaviour from here on is that $P$ increases linearly with $E_d$ up to a certain value. The graph then begins to curve over and eventually levels off. In the range of irradiance values where $P$ does not vary with $E_d$, photosynthesis is said to be light-saturated, $P$ now having the value $P_m$, the maximum specific photosynthetic rate, sometimes referred to as the *photosynthetic capacity*. With further increase in irradiance, $P$ begins to decrease again, a phenomenon referred to as *photoinhibition*. Fig. 10.1 shows two typical $P$ versus $E_d$ curves, one for marine phytoplankton,[410] the other for a mixed population of fresh-water diatoms.[36] $P$ versus $E_d$ curves for macrophytes have a similar shape but often do not show photoinhibition in full sunlight. Fig. 10.2 shows the variation of $P$ with irradiance in four fresh-water macrophyte species,[518] and in green, brown and red marine algal species.[250]

Because of the gradual onset of saturation as $E_d$ increases, it is difficult to pinpoint the irradiance value at which photosynthesis is just saturated. A more easily measured parameter by means of which the onset of saturation may be characterized[483] is the irradiance ($E_k$) at which the maximum rate, $P_m$, would be reached if $P$ were to continue to increase

Fig. 10.2. Photosynthetic rates of aquatic macrophytes as a function of irradiance of PAR. (*a*) Fresh-water macrophytes from lakes in Florida, USA (by permission, from Van, Haller & Bowes (1976) *Plant Physiology*, **58**, 761–8. The rates are limited by low, but typical *in situ*, $CO_2$ concentration (0.42 mg $l^{-1}$). Temperature is 30° C. L.S. = irradiance required for light saturation. L.C.P. = irradiance at the light compensation point. $\frac{1}{2}V_{max}$ = irradiance required for half-maximal photosynthetic rate. (*b*) Green, brown and red multicellular algae from the western Baltic Sea (plotted from data of King & Schramm, 1976). The samples of *Ulothrix speciosa* (eulittoral green), *Scytosiphon lomentaria* (eulittoral brown) and *Phycodrys rubens* (sublittoral red) were collected in the spring and the measurements were carried out at 10° C. The published irradiance values have been converted from mW $cm^{-2}$ to $\mu$einsteins (PAR) $m^{-2}$ $s^{-1}$.

linearly with $E_d$. The value of $E_k$ is that value of $E_d$ corresponding to the point of intersection between the extrapolated linear part of the curve and the horizontal line at $P_m$. This is illustrated on the idealized *P versus $E_d$* curve in Fig. 10.3. It can be seen that the slope, $\alpha$, of the linear part of the curve is equal to $P_m/E_k$: $\alpha$ is the rate of photosynthesis per unit biomass per unit of incident irradiance, and so is a measure of the efficiency with which the biomass utilizes light, at low intensities, to fix $CO_2$.

The light intensity required to saturate photosynthesis, and the compensation point, vary markedly from one species to another. Furthermore, as we shall see in a later section, these parameters also depend on the $CO_2$ concentration and the temperature. Thus, if measurements of the photosynthetic response as a function of light are to be ecologically meaningful, they must be measured under conditions approximating those in the aquatic ecosystem. In the case of measurements on phytoplankton the effect of photoinhibition, which develops to a greater extent during incubation in stationary bottles than in freely circulating cells (see later), is to lead to underestimation, both of the irradiance required for light saturation and of the maximum photosynthetic rate.[304] Thus, for phytoplankton *P versus $E_d$* data obtained in relatively short incubations are to be preferred: there is inevitably considerable uncertainty associated with results obtained from incubations of many hours duration, such as are required for unproductive oceanic waters. The ability of aquatic plants, including phytoplankton, to utilize light of any given intensity can be

Fig. 10.3. Idealized curve of specific photosynthetic rate ($P$) as a function of irradiance ($E_d$), illustrating the maximum photosynthetic rate, $P_m$, and the saturation onset parameter, $E_k$. The variation of $P/E_d$ (a measure of the efficiency of utilization of incident light) with irradiance value is also indicated ($\cdots\cdots$).

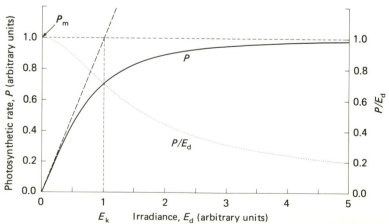

highly dependent on the light climate to which they were exposed during growth: it is therefore preferable, for ecological interpretation, if *P versus* $E_d$ curves are measured on naturally occurring, rather than laboratory-grown, plant material. Table 10.1 presents a selection of published data on the irradiance values which have been observed to correspond to light compensation, onset of light saturation and saturation, in a range of naturally occurring aquatic plants. The data do not permit any firm generalization with respect to differences in light saturation values between one algal class and another. This is partly because of the inherent difficulty in identifying the saturation irradiance in a *P versus* $E_d$ curve. Comparisons would be facilitated if the practice of always recording the more easily identified parameter, $E_k$, corresponding to onset of saturation, was generally adopted. One permissible generalization is that for any given algal species the light compensation point is lower in the winter or spring than in the summer or autumn: whether this is due simply to the differences in temperature or whether other factors are involved remains uncertain.[250] Within some species the greater the depth from which the sample was taken, the lower the irradiance required to saturate photosynthesis:[391] this is true of phytoplankton as well as macroalgae.[468] Adaptation to different ambient light levels is discussed more fully in Chapter 12. Within the phytoplankton there is evidence that the dinoflagellates have higher respiration rates, and therefore higher light compensation points, than the diatoms:[118, 449] this may be due to the energy required to sustain motility in the former group.

A number of attempts have been made to find mathematical expressions which give a reasonable fit to the empirical curves relating $P$ to $E_d$. Jassby & Platt (1976) tested eight different expressions which have at various times been proposed, against 188 *P versus* $E_d$ curves measured for marine phytoplankton in coastal Nova Scotia waters. The two which fitted best were

$$P = \frac{P_m \alpha E_d}{(P_m{}^2 + \alpha^2 E_d{}^2)^{\frac{1}{2}}} \tag{10.1}$$

originally (in a somewhat different form) proposed by Smith (1936), and

$$P = P \tanh(\alpha E_d / P_m) \tag{10.2}$$

proposed by Jassby & Platt (1976), the latter expression giving somewhat the better fit. These expressions describe the variation of $P$ with $E_d$ only up to the establishment of saturation; they do not encompass the decline in $P$ at higher values of $E_d$. Platt, Gallegos & Harrison (1980) have

Table 10.1. *Irradiance values required for saturation and light compensation of photosynthesis in various aquatic plants. Only data obtained for naturally occurring plant material, measured in natural water or its equivalent, have been used. Irradiance values published in other units have where necessary been converted to μeinsteins of PAR m⁻² s⁻¹ (where 1 μeinstein = 6.02 × 10¹⁷ quanta) with the use of appropriate conversion factors. In many cases saturation, $E_k$, or compensation irradiance values where not published explicitly have had to be estimated from authors' data: the uncertainty is particularly great in estimates of the irradiance required to saturate photosynthesis*

| Species or plant type | Location, season | Temperature (°C) | Irradiance, μeinstein (PAR) m⁻² s⁻¹ | | | Reference |
| | | | At saturation | At onset of saturation ($E_k$) | At compensation point | |
|---|---|---|---|---|---|---|
| **Freshwater phytoplankton** | | | | | | |
| *Diatoms* | | | | | | |
| *Asterionella formosa* | L. Windermere, England, spring | 5 | — | 28 | — | 482 |
| | | 10 | — | 50 | — | 482 |
| *Melosira italica* | L. Windermere, England, winter | 5 | — | 16 | — | 482 |
| | | | | | | |
| *Blue-green algae* | | | | | | |
| *Microcystis* etc. | L. George, Uganda | 27–34 | — | 135–323 | — | 137 |
| *Oscillatoria* sp. | L. Neagh, N. Ireland winter | 5 | — | 58 | — | 217 |
| | autumn | 18 | — | 232 | — | 217 |
| | | | | | | |
| **Freshwater macrophytes** | | | | | | |
| *Hydrilla verticillata* | Lakes, Fla, USA | 30 | 600 | — | 15 | 518 |
| *Ceratophyllum demersum* | Lakes, Fla, USA | 30 | 700 | — | 35 | 518 |
| *Myriophyllum spicatum* | Lakes, Fla, USA | 30 | 600 | — | 35 | 518 |
| *Cabomba caroliniana* | Lakes, Fla, USA | 30 | 700 | — | 55 | 518 |

| | | | | | | |
|---|---|---|---|---|---|---|
| *Vallisneria americana* | Lake, Wisc., USA summer | 25 | 140 | — | — | 510 |
| *Nuphar japonicum* | Japan | | | | | |
| floating leaf | | 20 | 400–600 | — | 7 | 197 |
| submerged leaf | | 21 | 75 | — | 3 | 197 |
| **Marine microalgae** | | | | | | |
| Oceanic phytoplankton (0 m) | Pacific (3° S) | 20–25 | 600 | — | — | 222 |
| Coastal phytoplankton (1–10 m) | Nova Scotia, Canada, all year | 0–15 | ~ 300 | 105 (av.) | 4 (av.) | 370 |
| Coastal phytoplankton (0 m) | Baltic, Denmark | | | | | |
| Feb. 3rd | | 1 | 400 | 200 | — | 469 |
| July 15th | | 17 | 1200 | 500 | — | 469 |
| Oct. 31st | | 12 | 800 | 300 | — | 469 |
| Intertidal benthic diatoms | Cape Cod, USA, summer | 27 | 57 | — | — | 496 |
| **Seagrasses** | | | | | | |
| *Cymodocea nodosa* | Malta | 25 | — | 158 | 17 | 97 |
| *Halophila stipulacea* | Malta | 25 | — | 83 | 8 | 97 |
| *Posidonia oceanica* | Malta | 17 | — | 108 | 17 | 97 |
| *Phyllospadix torreyi* | Calif., USA | 15 | — | 149 | 21 | 97 |
| *Zostera marina* | Calif., USA | 15 | — | 208 | 25 | 97 |
| *Zostera angustifolia* | Scotland | 10 | — | 133 | 12 | 97 |
| **Marine macroalgae** | | | | | | |
| *Chlorophyta* | | | | | | |
| *Enteromorpha intestinalis* | W. Baltic | | | | | |
| | spring | 10 | 450 | — | 6 | 250 |
| | summer | 20 | — | — | 15 | 250 |
| | Calif., USA | 21 | 245 | 56 | 6 | 9 |
| *Cladophora glomerata* | W. Baltic, summer | 20 | 700 | — | 8 | 250 |

Table 10.1 (*cont.*)

| Species or plant type | Location, season | Temperature (°C) | Irradiance, $\mu$einstein (PAR) m$^{-2}$ s$^{-1}$ | | | Reference |
| --- | --- | --- | --- | --- | --- | --- |
| | | | At saturation | At onset of saturation ($E_k$) | At compensation point | |
| *Acrosiphonia centralis* | W. Baltic, spring | 10 | 200 | – | 7 | 250 |
| *Ulothrix speciosa* | W. Baltic, spring | 10 | 700 | – | 6 | 250 |
| *Monostroma grevillei* | N. Baltic | 5 | 120 | – | – | 532 |
| *Ulva lactuca* | Woods Hole, Mass., USA summer | 23 | 250 | 160 | – | 391 |
| *Ulva lobata* | Calif., USA | 16 | 245 | 76 | 11 | 9 |
| *Ulva rigida* | Calif., USA | 21 | 412 | 50 | 9 | 9 |
| *Codium fragile* | Calif., USA | 21 | 346 | 50 | 9 | 9 |
| *Chaetomorpha linum* | Calif., USA | 21 | 418 | 82 | 10 | 9 |
| *Phaeophyta* | | | | | | |
| *Fucus serratus* | W. Baltic | | | | | |
| | winter | 5 | 350 | – | 5 | 250 |
| | autumn | 15 | 200 | – | 12 | 250 |
| *Laminaria saccharina* | W. Baltic | | | | | |
| | winter | 5 | 25 | – | 4 | 250 |
| | autumn | 15 | 700 | – | 18 | 250 |
| *Scytosiphon lomentaria* | W. Baltic, spring | 10 | 700 | – | 8 | 250 |
| *Ectocarpus confervoides* | W. Baltic, spring | 10 | 200 | – | 5 | 250 |
| *Dictyosiphon foeniculaceus* | N. Baltic | 14 | 300 | – | 18 | 532 |
| *Pilayella littoralis* | N. Baltic | 14 | 200 | – | 20 | 532 |
| | | 4 | 100 | – | – | 532 |

| | | | | | | |
|---|---|---|---|---|---|---|
| *Rhodophyta* | | | | | | |
| *Dumontia incrassata* | W. Baltic winter | 5 | 100 | — | 5 | 250 |
| | spring | 10 | 500 | — | 8 | 250 |
| *Phycodrys rubens* | W. Baltic spring | 10 | 200 | — | 5 | 250 |
| | summer | 20 | 200 | — | 14 | 250 |
| *Polysiphonia nigrescens* | W. Baltic spring | 10 | 400 | — | 7 | 250 |
| | autumn | 15 | 300 | — | 24 | 250 |
| *Ceramium tenuicorne* | N. Baltic | 11 | 100 | — | — | 532 |
| *Rhodomela confervoides* | N. Baltic | 4, 10 | 40 | — | — | 532 |
| *Chondrus crispus* | Woods Hole, Mass, USA summer | 23 | 120 | 60 | — | 391 |
| *Porphyra umbilicalis* | Woods Hole, Mass., USA summer | 23 | 250 | 90 | — | 391 |
| **Coral** | | | | | | |
| *Stylophora pistillata* | Sinai, Egypt | | | | | |
| High-light form | | 28 | 600–2000 | — | 350 | 117 |
| Low-light form | | 28 | 200 | — | 40 | 117 |

obtained an empirical equation which describes the photosynthetic rate of phytoplankton as a single continuous function of available light from the initial linear response up to and including photoinhibition.

### Photoinhibition

The inhibition of photosynthesis at high light intensities must be taken into account in ecological studies, since the intensities typically experienced in the surface layer of natural waters in sunny weather are in the range that can produce photoinhibition. Indeed if the depth profile of phytoplankton photosynthetic activity is measured by the suspended bottle method in inland or marine waters, a noticeable diminution in the specific photosynthetic rate or the rate per unit volume is commonly, although not invariably, observed near the surface. Fig. 10.4 shows examples of this surface inhibition of photosynthesis in a coastal and an inland water. With

Fig. 10.4. Depth profiles of phytoplankton photosynthetic rate per unit volume of water. The curves are for an inland water (Lake Windermere, England; plotted from data of Belay, 1981, assuming a photosynthetic quotient of 1.15), and a coastal water (Bedford Basin, NS, Canada; plotted from data of Marra, 1978).

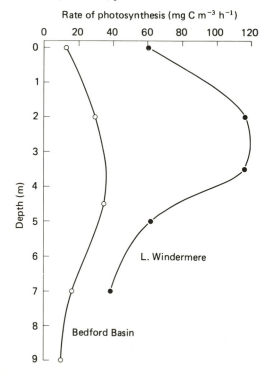

increasing depth and diminishing light intensity, photoinhibition lessens and the maximum, light-saturated but not inhibited, photosynthetic rate is achieved. With further increase in depth, irradiance falls to the point at which light intensity becomes limiting and from here on the photosynthetic rate diminishes roughly exponentially with depth approximately in parallel with irradiance.

Many, but not all, macrophyte species also show inhibition of photosynthesis at light intensities in the range of full sunlight. Ecologically, however, this phenomenon is of less significance since any given macrophyte species is generally to be found growing at a depth where the light intensity is one to which it is well adapted (Chapter 12), whereas phytoplankton are circulated within a range of depths by water movement. Marine macrophytic algae of the intertidal zone are intermittently exposed to very high light intensities. At irradiance values equivalent to full sunlight some of these species show no inhibition, but others are partially inhibited.[250]

Inhibition of photosynthesis by high light intensities takes time to develop. In the case of phytoplankton populations from Lake Ontario, Canada, the decline in photosynthetic activity began after about 10 min exposure.[178] Measurements of the time course of photosynthesis by populations of the diatom *Asterionella* in bottles suspended at the surface of a Welsh lake indicated that the inhibitory effect was small during the first hour but became significant during the second hour.[36] The higher the temperature at a given light intensity, the more rapidly inhibition ensues.[178] In the case of laboratory cultures of *Asterionella* grown at 18° C and 200 $\mu$einsteins m$^{-2}$ s$^{-1}$, exposure to 2000 $\mu$einsteins m$^{-2}$ s$^{-1}$ ($\sim$ full sunlight) for 1 h at 18 and 25° C reduced the subsequently measured photosynthetic rate by about 10 and 50%, respectively.[37]

Phytoplankton can recover from the inhibitory effects of intense light if they are transferred to a lower light intensity.[151] The longer the exposure to bright light, the longer the recovery takes. In the case of *Asterionella* populations from a Welsh lake, recovery from 2 h exposure to bright sunlight was complete after 4 h in low light intensity: after 6 h bright sunlight, which reduced photosynthetic rate by 70%, recovery took 20 h.[36]

The mechanism of photoinhibition has been studied in most detail in higher plants. Jones & Kok (1966) measured the action spectrum of photoinhibition of electron transport in spinach chloroplasts. The spectrum showed its main activity in the ultraviolet (uv) region with a peak at 250–260 nm. Photoinhibition also occurred in the visible region but with a very much lower quantum efficiency. Between 400 and 700 nm, the action spectrum followed the absorption spectrum of chloroplast pigments, with a distinct chlorophyll peak at 670–680 nm. The lesion appears primarily to affect the light reactions of photosynthesis possibly by damaging the reaction centre of photosystem II.[83, 227]

The shape of the action spectrum in the UV region suggests that plastoquinone or some other quinone functional in the reaction centre may be the sensitive molecule so far as UV inhibition is concerned. The shape of the action spectrum in the visible region indicates that at very high light intensities some of the energy absorbed by the photosynthetic pigments themselves is transferred to a sensitive site – not necessarily the same site as that affected by UV – where it causes damage.

Although detailed studies on the basis of photoinhibition of algal photosynthesis have not been carried out, the most plausible and economical hypothesis is that the mechanism is the same as in higher plants. For oceanic phytoplankton, the observed photoinhibition was found to vary linearly with the daily biological dose calculated using the Jones & Kok action spectrum.[445] In the surface layer of clear ocean waters, 50% of the photoinhibitory dose is at wavelengths less than 390 nm; in moderately productive waters (0.5 mg chl $a$ m$^{-3}$) at 10 m depth, 50% of the photoinhibitory dose is at wavelengths less than 430 nm.[442] Thus, for oceanic waters we may attribute about 50% of the photoinhibition to UV and about 50% to visible light. Aquatic yellow substances absorb strongly in the UV. We may therefore expect photoinhibition to be less apparent in the more coloured waters: this has been observed to be the case in highly productive tropical oceanic waters with high levels of gilvin.[269] By the same token it seems likely that photoinhibition in the more highly coloured (i.e. most inland) waters is caused mainly by the visible (400–700 nm) component of the solar radiation.

Depth profiles of phytoplankton photosynthesis, such as those in Fig. 10.4, determined by the suspended bottle method, tend to overestimate the extent to which photoinhibition diminishes primary production. In nature the phytoplankton are not forced to remain at the same depth for prolonged periods. Some, such as dinoflagellates and blue-green algae, can migrate to a depth where the light intensity is more suitable (see §12.6). Even the non-motile algae will only remain at the same depth for extended periods under rather still conditions. Wind blowing across a water surface induces circulatory currents known as Langmuir cells, after the eminent physical chemist, Irving Langmuir, who first studied them.[277] Langmuir cells are horizontal tubes (roll vortices) of rotating water, their axes aligned approximately parallel to the wind direction (Fig. 10.5). Adjacent tubes rotate in opposite directions and tubes of varying diameter can be present at the same time. The simultaneous occurrence of both wind and waves is necessary for the generation of these roll vortices,[120] but even a light wind over small-amplitude waves can set them going. Cells can have diameters ranging from a few centimetres to hundreds of metres: with a wind speed of 5 m s$^{-1}$, a typical cell might have a diameter of 10 m and a surface speed of 1.5 cm s$^{-1}$.[116]

Thus it will very commonly be the case that phytoplankton are not held in the intensely illuminated surface layer but are slowly circulating throughout the mixed layer. Harris & Piccinin (1977) point out that on the Great Lakes of North America the average monthly wind speed throughout all the winter, and most of the summer period, is sufficient to generate Langmuir cells, and that the residence time at the surface under such conditions will not be long enough for photoinhibition to set in.

*algae are rarely limited to one level +: specific irradiance*

Fig. 10.5. Wind-induced circulatory currents (Langmuir cells) in a water body.

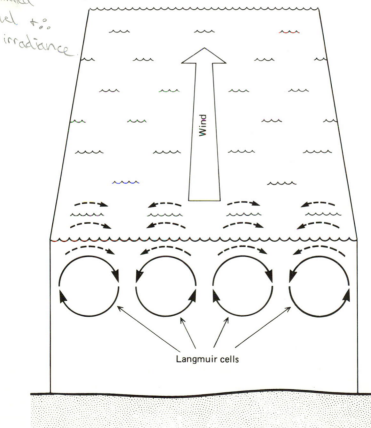

Wind

Langmuir cells

Within any given month of course, although the average wind speed may be enough to ensure Langmuir circulation, there will be calm periods in which it does not occur.

There is direct experimental evidence that the absence of vertical circulation associated with the fixed-bottle method does lead to underestimates of primary production. In the case of phytoplankton populations from Lake Ontario, Harris & Piccinin (1977) found that when 4 h incubations with $^{14}CO_2$ were used, the usual surface photoinhibition was observed, whereas when a method involving incubations for only 10 min was used (based on $O_2$ liberation by concentrated suspensions), no inhibition at surface intensities occurred. In the latter case the exposure time was too short for photoinhibition to set in. In four measurements over a 3-wk period, the 10-min method gave values of areal (integral) photosynthesis ranging from 20 to 83% (av. 41%) higher than those obtained by the 4-h method. Marra (1978), using $^{14}CO_2$ with 4-h incubations, in a Canadian coastal inlet, found that if some of the bottles were vertically cycled (1–2 m every 15–20 min) then, on days when there was surface photoinhibition in stationary bottles, the estimates of areal photosynthesis from the cycled bottles were 19–87% (av. 46%) higher than those from the fixed bottles: on days of no surface inhibition there was good agreement. On the other hand, Jewson & Wood (1975) found that in Lough Neagh (N. Ireland) there was quite good agreement (not more than 10% discrepancy in 4 of 7 trials) between estimates of areal photosynthesis obtained with stationary bottles and with water samples circulating in a tube through depths roughly corresponding to the euphotic zone. Such variations from one water body to another are not necessarily surprising since different waters will have different concentrations of background UV-absorbing material, and also different phytoplankton species are likely to vary in their susceptibility to photoinhibition. As we shall discuss more fully in the next chapter, circulation through too great a depth can inhibit total photosynthesis by keeping the cells for significant periods in light intensities too low for photosynthesis.

On balance we may reasonably conclude that photoinhibition of photosynthesis in the surface layer, although it exists, is not as frequent a phenomenon as was originally thought. It can significantly reduce areal photosynthesis under sunny, still, conditions, but is likely to be of small or no significance when there is even a light wind. Underestimates of primary production resulting from the use of stationary bottles are likely to be more serious in oligotrophic waters requiring long incubation times, than in productive waters.

## 10.2 Efficiency of utilization of incident light energy

Of the light energy incident on the water surface, only a small fraction is converted to chemical energy in the form of aquatic plant biomass. We shall now consider the reasons why this is so. Efficiency of utilization of solar energy by phytoplankton is also discussed by Morel (1978) and Talling (1982).

The first mode by which energy is lost is reflection at the surface. As we saw earlier, however, (§2.5, Table 2.1) such losses are small. For that range of solar angles at which most aquatic primary production takes place, only 2–6% of the incident light is lost by surface reflection. Thus the main causes of inefficiency of light utilization are to be found beneath the water surface.

In shallow water bodies (very shallow in coloured and/or turbid waters; moderately shallow in clear waters) substantial amounts of light reach the bottom. Some is absorbed, some reflected, the proportion depending on the optical characteristics of the substrate. Of the bottom-reflected light, a fraction will succeed in passing up through the water column again and escaping through the surface. Thus in shallow waters, bottom absorption and bottom reflection, followed by surface escape, are mechanisms preventing utilization of some of the light in photosynthesis. The light lost can be anything from a trivial proportion up to nearly 100% in, say, very shallow clear water over a white sandy bottom. Our main concern here, however, is with optically deep waters in which the fraction of the incident light which penetrates to the bottom is negligible. In such waters, most of the light which penetrates the surface is absorbed within the aquatic medium. A fraction of the light, however, usually small, is backscattered upwards within the water (see §6.4) and some of this succeeds in passing up to, and out through, the surface. Combining data for the irradiance reflectance just beneath the surface (§6.4) with the fact that about half the upwelling flux is reflected down again at the water/air boundary,[16] we may conclude that the amount of incident PAR lost in this way is 1–2.5% in oceanic waters, 1–10% in inland waters of low to high turbidity, and as little as 0.1–0.6% in waters with intense colour but low scattering.

### *Proportion of incident light captured by phytoplankton*

A major factor limiting conversion of solar energy to chemical energy by phytoplankton is, as was pointed out by Clarke (1939), the competition for radiant energy by all the non-living components of the water. We saw in Chapter 3 that the different components of the aquatic medium – water, soluble colour, tripton and phytoplankton – each account for a proportion of the total light absorbed by the water body. We also saw that to obtain an accurate estimate of the amount of PAR captured by each component separately, calculations should be carried out using the absorption coefficients for a series of narrow wavebands followed by summation across the photosynthetic spectrum. A useful approximation, however, is to first consider the total photosynthetically available radiation and then assume that the relative rates of absorption of light by the different components of the aquatic medium are in proportion to their individual contributions to the total vertical attenuation coefficient for downward irradiance of PAR, in accordance with eqn 9.8

$$K_d(PAR) = K_W + K_G + K_{TR} + K_{PH}$$

For many, perhaps most, waters of interest to limnologists and marine biologists, this assumption will not be intolerably far from the truth but it does presuppose that absorption rather than scattering is the dominant contributor to each of the partial attenuation coefficients on the right-hand side of eqn 9.8. For $K_W$ (water), $K_G$ (gilvin) and $K_{PH}$ (phytoplankton) this will be true, and when the tripton fraction is strongly coloured (e.g. by insoluble humic material) it will also be true for $K_{TR}$. If, however, the tripton fraction is high in concentration so that $K_{TR}$ is high, but consists of mineral particles low in intrinsic colour so that $K_{TR}$ is made up mainly of the scattering contribution (see §6.7, eqn 6.9), then the assumption that relative rates of absorption are proportional to the partial vertical attenuation coefficients will be seriously in error.

Nevertheless, for waters other than the kind we have just described, the fraction of the total absorbed light which is captured by the phytoplankton is, with this approximate treatment, given by

$$\frac{K_{PH}}{K_d(PAR)} = \frac{B_c k_c}{K_d(PAR)} = \frac{B_c k_c}{K_W + K_G + K_{TR} + B_c k_c} = \frac{B_c k_c}{K_{NP} + B_c k_c}$$

$$(10.3)$$

where $K_{NP}$ (which is equal to $K_W + K_G + K_{TR}$) is the vertical attenuation coefficient due to all non-phytoplankton components, $B_c$ is the phytoplankton concentration (mg chl $a$ m$^{-3}$) and $k_c$ is the specific vertical attenuation coefficient per unit phytoplankton concentration. Thus the extent to which the phytoplankton succeed in competing with other components of the medium for the available quanta depends on the relative size of $B_c k_c$ and $K_{NP}$. The range of possibilities is limitless but we shall consider a few specific examples. We shall take 0.014 m² mg$^{-1}$ as a typical mid-range value of $k_c$ (see Table 9.1). Table 10.2 lists some values for the proportion of absorbed PAR captured by phytoplankton in a number of hypothetical (but typical) water bodies ranging from very pure oceanic water with $K_{NP}$ not much greater than that due to pure water

($K_W = 0.03$–$0.06$ m$^{-1}$), to a quite highly coloured, but productive, inland water. Very approximate though these calculations are, they do show that the share of the available quanta collected by the phytoplankton can vary from a few per cent in the less productive waters, to well over 50% in highly productive systems. They also emphasize the point that quite dilute algal populations can collect a substantial proportion of the quanta, provided the background absorption is low.

Some calculations of this type have been carried out for real water bodies. Dubinsky & Berman (1981) estimated that in the eutrophic Lake Kinneret (Sea of Galilee) the proportion of the absorbed quanta captured by phytoplankton (mainly the dinoflagellate *Ceratium*) varied from about 4 to 60% as the algal concentration rose from about 5 to 100 mg chl *a* m$^{-3}$. For the eutrophic, blue-green-algal-dominated Halsted Bay, Lake Minnetonka, USA, Megard, Combs, Smith & Knoll (1979) calculated that the proportion of absorbed PAR collected by the algae rose from about 8 to

Table 10.2. *Proportion of total absorbed PAR captured by phytoplankton in idealized water bodies of different types. Calculations carried out using eqn 10.3 and assuming* $k_c = 0.014$ *m*$^2$ *mg chl* a$^{-1}$. $K_{NP}$ *is vertical attenuation coefficient due to all non-phytoplankton material*

| Type of water body | $K_{NP}$ (m$^{-1}$) | Phyto-plankton (mg chl *a* m$^{-3}$) | Proportion of absorbed PAR captured by phytoplankton (%) | Proportion of absorbed PAR captured by non-phytoplankton material (%) |
|---|---|---|---|---|
| Clear oceanic | 0.08 | 0.2 | 3.4 | 96.6 |
| | | 0.5 | 8.0 | 92.0 |
| | | 1.0 | 14.9 | 85.1 |
| Coastal | 0.15 | 1.0 | 8.5 | 91.5 |
| | | 2.0 | 15.7 | 84.3 |
| | | 4.0 | 27.2 | 72.8 |
| Clear lake, limestone catchment | 0.4 | 4.0 | 12.3 | 87.7 |
| | | 8.0 | 21.9 | 78.1 |
| | | 12.0 | 29.6 | 70.4 |
| Productive lake, coloured water | 1.0 | 8.0 | 10.1 | 89.9 |
| | | 16.0 | 18.3 | 81.7 |
| | | 32.0 | 30.9 | 69.1 |
| | | 64.0 | 47.3 | 52.7 |
| Oligotrophic lake, coloured water | 2.0 | 1.0 | 0.7 | 99.3 |
| | | 2.0 | 1.4 | 98.6 |
| | | 4.0 | 2.7 | 97.3 |

80% as the phytoplankton concentration increased from about 3 to 100 mg chl $a$ m$^{-3}$. From the data of Talling (1960) these workers estimated that for Lake Windermere (*Asterionella*-dominated), England, which has relatively low background absorption, the proportion of absorbed PAR collected by the algae rose from about 5 to 25% over the population density range of 1 to 7 mg chl $a$ m$^{-3}$. For the eutrophic, shallow (and therefore turbid) Lough Neagh, Ireland, Jewson (1977) estimated that phytoplankton accounted for about 20% of the absorbed light at the lowest population level (26.5 mg chl $a$ m$^{-3}$) and 50% at the highest (92 mg chl $a$ m$^{-3}$).

### Efficiency of conversion of absorbed light

Once the light energy is absorbed by the chloroplast pigments of the phytoplankton or aquatic macrophytes it is used, by means of the photosynthetic fixation of $CO_2$, to generate useful chemical energy in the form of carbohydrate. We shall now consider the efficiency of this conversion of excitation energy to chemical energy.

An upper limit to the efficiency is imposed by the nature of the physical and chemical processes that go on within photosynthesis. We saw in Chapter 8 that the transfer of each hydrogen atom from water down the electron transport chain to NADP requires two photons, each driving a distinct photochemical step. The reduction of one molecule of $CO_2$ to the level of carbohydrate uses four hydrogen atoms ($2 \times NADPH_2$) and so requires eight photons. Putting it another way, to convert one mole of $CO_2$ to its carbohydrate equivalent (one-sixth of a mole of a glucose unit incorporated in starch) requires not less than eight molar equivalents, i.e. 8 einsteins, of light, where an einstein is Avogadro's number ($\sim 6 \times 10^{23}$) of photons. The energy in a photon varies with wavelength ($\varepsilon = hc/\lambda$) and so we shall obtain an average value by taking advantage of the observation by Morel & Smith (1974) that for a wide range of water types $2.5 \times 10^{18}$ quanta of underwater PAR $\equiv 1$ J, with an accuracy of better than 10%. We may thus regard typical underwater light as containing 0.24 MJ (megajoules) of energy per einstein and so 8 einsteins is equivalent to 1.92 MJ. The increase in chemical energy associated with the photosynthetic conversion of one mole of $CO_2$ to its starch equivalent is 0.472 MJ. Thus, of the light energy absorbed and delivered to the reaction centres, about 25% is converted to chemical energy as carbohydrate and this is the maximum possible efficiency.

To equate the plant biomass to carbohydrate is an oversimplification, since the aquatic plants also contain protein, lipids and nucleic acids, none of which conform closely in their overall composition to $CH_2O$. The biosynthesis of these substances requires additional photosynthetically generated reducing power and chemical energy in the form of $NADPH_2$ and ATP, and so requires additional light quanta per $CO_2$ incorporated. The true minimum quantum requirement per $CO_2$ for growing cells is likely to be about 10–12 rather than 8[390] which brings the maximum efficiency down to 16–20%. Thus the best efficiency we can hope for in the conversion of absorbed light energy to chemical energy in the form of new aquatic plant biomass is about 18%.

The conversion efficiency actually achieved by a given phytoplankton population or macrophyte can be determined from measurements of the photosynthetic rate and the irradiance, provided that information on the light absorption properties of the plant material is available. Using the absorption spectrum, 400–700 nm, rates of light absorption for a series of wavebands can be calculated and summed. An alternative, more approximate, procedure applicable to phytoplankton populations is to make use of an expression derived by Duntley (1963) for the rate of absorption of light energy per unit volume of medium at a given depth

$$\frac{d\Phi(z)}{dv} = K_d E_d(z)(1 - R(z)) \tag{10.4}$$

where $R(z)$ is irradiance reflectance at depth $z$. Since the fraction of the total absorbed PAR captured by the phytoplankton is approximately $B_c k_c / K_d$(PAR), then the rate of absorption of light energy by phytoplankton per unit volume at depth $z$ is

$$\frac{d\Phi_p(z)}{dv} = B_c k_c E_d(z)(1 - R(z)) \tag{10.5}$$

which may usually be further approximated to

$$\frac{d\Phi_p(z)}{dv} = B_c k_c E_d(z) \tag{10.6}$$

when the reflectance is only a few per cent, as will in fact be the case in all except rather turbid waters.

We saw in the previous chapter that because of the change in spectral distribution of underwater light with depth, $k_c$ for the whole photosynthetic waveband can change significantly with depth. A more accurate formulation of eqn 10.6 would be

$$\frac{d\Phi_p(z)}{dv} = B_c \int_{400}^{700} k_c(\lambda, z) E_d(\lambda, z) \, d\lambda \tag{10.7}$$

in which $k_c(\lambda, z)$ is the specific vertical attenuation coefficient for phytoplankton at wavelength $\lambda$ and depth $z$ m, and $E_d(\lambda, z)$ is the downward irradiance per unit bandwidth ($nm^{-1}$) at wavelength $\lambda$ and depth $z$ m. This formulation would only be required when a particularly accurate estimate of efficiency was needed, e.g. when attempting to determine if a particular *in situ* efficiency had reached the theoretical maximum. Ideally the variation of $k_c(\lambda, z)$ both with wavelength (in which it would follow the absorption spectrum of the algae) and with depth (due to the changing angular structure of the field) should be taken into account. In practice it would be hard to obtain all the information necessary to make use of eqn 10.7. A compromise would be to use eqn 10.5 but with a $k_c$ value approximately corrected for depth variation using the data of Morel (1978) or Atlas & Bannister (1980).

To obtain the energy conversion efficiency at a given depth we divide the rate of accumulation of chemical energy per unit volume at that depth by the rate of absorption of light energy by phytoplankton per unit volume. Given a specific photosynthetic rate of $P(CO_2)$ moles $CO_2$ fixed mg chl $a^{-1}$ $h^{-1}$, and an increase in chemical energy of 0.472 MJ associated with the fixation of each mole of $CO_2$, then the rate of accumulation of chemical energy per unit volume is 0.472 $B_c P(CO_2)$ MJ m$^{-3}$ h$^{-1}$. Dividing by $B_c k_c E_d(z)$ we obtain the conversion efficiency

$$\varepsilon_c = \frac{0.472 P(CO_2)}{k_c E_d(z)} \tag{10.8}$$

$E_d(z)$ being expressed in MJ m$^{-2}$ h$^{-1}$ (quanta measurements can be converted using $2.5 \times 10^{24}$ quanta $\equiv 1$ MJ, see above). Alternatively, if $P$ is expressed in terms of mg carbon fixed mg chl $a^{-1}$ h$^{-1}$ ($P(C)$), then, since there is an increase in chemical energy of $3.93 \times 10^{-5}$ MJ associated with the fixation of 1 mg C, the conversion efficiency is given by

$$\varepsilon_c = \frac{3.93 \times 10^{-5} P(C)}{k_c E_d(z)} \tag{10.9}$$

Another way of expressing the efficiency of conversion of absorbed light energy to chemical energy by aquatic plants is the quantum yield, $\phi$. This is defined to be the number of $CO_2$ molecules fixed in biomass per quantum of light absorbed by the plant. Given the quantum requirement per $CO_2$ fixed, imposed by the mechanism of photosynthesis (see above), it follows that the quantum yield could never be greater than 0.125, and for growing cells, even under ideal conditions, is unlikely to exceed about 0.1. Quantum yield and percent conversion efficiency are, of course, linearly related. Allowing 0.472 MJ chemical energy per $CO_2$ fixed and 0.24 MJ per einstein of underwater PAR, we arrive at

$$\varepsilon_c = 1.97 \phi \tag{10.10}$$

Equations corresponding to 10.8 and 10.9 can be written for the calculation of quantum yield from $E_d$, $k_c$, and specific photosynthetic rate

$$\phi = \frac{P(CO_2)}{k_c\,E_d(z)} \tag{10.11}$$

$$\phi = \frac{P(C)}{12\,000\,.\,k_c\,E_d(z)} \tag{10.12}$$

$E_d(z)$ in these equations has the units einsteins $m^{-2}\,h^{-1}$.

The quantum yield attained by an aquatic plant is a function of the light intensity to which it is exposed. That this is so is apparent from the variation of specific photosynthetic rate with irradiance (Fig. 10.3). Ignoring for the moment any changes in chloroplast shape or position with light intensity, we may assume that the rate of absorption of quanta is proportional to the incident irradiance. Thus, at any point in the photosynthesis *versus* irradiance curve the value of $P/E_d$ (see Fig. 10.3) is proportional to the quantum yield.

For a plant to be able to make efficient use of light quanta being absorbed at a given rate, the activity of the electron transfer components in the thylakoid membranes and of the enzymes of $CO_2$ fixation in the stroma must both be high enough to ensure that the excitation energy collected by the light-harvesting pigments is utilized as fast as it arrives at the reaction centres. If this situation exists, then a moderate increase in light intensity, causing a proportionate increase in quantum absorption rate, leads to a corresponding increase in the specific rate of photosynthesis. In the initial, linear, region of the $P$ *versus* $E_d$ curve this is what is happening. Over this range of intensity, $P/E_d$ is constant and has its highest value, indicating that the plants are achieving their highest conversion efficiency and quantum yield. If the absorption characteristics of the plant are known, then this maximum value of $P/E_d$ can be used to calculate $\phi_m$, the maximum quantum yield.

As the incident light intensity is further increased, the rate of absorption of quanta reaches the point at which excitation energy begins to arrive at the reaction centres faster than it can be made use of by the electron transfer components and/or the $CO_2$ fixation enzymes. At this stage some of the additional absorbed quanta (over and above those the system can readily handle) are utilized for photosynthesis, and some are not, the energy of the latter being eventually dissipated, mainly as heat. For this reason, in this range of light intensity, increments in $E_d$ are accompanied by less than commensurate increases in $P$, i.e. the slope of the curve progressively diminishes, until eventually the point is reached at which

$\Delta P/\Delta E_d$ becomes zero. In this light-saturated state, the electron transfer and/or $CO_2$ fixation enzymes (most likely, the latter) are working as fast as they are capable, and so any additional absorbed quanta are not used for photosynthesis at all. From the end of the linear region through to the light-saturated region, since photosynthetic rate does not increase in proportion to irradiance ($P/E_d$ steadily falls − Fig. 10.3) the quantum yield and conversion efficiency necessarily undergo a progressive fall in value. This is accentuated further if, at even higher light intensities, photoinhibition sets in.

The consequence of this in natural water bodies is that quantum yield and conversion efficiency, depending as they do on light intensity, vary markedly with depth, the general tendency being, as might be expected, for $\phi$ and $\varepsilon_c$ to increase with depth.[99, 328] Quantum yields in the surface layer in the middle period of the day, when irradiance values are generally above the range corresponding to the linear part of the $P$ *versus* $E_d$ curve, are usually below $\phi_m$.

Morel (1978) calculated the quantum yield at a series of depths from $^{14}CO_2$ fixation, chlorophyll and light data for a variety of oceanic waters from the highly oligotrophic Sargasso Sea to the productive waters of the Mauritanian upwelling area. In most cases $\phi$ increased with depth, i.e. with decreasing irradiance. On average, the $\phi$ values for green eutrophic waters were higher than those observed in blue, oligotrophic waters. In the surface layers the $\phi$ values were mainly in the range 0.003–0.012 (equivalent to $\varepsilon_c$ values of 0.6–2.4%. At some stations the apparent value of quantum yield near the bottom of the euphotic zone was in the region of 0.1–0.125 ($\varepsilon_c = 20$ to $25\%$), the maximum possible on theoretical grounds. For reasons given below, however, there are inherent limitations on the accuracy with which *in situ* quantum yield can be determined. Koblents-Mishke (1979) estimated that conversion efficiency ($\varepsilon_c$) by phytoplankton in the oceans does not exceed $2\%$ near the surface, but approaches $20\%$ near the bottom of the euphotic zone. Dubinsky & Berman (1981) estimated that during the spring *Peridinium* bloom in Lake Kinneret (Sea of Galilee), $\varepsilon_c$ rose from about $5\%$ at the surface to $8.5\%$ at 3 m. In late summer, with a much lower population of different (green) algae, $\varepsilon_c$ was $2.5\%$ at the surface but rose to about $12\%$ at 5–7 m.

In contrast to laboratory studies in which all the essential parameters can be accurately determined, estimates of quantum yield for real water bodies are beset with inescapable uncertainties. Probably the most important of these is the true *in situ* value of $k_c$, the specific vertical attenuation coefficient for the phytoplankton population present. We saw in the last chapter (Table 9.1) that $k_c$ can vary up to four-fold, being a function of the type of phytoplankton present, the background colour of the water, and also varying with depth. Thus in any real-world estimates of the value of $\phi$, the selection of the $k_c$ value used must be carried out with particular care. Values based on measurements of $K_d$ as a function of phytoplankton concentration within the water body of interest, or on the measured absorption spectrum of a phytoplankton sample from the water, are to be preferred. These uncertainties do not invalidate the qualitative observations of the marked variation of $\phi$ with depth, but do make it hard to arrive at firm estimates of the maximum quantum yield, $\phi_m$, a parameter of considerable theoretical interest.

On the assumption (see above) that in the linear region of the $P$ *versus* $E_d$ curve the cells are achieving their maximum efficiency, estimates of $\phi_m$ are normally obtained from the observed slope, $P/E_d$, commonly referred to as $\alpha$, in this region of the curve. From any of the equations for $\phi$ it follows that $\phi_m$ should be proportional to $\alpha$, e.g. from eqn 10.11, $\phi_m = \alpha/k_c$. The photosynthetic process imposes by its essential nature a maximum value for $\phi_m$ of 0.1–0.125, for all photosynthetic systems. The fact that this $\phi_m$ value is one to which any plant species might reasonably aspire, and also that $\alpha$ is linearly related to $\phi_m$ (since $\alpha = k_c \phi_m$), has aroused expectations that $\alpha$ should not vary markedly for a given species in different environments, or from one species to another. Such expectations, however, ignore the extent to which the proportionality factor, $k_c$, can vary. We saw earlier (Chapter 9) how markedly $k_c$ varies with the size and shape of the cells or colonies, for example decreasing as the absorbing units become larger or more intensely pigmented. Taguchi (1976), from studies on seven species of marine diatom, found, as might be predicted, that $\alpha$ decreased with increasing size of the cells. Platt & Jassby (1976) found that the $\alpha$ values of natural phytoplankton populations in Nova Scotia coastal waters tended to be lower in deeper water, and also in the winter: a possible explanation would be an increased pigment content (and therefore, lower $k_c$) resulting from adaptation by the phytoplankton (or changes in the species present) at lower light intensities. The total variation in $\alpha$ was as much as 20-fold, although most values occurred within a five-fold range. Welschmeyer & Lorenzen (1981), in a comparison of six species of marine phytoplankton growing exponentially under identical conditions, found that there were no significant differences in $\phi$, but there were significant differences in $\alpha$: these they attributed to differences in the light absorption efficiency per unit of chlorophyll.

Thus, while we might reasonably expect $\phi_m$ not to vary dramatically with species or environment, we should not expect the same constancy of $\alpha$, because of the variability of $k_c$. Unfortunately this variability in $k_c$, and the consequent uncertainty in its value, make the determination of $\phi_m$ in natural populations, from $\alpha$, difficult.

Some typical values of $\alpha$ for natural phytoplankton populations, in the units mg C mg chl $a^{-1}$ h$^{-1}$ (W m$^{-2}$)$^{-1}$, are: 0.21 (range 0.03–0.63) for Nova Scotia coastal waters,[370] 0.25 for nanoplankton ($< 22$ $\mu$m) in the lower Hudson estuary, USA,[299] and 0.10 for diatoms and 0.14 for blue-green algae in eutrophic Lough Neagh, N. Ireland.[217] To convert these values to a basis of $\mu$einsteins m$^{-2}$ s$^{-1}$, rather than W m$^{-2}$, they can be divided by 4.15.

### Areal and volumetric efficiencies

We have seen that the efficiency of utilization of the light incident on the aquatic ecosystem for primary production is determined by two main factors: the extent to which the aquatic plants succeed in competing with the other components of the system for the quanta in the underwater light field, and the efficiency with which the absorbed light energy is converted to chemical energy. We shall now consider the overall efficiency which results from the simultaneous operation of these two factors.

The most common and generally useful way of expressing the overall efficiency is as the proportion of the light energy (400–700 nm) incident per m$^2$ of water surface which is photosynthetically stored as chemical energy in plant biomass throughout the water column. This we shall refer to as the areal efficiency, $\varepsilon_A$: it is obtained by dividing the areal (integral) photosynthetic rate expressed in energy units (MJ equivalent of photosynthetic assimilate) per m$^2$ per unit of time (hour or day), by the total PAR (in MJ) incident per m$^2$ of water surface in the same time

$$\varepsilon_A = 0.472\, P_A(CO_2)/E_d(\text{on surface}) \tag{10.13}$$

For oceanic waters, $\varepsilon_A$ values vary by a factor of 100–300 from the least to the most productive. Koblents-Mishke (1979) reviewing Russian and other work reported that marine $\varepsilon_A$ values worldwide ranged from 0.02 to 5%. Morel (1978) reported $\varepsilon_A$ values (calculated on a daily basis) of about 0.02% for the very oligotrophic Sargasso Sea, 0.02–0.07% for the Caribbean Sea (oligotrophic), 0.06–0.25% for the moderately productive eastern equatorial Pacific, and 0.4–1.66% for the eutrophic Mauritanian upwelling area in the eastern tropical Atlantic.

Brylinsky (1980) analysed the collected International Biological Programme data for lakes and reservoirs throughout the world. The calculated $\varepsilon_A$ values for the whole growing season ranged from about 0.002 to 1.0%, most of the values being in the range 0.1 to 1.0%. Talling, Wood, Prosser & Baxter (1973) found high efficiencies in two very productive Ethiopian soda lakes dominated by blue-green algae: $\varepsilon_A$ based on half-hour incubations at high incident irradiances ranged from 0.5 to 1.6% in Lake Kilotes and from 1.2 to 3.3% in Lake Aranguadi. In the Sea of Galilee (Lake Kinneret), Dubinsky & Berman (1981) found $\varepsilon_A$ (measured 09.00–12.00 h) to vary from 0.3% in August when the phytoplankton consisted mainly of small chlorophytes, to a maximum of 4% in April during the *Peridinium* (dinoflagellate) bloom.

From literature data for eight lakes covering a wide range of latitude and trophic status, Tilzer, Goldman & De Amezaga (1975), calculated values of $\varepsilon_A$ ranging from 0.035% in the very oligotrophic, high-altitude Lake Tahoe (California, USA) up to 1.76% in the eutrophic Loch Leven (Scotland). Areal efficiency was strongly correlated with the concentration of algal biomass per unit volume: Tilzer *et al.* considered that the key factor responsible for the variation in $\varepsilon_A$ in these lakes was the proportion of the total incident light captured by the phytoplankton.

A way of expressing the overall efficiency of light utilization which can provide information about its variation with depth through the water column is the volumetric efficiency, $\varepsilon_V$. We may define this as the proportion of the downwelling light energy (400–700 nm) incident upon the upper surface of any unit volume within a water body which is photosynthetically stored as chemical energy in plant biomass within that volume. A more all-embracing definition of $\varepsilon_V$ should take into account the light incident upon the unit volume from all directions, i.e. scalar irradiance, $E_0$, rather than downward irradiance, $E_d$. The definition based on $E_d$ is, however, more convenient and serves the purpose well enough: $\varepsilon_V$ is given by

$$\varepsilon_V = \frac{0.472 \, B_c \, P(CO_2)}{E_d(z)} \tag{10.14}$$

$P(CO_2)$ being expressed in moles $CO_2$ mg chl $a^{-1}$ h$^{-1}$, and $E_d(z)$ in MJ m$^{-2}$ h$^{-1}$. Unlike $\varepsilon_C$ and $\varepsilon_A$, $\varepsilon_V$ is not dimensionless, since it has the units m$^{-1}$. Platt (1969) pointed out that volumetric efficiency at a specific depth has the same dimensions, m$^{-1}$, as the vertical attenuation coefficient for irradiance, and that it is in fact equivalent to that part of the total vertical attenuation coefficient for downward irradiance which is due to removal of light by photosynthetic conversion to chemical energy

$$K_d(\text{total}) = K_d(\text{photosynthetic}) + K_d(\text{physical})$$

$K_d(\text{physical})$ being that part of the total vertical attenuation coefficient which is due to removal of light by all processes other than photosynthetic conversion to chemical energy: it thus includes that part of the light absorption by phytoplankton which fails to result in photosynthesis.

$K_d$(photosynthetic) is equivalent to $\varepsilon_V$. Platt's data for St Margaret's Bay, Nova Scotia, Canada, indicate a general tendency for $\varepsilon_V$ to increase with depth (average $\varepsilon_V$ was 0.07% m$^{-1}$ at 1 m and 0.21% m$^{-1}$ at 10 m). This is to be expected since, as we have already noted, conversion efficiency, $\varepsilon_C$, increases with depth, and $\varepsilon_V = B_c k_c \varepsilon_C$.

Having noted the equivalence of $\varepsilon_V$ with $K_d$(photosynthetic), we can now go on to define one more efficiency parameter, namely, the proportion of the total light energy *absorbed* within unit volume of medium which is photosynthetically stored as biomass chemical energy. This is equal to $K_d$(photosynthetic)/$K_d$(total) and, following Morel (1978), we shall refer to it as the radiation utilization efficiency, and give it the symbol, $\varepsilon$

$$\varepsilon = \varepsilon_V / K_d \qquad (10.15)$$

From the definitions of $\varepsilon_C$ and $\varepsilon_V$ it follows that

$$\varepsilon = \frac{B_c k_c}{K_d} \cdot \varepsilon_C \qquad (10.16)$$

i.e. $\varepsilon$ is the (dimensionless) product of the fraction of the total absorbed light which is captured by phytoplankton and the energy conversion efficiency of the phytoplankton. Thus, $\varepsilon$ combines directly the two factors controlling the efficiency with which the aquatic ecosystem utilizes incident light energy for photosynthesis: $\varepsilon$ varies with depth and the integral of $\varepsilon(z)$ with respect to depth over the whole euphotic zone gives $\varepsilon_A$, the areal efficiency.

Given the tendency of the conversion efficiency, $\varepsilon_C$, to increase with depth we might expect $\varepsilon$ also to increase with depth, and this generally seems to be the case.[328] However, variation in $\varepsilon$ with depth depends also on the variation of $k_c$ with depth resulting from changes in the spectral distribution. From one water body to another $\varepsilon$ will increase as the phytoplankton biomass increases ($B_c$ in eqn 10.16) but decrease as the background attenuation due to dissolved colour etc. (increasing $K_d$ in eqn 10.16) rises. For the oligotrophic Sargasso Sea and Caribbean Sea Morel (1978) found $\varepsilon$ to increase with depth from about 0.01 to 0.1%. In the somewhat more productive waters of the equatorial eastern Pacific the range was from about 0.01% near the surface to approaching 1% at low light levels. In the productive waters of the Mauritanian upwelling, $\varepsilon$ rose from 0.1–0.4% near the surface to 2–7% near the bottom of the euphotic zone. In Lake Kinneret (Sea of Galilee), Dubinsky & Berman (1981) found that during the *Peridinium* bloom (428 mg chl *a* m$^{-2}$), $\varepsilon$ rose from just under 1% at the surface to 6.5% at 3 m. After the *Peridinium* bloom had collapsed, to be replaced with a much lower biomass (50 mg chl *a* m$^{-2}$, chlorophytes), $\varepsilon$ was $\sim 0.3$% at the surface, and rose to about 2% at 10 m.

## 10.3 Photosynthesis and wavelength of incident light

The spectral composition of underwater light in a given water body varies markedly with depth and at any specified depth it varies with the optical properties of the water (Chapter 6). Thus, in order to assess the suitability of a given underwater light field for photosynthesis by different kinds of aquatic plant, we need to know in what ways the photosynthetic rates of the various plant types depend upon the wavelengths of the light to which they are exposed. Some of the information we need can be provided by an action spectrum of photosynthesis: this is the curve obtained by plotting the photosynthetic response of a plant per unit incident irradiance (quanta m$^{-2}$ s$^{-1}$) of incident light at a series of wavelengths across the photosynthetic range. Action spectra are normally measured at light intensities low enough to ensure that the response is proportional to the incident irradiance, i.e. they are measured in the initial, linear, region of the *P versus $E_d$* curve. The rate of photosynthesis at any wavelength under these conditions will be equal to the rate of photon absorption times the quantum yield, $\phi$. In the case of a macrophyte thallus or leaf, the rate of light absorption at a given wavelength is proportional to the absorptance at that wavelength. Thus the action spectrum will have a shape similar to that of the absorptance spectrum, modified by any variation in quantum yield with wavelength that may occur (see later).

The action spectrum of macrophytes can usually be measured on a single leaf or thallus or piece thereof, and can reasonably be taken as representative of the ability of the original plant *in situ* to make use of light of different wavelengths. In the case of phytoplankton, however, action spectra must for practical reasons be measured on suspensions containing many cells or colonies: sometimes, if reasonable rates are to be achieved, on concentrated suspensions. How does an action spectrum determined on a suspension relate to the action spectrum we wish to know, namely that of the individual cell or colony? The action spectrum of a cell or colony would be identical to the spectral variation in rate of absorption of quanta, modified as usual by any variation in quantum yield. The average rate of absorption of quanta per individual cell or colony, from a light beam, is the product of the irradiance and the mean absorption cross-section $(\overline{\alpha_p A_p})$ of the cells or colonies (§9.4). Thus the action spectrum of a single cell or colony would have the same shape (allowing for variations in $\phi$) as the spectral variation in $\overline{\alpha_p A_p}$, and this is fixed, being determined only by the absorption characteristics of the individual cells or colonies.

The action spectrum of the phytoplankton suspension measured in the laboratory will, by contrast, have the same shape (allowing for variations in $\phi$), as the absorptance spectrum of the whole suspension. The shape of this spectrum is not fixed, being a function of the number of cells/colonies per unit area of the illuminating beam. Since absorptance can only approach 1.0 and can never exceed it, the absorptance at weakly absorbed wavelengths (e.g. in the green in the case of chlorophytes, diatoms etc.) progressively approaches absorptance at strongly absorbed wavelengths (e.g. at the chlorophyll red peak) as the concentration of cells or colonies increases.

From its definition (§3.2) it follows that the *absorbance* of the suspension, $D_{sus}$, is equal to $-0.434 \ln (1 - A_{sus})$ where $A_{sus}$ is the absorptance of the suspension. From this it in turn follows (since $\ln (1+x) \simeq x$ when $x \ll 1$) that for low values of $A_{sus}$, $D_{sus} \simeq 0.434 A_{sus}$. We saw earlier (§9.2) that $D_{sus} = 0.434 \, n \overline{\alpha_p A_p}$, and therefore at low values of $A_{sus}$, $A_{sus} \simeq n \overline{\alpha_p A_p}$, where $n$ is the number of cells or colonies per ml. The true action spectrum of the individual cells or colonies follows (allowing for variations in $\phi$) the spectral variation in $\overline{\alpha_p A_p}$ (see above). Therefore, the action spectrum of the suspension will have approximately the same shape as the action spectrum of the individual cells or colonies provided that it is measured on suspensions with low values of absorptance. The discrepancies are about 5, 11 and 19% when $A_{sus}$ is 0.1, 0.2 and 0.3, respectively. The general effect of the error is that the action spectrum of the suspension will be flattened (i.e. peaks lowered relative to the valleys) compared to the true action spectrum of the cells or colonies.

The earliest measurements of photosynthetic action spectra were those of Engelmann (1884). He illuminated pieces of algal tissue with a spectrum and then observed the migration of oxygen-requiring bacteria to the different parts of the tissue. He used the concentration of bacteria around different parts of the tissue as an indication of the relative ability of different colours of light to elicit $O_2$ evolution. Approximate though this method was, the results did indicate that other pigments in addition to chlorophyll participated in photosynthesis. Nowadays, laboratory determination of action spectra is most commonly carried out using a high-intensity monochromator as a light source, and a platinum electrode to measure the rate of $O_2$ evolution.[181] Manometry has also been used,[110, 111] as has $^{14}CO_2$ fixation.[200]

When the action spectrum is measured by determining the rate at a series of different wavelengths, then the curve of photosynthetic rate per unit irradiance obtained is generally found to be somewhat similar, but not identical, to the curve of absorptance for the tissue or cell suspension. Fig. 10.6 shows the action and corresponding absorptance spectra measured on multicellular green, brown and red algae,[181] and on the unicellular green

Fig. 10.6. Action spectra of photosynthesis in multicellular and unicellular algae. In (a)–(d) the action spectrum (photosynthetic rate per unit incident irradiance) has been plotted so that it coincides with the absorptance spectrum at some appropriate wavelength. (a), (b) and (c) – bare platinum electrode method (after Haxo & Blinks, 1950). (d) – manometric method (after Emerson & Lewis, 1943); the spectra in this case were obtained with dense suspensions and so do not accurately correspond to the absorption and action spectra of individual cells (see text). (e) – $^{14}CO_2$ method (plotted from data of Iverson & Curl, 1973). (a) Ulva taeniata (multicellular green). (b) Coilodesme californica (multicellular brown). (c) Delesseria decipiens (multicellular red, from shaded habitat). (d) Chlorella pyrenoidosa (unicellular fresh-water chlorophyte). (e) Skeletonema costatum (marine diatom).

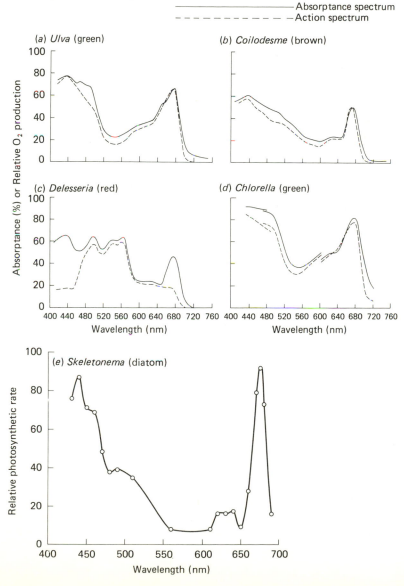

alga *Chlorella*,[111] together with an action spectrum for the marine diatom *Skeletonema*.[200]

It will be observed that the action spectrum falls below the absorption spectrum in certain spectral regions. This means that the amount of photosynthesis achieved per unit light absorbed is lower, i.e. the quantum yield is lower, at these wavelengths than at others. Fig. 10.7 shows the quantum yield as a function of wavelength in the blue-green alga *Chroococcus*, the green alga *Chlorella*, and the diatom *Navicula*.[110, 111, 493] A fall in quantum yield is observed on the long-wavelength side (680–710 nm) of the chlorophyll *a* red absorption peak, and also in the region of carotenoid absorption (440–520 nm), this latter fall being particularly marked in the blue-green alga and significant in the green alga but only slight in the diatom. Such observations led to the belief, initially, that while light absorbed by the carotenoid fucoxanthin in diatoms is efficiently used for photosynthesis, light absorbed by carotenoids in blue-green and green algae is used with much lower efficiency. While there can indeed be some variation in the efficiency with which light absorbed by different pigments is utilized for photosynthesis, it is now thought that this is not the major factor involved in the fall in quantum yield. Furthermore, the explanation originally suggested for the decrease in $\phi$ in the far-red – namely, that the quanta are insufficiently energetic to bring about photosynthesis – is also incorrect.

The variation in quantum yield with wavelength is now considered to be primarily due to the fact that there are two light reactions, both of which

Fig. 10.7. Quantum yield as a function of wavelength in three unicellular algae. (——— *Navicula minima*, diatom (after Tanada, 1951); ----- *Chlorella pyrenoidosa*, green (after Emerson & Lewis, 1943); ...... *Chroococcus* sp., blue-green (after Emerson & Lewis, 1942).)

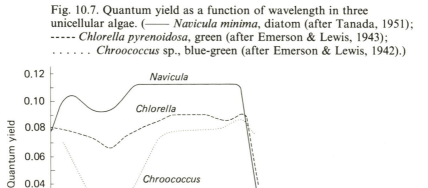

must be energized for photosynthesis to occur, and which have – to an extent varying with the algal type – different light-harvesting pigment arrays with differing absorption spectra. Thus, during the measurement of an action spectrum, when the wavelength of the monochromatic light provided is strongly absorbed by one photosystem but weakly by the other, then the two light reactions do not operate at the same rate and so photosynthetic efficiency is low. For example, 680–710 nm light is absorbed quite well by photosystem I but poorly by photosystem II. Consequently when cells are illuminated with far-red light, the rate of photosynthesis is low because photosystem II cannot keep up with photosystem I.

Emerson (1958) showed that if *Chlorella* cells illuminated with far-red light were simultaneously exposed to light absorbed by chlorophyll *b* (which transfers most, perhaps all, its energy to photosystem II), then the quantum yield increased above that expected on the basis of the additive contributions of the two kinds of light alone. This phenomenon, of a synergistic effect of two wavebands of light, is referred to as *enhancement* and can plausibly be explained in terms of the existence of the two photosystems. Invoking the photosystems to explain the dependence of quantum yield on wavelength immediately raises problems of its own however. The virtually constant and high value of $\phi$ between 560 and 680 nm (Fig. 10.7) indicates that both photosystems are working at the same rate throughout this wavelength range, which is rather surprising in view of the differing pigment complements of the two photosystems. The generally accepted explanation of this phenomenon, originally proposed by Myers & Graham (1963), is that when photosystem II is absorbing light faster than photosystem I, some of its excitation energy is transferred to photosystem I, so that both systems can operate at the same rate. This is known as the 'spillover' hypothesis, and there is now good evidence for its validity:[72] transfer of energy from photosystem I to II does not appear to take place. Put simply, light absorbed by photosystem II can be used to drive both light reactions; light absorbed by photosystem I drives only light reaction 1. We may therefore take the fall-off in quantum yield in certain parts of the spectrum as indicating that in those spectral regions absorption is mainly due to photosystem I pigments.

It follows from all this that action spectra measured with monochromatic light are going to be biased heavily in their shape towards the absorption spectrum of photosystem II, and may therefore be a poor guide to the ability of a particular plant to make use of the mixture of wavelengths (however enriched this may be in one spectral region or another) that is invariably present in underwater light fields. A more meaningful action spectrum, for ecological purposes, may be obtained by carrying out the measurements with two simultaneously applied light sources; one with wavelength changing progressively to cover the whole photosynthetic range, the other constant at a wavelength absorbed by photosystem II. In this way, even when the variable source is set to spectral regions where absorption is mainly due to photosystem I, the fixed-wavelength background source ensures that photosystem II continues to operate. Fig. 10.8 shows, for a red alga, the difference between action spectra measured[126] in the absence or the presence of background green (546 nm) light, absorbed by phycoerythrin, which is known to supply its energy directly to photosystem II. When photosystem II is energized in this way, there is a great increase in the photosynthetic response to light in the 400–480 nm and 600–700 nm regions. It appears likely that most of the chlorophyll *a* (400–450 nm, 650–700 nm) and carotenoids (400–500 nm) in this alga are part of photosystem I. The discrepancy between action spectra measured with and without background photosystem II light is particularly great in biliprotein-containing algae, because in such algae it seems to be generally true that the biliproteins, which constitute such a major but spectrally confined part of the light-harvesting apparatus, transfer all their energy directly to photosystem II, whereas the chlorophyll *a* directs most of its energy to photosystem I. In algae such as the diatoms, brown algae and green algae, where the pigment arrays of the two photosystems are not so different, the difference between action spectra measured with and without background photosystem II light is not so marked.

Fig. 10.8. Photosynthetic action spectra of the red alga *Cryptopleura crispa* measured with and without background green (546 nm) light, together with the absorptance spectrum (from Fork, 1963).
Reproduced from *Photosynthetic mechanisms in green plants*, National Academy Press, Washington DC, 1963.

It can be observed in Fig. 10.8 that the action spectrum measured with background photosystem II light is fairly close to the absorptance spectrum of the thallus. A similar observation has been made for other red algae.[399] In the absence of action spectra data measured with background photosystem II light, the absorptance spectrum may be used as an approximate guide to the ability of an aquatic plant not only to capture but also to use light of different wavelengths. The discrepancies between the absorptance spectra and the photosystem II-supplemented action spectra of aquatic plants are nevertheless significant and of interest. It has been observed for a number of red algae,[126, 399] and can be seen in Fig. 10.8, that the green-supplemented action spectrum falls significantly below the absorptance spectrum in the 400–520 nm region, indicating that some of the pigment molecules absorbing in this region are not transferring their excitation energy efficiently to the reaction centres. The molecules in question are probably carotenoids. It seems unlikely that the carotenoids in the red algae are completely inactive: if they were inactive we would expect a much greater fall in the green-supplemented action spectrum between 450 and 500 nm. The photosystem II-supplemented action spectrum of the blue-green alga *Anacystis* also indicates that carotenoids contribute excitation energy for photosynthesis.[228]

Thus, in the red and blue-green, and possibly also green and xanthophycean algae[150] the absorption spectrum is likely, because of the inactivity of some of the xanthophyll molecules, partly to overestimate the ability of the cells to utilize light in the blue region of the spectrum. In those algal groups (diatoms, phaeophytes, dinoflagellates, siphonous green algae) which rely heavily on specialized carotenoids (fucoxanthin, peridinin, siphonaxanthin) for light harvesting, however, virtually all the carotenoid molecules transfer their absorbed energy to the reaction centres as efficiently as does chlorophyll. In these algae we may reasonably assume that the absorption spectrum (absorptance or absorption cross-section, as appropriate) is quite an accurate guide to the ability of the cells to utilize light in all regions of the photosynthetic spectrum.

# 11

Photosynthesis in the aquatic
environment

Having considered the photosynthetic response of aquatic plants
to light of different intensities and spectral qualities, we shall now examine
how the availability of light influences where, when and how much
photosynthesis takes place in aquatic ecosystems, and also the extent to
which other parameters of the environment can limit photosynthesis.
Aquatic production ecology is an enormous field: a comprehensive
account will not therefore be attempted. Rather, the broad principles
governing the controlling influence of light and other parameters will be
outlined and illustrated by a few examples. More detailed accounts and
extensive bibliographies can be found in the final International Biological
Programme Report on *The functioning of freshwater ecosystems* edited by
Le Cren & Lowe-McConnell (1980), in the book by Raymont (1980) on
*Plankton and productivity in the oceans*, and in the multi-author work edited
by Morris (1980) on *The physiological ecology of phytoplankton*.

## 11.1 Circulation and depth

We saw in the previous chapter that except under very still
conditions with virtually no wind or waves, there is always circulation of
water in the upper layer. We also saw that this can be an advantage to
the phytoplankton insofar as, by ensuring that they are not exposed to the
intense light just below the surface for very long, they avoid photoinhibition.
This circulation can, however, also be a disadvantage to the phytoplankton
if, in the lower reaches of the mixed layer, the light intensity is too low
for net photosynthesis to be achieved. As the depth of the mixed layer
through which the phytoplankton is circulating increases, so the average
light intensity to which the cells are exposed decreases, and consequently
the total rate of photosynthesis by the whole phytoplankton population
throughout the water column decreases. The rate of respiration of the

whole phytoplankton population, on the other hand, will be essentially constant whatever the mixing depth. Thus, as was first pointed out by Braarud & Klem (1931), there exists a mixed-layer depth – the *critical depth*, $z_c$ – beyond which respiratory carbon loss by the whole population exceeds photosynthetic carbon gain, and so net phytoplankton growth cannot occur. Even when the critical depth is not exceeded, increases in mixing depth tend to reduce total photosynthesis.

The depth through which circulation can occur is limited either by the depth of the bottom or (from spring to autumn in temperate latitudes and throughout the year in tropical oceans) by the presence of a shallow thermocline. Thus, one answer to the question, where does aquatic photosynthesis take place is that it takes place best in shallow waters or in waters in which circulation is confined to a shallow layer by thermal stratification. The converse is that net photosynthesis occurs poorly, or not at all, in waters in which the phytoplankton is circulated through great depths. For a series of Japanese lakes covering a wide range of depths, Sakamoto (1966) found a general tendency for deeper lakes to be less productive than shallow ones. Even when differences in nutrient supply were allowed for, the inhibiting effect of depth remained. In the marine environment too, we may plausibly attribute the greater productivity of shallow coastal waters (relative to the deeper waters) in part (i.e. in addition to nutrient differences) to their lesser depth. Isolated shallow areas within the oceans such as the Faroebank (100 m depth) west of the Faroe Islands are also more productive than the surrounding oceanic waters.[467]

The higher productivity of shallow water applies to the benthic flora also. Any surface within the euphotic zone of a water body is usually found to support a productive plant community. This applies not only to the more obvious macrophyte communities such as kelp forests, seagrass beds, brown and red algal associations on underwater rocks, but also to seemingly bare sand and mud surfaces which usually harbour a dense microflora amongst the grains. This microflora is likely to consist of specialized benthic diatoms in temperate mud flats and sand, or of symbiotic algae within Foraminifera in the sands of tropical lagoons.[452] Moving in a seawards direction, as depth increases so the standing crop of biomass (macrophytes and microflora) per unit area, and the rate of primary production per unit area, both decrease until eventually a depth is reached beyond which the light intensity is too low to support any plant growth. For the important group of large sublittoral brown algae, loosely referred to as the kelps, the lower limit occurs at depths where the downward irradiance is 0.7–1.4% of that penetrating the surface: this depth limit ranges from as little as 8 m in the turbid water around the island of Helgoland (North Sea) to about 95 m in the very clear waters of the central Mediterranean.[296] For the coralline crustose red algae – commonly the most deeply occurring algal type – the lower depth limit occurs where the irradiance is 0.05–0.1% of the subsurface value: this corresponds to about 15 m depth off Helgoland but to as much as 175 m in the Caribbean.[296] Lüning & Dring (1979) found for the sublittoral region off Helgoland that the total downwelling light (400–700 nm) received *per year* was about 15 MJ m$^{-2}$ or 70 einsteins m$^{-2}$ at the lower kelp limit (8 m) and 1 MJ m$^{-2}$ or 6 einsteins m$^{-2}$ at the lower red-algal limit (15 m): these may be regarded as the approximate minimum yearly requirements of PAR for growth of these algal types in any waters.

The macrophyte *Najas marina*, from the Yarkan River in Israel, has a light compensation point of $\sim 5\ \mu$einsteins (PAR) m$^{-2}$ s$^{-1}$, and this corresponds to the mid-day (August) irradiance value at 1.7 m depth in the river. This is in reasonable agreement with the maximum depth – 1.5 m – at which the plant is found growing: plants transplanted to greater depths (2.0 or 2.5 m) deteriorate.[3] Bodkin, Posluszny & Dale (1980) found the macrophyte *Hippuris vulgaris* growing down to a depth of 6 m in two clear-water Scottish lakes. In midsummer the irradiance at this depth was about 100 $\mu$einsteins (PAR) m$^{-2}$ s$^{-1}$, and the temperature was 15–18° C. In the laboratory, shoots of *H. vulgaris* at 15° C continued to grow at 140 $\mu$einsteins m$^{-2}$ s$^{-1}$ but ceased to grow at 60 $\mu$einsteins m$^{-2}$ s$^{-1}$. Bodkin *et al.* concluded that the depth limits of *H. vulgaris* in the summer are governed by the decrease in light intensity with depth.

It is instructive to consider under what conditions phytoplankton primary production may be prevented altogether as a result of the mixed depth's exceeding the critical depth. Using Talling's (1957*b*) model for calculating integral photosynthesis (see § 11.5) we can derive an approximate expression for the critical depth

$$z_{\mathrm{c}} = \frac{N}{24\,K_{\mathrm{d}}\rho} \ln\!\left(\frac{\overline{E_{\mathrm{d}}(0)}}{0.5\,E_{\mathrm{k}}}\right) \tag{11.1}$$

where $\rho$ is the ratio of respiration rate to light-saturated photosynthetic rate in the phytoplankton, $K_{\mathrm{d}}$ is the vertical attenuation coefficient for downwards irradiance of PAR, $N$ is daylength in hours (readily calculated from eqn 2.11), $\overline{E_{\mathrm{d}}(0)}$ is the average value of downward irradiance just below the surface during daylight hours, and $E_{\mathrm{k}}$ is the irradiance value defining the onset of saturation (§ 10.1). We can see from eqn 11.1 that the likelihood of the mixed layer depth exceeding the critical depth increases (i.e. $z_{\mathrm{c}}$ decreases) as respiration rate (relative to photosynthetic rate), attenuation by the water or the light-saturation parameter increase: an increase in incident irradiance or in daylength has the opposite effect.

As an example of a temperate-zone lake with relatively low attenuation we shall take Lake Windermere, England, and use the data of Talling (1957a). For a day in late spring (May 1st, 1953), with $\overline{E_\mathrm{d}}(0) \simeq 170$ W m$^{-2}$, daylength $\simeq 15$ h, $K_\mathrm{d} \simeq 0.43$ m$^{-1}$, and the phytoplankton (predominantly the diatom, *Asterionella*) having $E_\mathrm{k} = 6.3$ W m$^{-2}$, and $\rho = 0.033$, we may calculate that the critical depth was about 177 m. Since the lake itself (North basin) has a maximum depth of only about 63 m, there is clearly no possibility of net primary production being prevented altogether. Even if the lake were much deeper, once the spring–summer thermal stratification had set in, the comparatively shallow depth of the epilimnion (the layer, commonly 5–20 m deep, above the thermocline, within which wind-induced circulation occurs) would prevent the critical depth's being exceeded. For Lake Windermere in midwinter, on the other hand, with daily incident light down to a small fraction of the summer value and circulation through the full depth of the lake, Talling (1971) estimated that phytoplankton growth was impossible. In the much shallower Esthwaite Water ($z_\mathrm{max} \simeq 15$ m), in contrast, he concluded that phytoplankton growth should be possible throughout the year.

Even in shallow water bodies, if the attenuation is high enough the circulation depth can exceed the critical depth. For turbid Lough Neagh (N. Ireland), with a mean depth of 8.6 m and $K_\mathrm{d}$ values commonly 1.5–3.5 m$^{-1}$, Jewson (1976) estimated that the critical depth would be exceeded in winter, thus precluding growth. In spring, despite the lack of stratification in this well-mixed lake, the increase in illumination was sufficient to ensure that $z_\mathrm{c}$ was not exceeded.

When high attenuation is combined with a requirement for high light intensity for saturation, then the critical depth can be very shallow. In Lake George, Uganda, on the equator, Ganf (1975) observed $K_\mathrm{d}$ values in the region of 9 m$^{-1}$, and the blue-green algal population had a light-saturation onset parameter ($E_\mathrm{k}$) of about 55 W m$^{-2}$, on a day (12th April, 1968) when the average incident irradiance ($\overline{E_\mathrm{d}}(0)$) was 360 W m$^{-2}$. Assuming $\rho$ to be about 0.1 and the daylength to be 12 h, eqn 11.1 gives a critical depth of 1.4 m. Thus, despite the high irradiance incident on this tropical lake, the mixed depth could often exceed the critical depth. In productive lakes of this type a large part of the attenuation is due to the phytoplankton itself and so the situation is self-correcting. Cessation of growth is likely to be followed by breakdown of the algae, following which attenuation falls and photosynthesis and growth can recommence.

In the marine environment, the relation between the mixed depth and the critical depth is of particular importance, and indeed the first attempt to characterize this relation quantitatively was carried out by Sverdrup

(1953) for the Norwegian Sea. In middle and high latitudes the mixed layer is deep at the end of the winter (varying from 100–400 m in March at 66°N 2°E), but in the spring a shallow mixed layer (25–100 m) develops due to thermal stratification. For the 66°N station that Sverdrup studied, he estimated that until the last week of April the mixed layer was much deeper than the critical depth, thus precluding phytoplankton growth, but that after the middle of May the mixed-layer depth was smaller than the critical depth so that growth was possible.

Strictly speaking, these various interactions between mixed depth and critical depth apply to non-motile forms such as diatoms. Motile algae such as dinoflagellates can to some degree escape these effects by migrating again up to regions of higher irradiance if the circulating water moves them down. Nevertheless the evidence is clear that for the total phytoplankton population, circulation of water within the mixed layer is a major factor influencing primary production, and frequently determines whether such production takes place at all.

In the sea, the mixed layer, once thermal stratification has been established (spring to autumn in moderate and high latitudes, all year in the tropics), is typically 20–100 m deep. Within and below the thermocline there is comparatively little circulation of the water. Since the euphotic zone will frequently be deeper than the mixed layer (for marine waters with $K_d$ values of 0.03, 0.05, 0.11, 0.16 m$^{-1}$ (Table 6.1), $z_{eu} \simeq 153, 92, 42$ and 29 m, respectively), there is, throughout much of the oceans and coastal seas, a substantial layer of water which combines sufficient light intensity to support photosynthesis, with a stable water column.

When the vertical profile of phytoplankton chlorophyll is measured in the (stratified) ocean, a distinct peak of chlorophyll concentration is normally found close to the bottom of the (nominal) euphotic zone. An example of this *deep chlorophyll maximum* as it is often called, in the northeast Pacific Ocean is shown in Fig. 11.1. The peak of the chlorophyll distribution is either a little above or a little below (as in Fig. 11.1) the depth at which $E_d$ is 1% of the subsurface value. It will be noted that there is a layer of increased photosynthetic activity in this region indicating that the algal cells are active, not moribund. This layer of deep phytoplankton, commonly about 10 m thick, is very widespread in the world's oceans: it appears, for example, to extend right across the Pacific.[429] It seems likely that the increase in chlorophyll concentration in this deep layer, relative to that in the mixed layer, is due to an increase in chlorophyll concentration within the cells as well as to an increased population of phytoplankton.[7, 246, 353] Kiefer, Olson & Holm-Hansen (1976) found the chlorophyll per unit biomass in the deep phytoplankton layer to be about twice the value near the surface.

The mechanism (apart from shade adaptation) by which the deep chlorophyll maximum becomes established is uncertain. A commonly held view is that it is related to the distribution of nutrients. In the well-illuminated mixed layer the phytoplankton rapidly consumes the nutrients in the spring, and the population then decreases. Below the thermocline nutrient concentrations are much higher. Furthermore, additional nutrients can diffuse upwards from the deep water below. The continued availability of nutrients in this region, possibly combined with a diminution in the sinking rate of the phytoplankton once it reaches these nutrient-rich waters, may account for the establishment and maintenance of this phytoplankton layer.[7, 202, 465] The deep phytoplankton layer makes a significant contribution to total primary production in those waters where it occurs.

In most inland waters the euphotic depth is less than the mixed depth, and so the conditions for the development of a deep chlorophyll maximum do not occur. However, Fee (1976) found a very narrow layer with very high phytoplankton chlorophyll concentration in the region of, or below, the thermocline in a series of lakes in northwestern Ontario (Canada), which were both clear and thermally stratified. This layer was found between 4 and 10 m depth (depending on the lake) where the irradiance was 0.3–3.5% of the subsurface value. Phytoplankton samples from this

Fig. 11.1. The deep chlorophyll maximum in the Northeast Pacific Ocean (after Anderson, 1969). The temperature profile shows that the mixed layer is about 25 m deep.

layer were always dominated by large colonial chrysophycean flagellates of the genera *Dinobryon, Synura, Uroglena* and *Chrysosphaerella*, just one species normally being dominant in the deep chlorophyll layer of any given lake. In the very clear, colourless waters of Crater Lake and Lake Tahoe, comparable optically to ocean water, a deep chlorophyll maximum develops at considerable depth ($\sim 75$ m) in the summer.[279, 504]

## 11.2    Optical characteristics of the water

We saw in the previous chapter that a major factor in limiting efficiency of utilization of incident light in aquatic ecosystems is the removal of a large proportion of the light energy by the aquatic medium. This occurs in waters in which the vertical attenuation of PAR by non-phytoplanktonic material is high (see Table 10.1). We would therefore expect, for example, brown-water lakes with a high concentration of gilvin to be on the average less productive than lakes with low background colour. Observation suggests that this is indeed the case although no wide-ranging survey appears yet to have been carried out. However, with the ready availability nowadays of spectrophotometric techniques for determining the absorption properties of natural waters, it is to be hoped that the relation between productivity and background water colour will be studied quantitatively for a variety of water-body types. It should be noted that a high total vertical attenuation coefficient need not necessarily be associated with low productivity since that high coefficient may be due to a high concentration of phytoplankton: it is high attenuation by non-algal material that we expect to lower the productivity.

Turbid waters in which the turbidity is due to large numbers of mineral particles which are themselves of low intrinsic colour, and in which dissolved colour is also low, represent a special case. Because of the intense scattering, the total vertical attenuation coefficient ($K_d(PAR)$) may be quite high so that the euphotic zone is shallow. This fact alone does limit production by reducing the volume of medium which the phytoplankton can exploit photosynthetically. Nevertheless, within the euphotic zone the relative ability of the phytoplankton to collect light is dependent on its absorption properties compared to the other constituents of the medium, and since in the present case the rest of the medium does not absorb the light strongly despite its intense scattering, the phytoplankton is well placed to compete for the available photons. In a medium of this type, the assumption, contained in eqn 10.3, that the relative amounts of the absorbed light captured by phytoplankton and by the rest of the medium are in proportion to $B_c k_c$ and $K_{NP}$, is no longer even approximately valid.

We saw in §10.2 that this assumption can in fact only be made with reasonable accuracy for waters in which vertical attenuation is absorption dominated.

The kind of mechanism we have envisaged, namely effective competition by phytoplankton for the photons within the shallow euphotic zone, may in part account for the surprisingly high productivity of some waters with high inorganic turbidity.[347] The shallowness of such waters (the turbidity often being due to wind resuspension of bottom sediments) will further promote their productivity by ensuring that the circulating cells do not for long remain below the illuminated layer. In addition, the ratio of scalar to downward irradiance is high in turbid waters with a high ratio of scattering to absorption (see Fig. 6.9), and so there is substantially more light available for photosynthesis than the values of downward irradiance alone would indicate.

## 11.3 Other limiting factors

To fully understand the extent to which light availability limits primary production it is necessary to be aware of the limitations which are imposed at the same time by other environmental factors. Consider, for example, a diatom population photosynthesizing in lake water at a saturating light intensity of $\sim 400$ $\mu$einsteins m$^{-2}$ s$^{-1}$, as illustrated in Fig. 10.1. A small increase in light intensity will bring about no change in photosynthetic rate and a large increase will even lead to photoinhibition. What is happening within the cells is that the rate of formation of the products of the light reactions of photosynthesis, namely NADPH$_2$ and ATP, is so high that the enzyme system responsible for the dark reactions

is saturated so far as NADPH$_2$ and ATP are concerned. Any increase in their steady-state concentration resulting from an increase in light intensity is therefore not accompanied by an increase in the rate of CO$_2$ fixation. This does not mean, however, that the system is necessarily working at its maximum possible rate. The CO$_2$ concentration may be too low to saturate the first enzyme, ribulose diphosphate (RuDP) carboxylase, in the dark reaction sequence so that it is not functioning at its maximum capacity. Alternatively, or in addition, the temperature may be too low for maximum activity: at a higher temperature the enzymes of the whole dark reaction system might operate more rapidly and thus be able to consume the $NADPH_2$ and ATP at a faster rate.

### Carbon dioxide

We shall first consider the extent to which $CO_2$ availability limits the overall rate. Given that $CO_2$ is a substrate which is used by an enzyme (system), then we may plausibly suppose that the photosynthetic rate at any given light intensity will vary with $CO_2$ concentration approximately in accordance with the well-known Michaelis–Menten equation for enzyme kinetics

$$v = \frac{Vs}{K_m + s}$$

where $v$ is the rate of the enzyme reaction at substrate concentration $s$, $V$ is the maximum rate obtainable at saturating substrate concentrations, and $K_m$ is the dissociation constant of the enzyme–substrate complex, but is also equivalent to the substrate concentration which gives half the maximum rate. Re-expressing the equation in photosynthetic terms we obtain

$$P = \frac{P_m[CO_2]}{K_m(CO_2) + [CO_2]} \tag{11.2}$$

Fig. 11.2 shows an idealized curve of rate *versus* substrate concentration

Fig. 11.2. Idealized curve of rate of an enzymic reaction as a function of substrate concentration. Calculated from Michaelis–Menten equation.

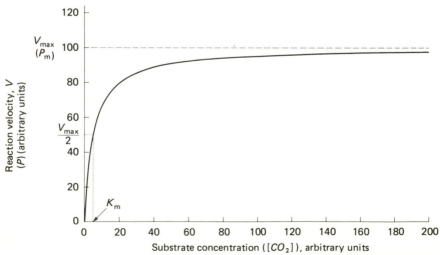

for an enzymic reaction, in accordance with the Michaelis–Menten equation. Observed curves of photosynthetic rate *versus* $CO_2$ concentration for aquatic plants are approximately of this type, and so by obtaining the $K_m$ values from such curves and comparing them with the *in situ* $CO_2$ concentrations we may be able to assess to what extent $CO_2$ availability limits photosynthesis in natural waters.

In the isolated state, ribulose diphosphate carboxylase has a $K_m$ value for $CO_2$ of about 20 $\mu$M. The $K_m(CO_2)$ values for living plants, however, are higher because of diffusive resistance to entry of carbon dioxide into the plant. Some modern values for $K_m(CO_2)$ determined for fresh-water phytoplankton and macrophyte species are listed in Table 11.1: they range from 20 to 170 $\mu$M. Fresh water in equilibrium with the atmosphere ($\sim 0.034$ vol% $CO_2$) at 15° C contains dissolved $CO_2$ at about 14 $\mu$M concentration. Calculations for sea water at 26° C indicate a free $CO_2$ concentration of about 12 $\mu$M.[55] We might therefore expect it generally to be the case that aquatic plant photosynthesis in natural waters is undersaturated with respect to $CO_2$ concentration, and should respond to an increase in $CO_2$ concentration with an increase in photosynthetic rate.

Table 11.1. *Apparent half-rate constants* ($K_m$) *for* $CO_2$ *for photosynthesis in certain fresh-water phytoplankton and macrophyte species*

| Plant species | $K_m(CO_2)$ ($\mu$M) | Reference |
|---|---|---|
| *Phytoplankton* | | |
| *Pediastrum boryanum* (Chlorococcales) | 40[a] | |
| *Cosmarium botrytis* (desmid) | 170[a] | |
| *Anabaena cylindrica* | 60[a] | Allen & Spence (1981) |
| *Macrophytes* | | |
| *Nitella flexilis* | 100[a] | |
| *Eurhynchium rusciforme* | 80[a] | |
| *Fontinalis antipyretica* | 170[a] | |
| *Elodea canadensis* | 30[a] | |
| *Potamogeton crispus* | 20[a] | |
| *Hydrilla verticillata* | 170[b] | |
| *Myriophyllum spicatum* | 150[b] | Van, Haller & Bowes (1976) |
| *Ceratophyllum demersum* | 165[b] | |

*Values determined*: [a]at pH 5.5 and 560 $\mu$einsteins PAR m$^{-2}$ s$^{-1}$; [b]at pH 4.0 and 1000 $\mu$einsteins PAR m$^{-2}$ s$^{-1}$ (full noon summer sunlight is $\sim 2000$ $\mu$einsteins m$^{-2}$ s$^{-1}$)

This need not apply only at saturating light intensity: the assumption that photosynthetic rate depends on $CO_2$ concentration in accordance with the Michaelis–Menten relation implies that even at low light intensities an increase in $CO_2$ concentration above a typical starting value of 12–14 $\mu$M would lead to an increase in photosynthetic rate. This means that at subsaturating light intensities, photosynthesis by plants within a water body is likely to be simultaneously limited by availability of both light and $CO_2$. The aquatic moss, *Fontinalis antipyretica*, and the aquatic higher plant, *Cabomba caroliniana*, both show marked increase in photosynthetic rate with increasing $CO_2$ concentration in the range 10–25 $\mu$M at low, subsaturating, as well as at high, light intensities.[170, 435] Fig. 11.3 shows Smith's (1938) data for *Cabomba*.

Sea water and most inland waters contain much more inorganic carbon in the form of bicarbonate ion, $HCO_3^-$, than in the form of $CO_2$. The different forms of inorganic carbon are interconverted in accordance with

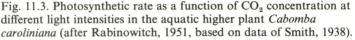

$$H_2O + CO_2 \rightleftharpoons H_2CO_3 \rightleftharpoons H^+ + HCO_3^- \rightleftharpoons 2H^+ + CO_3^{2-}$$

The higher the pH of the water, the more this equilibrium shifts to the right

Fig. 11.3. Photosynthetic rate as a function of $CO_2$ concentration at different light intensities in the aquatic higher plant *Cabomba caroliniana* (after Rabinowitch, 1951, based on data of Smith, 1938).

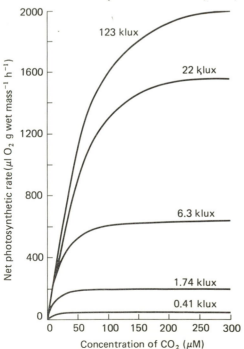

with $HCO_3^-$ constituting more than $50\%$ of the total from pH 6.2 to 9.3, and more than $80\%$ between pH 6.7 and 8.8. Given that $CO_2$ tends, as we have seen, to be present at suboptimal concentration, it would clearly be an advantage in many waters for aquatic plants to be able to use bicarbonate as a carbon source for photosynthesis, and in fact many, but not all, species do. This topic has been reviewed by Raven (1970), and has also more recently been discussed by Hutchinson (1975), Talling (1976) and Allen & Spence (1981). The inorganic carbon species used by the carboxylase enzyme is always $CO_2$. In the plants which can utilize bicarbonate, the $HCO_3^-$ ions are transported into the cell where they give rise to $CO_2$ by the reversal of the first two of the above interconversion reactions. The reversible dehydration of $H_2CO_3$ to give $CO_2$ is catalyzed by carbonic anhydrase, a chloroplast enzyme. The liberated $CO_2$ is then used in photosynthesis. The combination of $HCO_3^-$ ions with $H^+$ to give $H_2CO_3$ leads to a corresponding accumulation of $OH^-$ ions (from $H_2O \rightarrow H^+ + OH^-$) which are excreted from the cells to balance the uptake of $HCO_3^-$.

So far as marine ecosystems are concerned, photosynthetic utilization of bicarbonate has been shown for a number of seaweeds[395] – *Ulva* and *Enteromorpha* (green), *Sargassum muticum* and *Fucus* (brown), *Iridaea* and *Gigartina* (red) – and we may reasonably suspect that it occurs in most marine multicellular algae. Amongst the marine phytoplankton, $HCO_3^-$ utilization has been shown in the coccolithophorid *Coccolithus huxleyi*,[354] but Steemann Nielsen (1975) believes that the coccolithophorids may be an exception: studies on other marine phytoplankton species are needed.

In fresh waters, according to Hutchinson's (1975) review, most aquatic higher plants can utilize bicarbonate for photosynthesis. The few species which cannot utilize bicarbonate typically occur in soft waters in which bicarbonate concentrations are low. Aquatic mosses seem to lack the ability to utilize $HCO_3^-$. Most charophyte species (other than those from soft waters) can utilize $HCO_3^-$ as can benthic filamentous algae such as *Cladophora*. Amongst the fresh-water phytoplankton, the ability to utilize $HCO_3^-$ for photosynthesis varies widely. *Chlorella emersonii, Scenedesmus quadricauda* (green), *Ceratium hirundinella* (dinoflagellate), *Fragilaria crotonensis* (diatom), *Microcystis aeruginosa* and *Anabaena cylindrica* (blue-green) can all effectively use $HCO_3^-$, but *Chlorella pyrenoidosa* (green), *Asterionella formosa* and *Melosira italica* (diatoms) cannot.[5, 395, 487]

Despite the wide occurrence of ability to utilize bicarbonate in the aquatic plant kingdom, it is generally true that, with the possible exception of phytoplankton species adapted to alkaline waters, free $CO_2$ if available is the preferred, i.e. more effectively used, carbon source. For example, the fresh-water macrophyte *Myriophyllum spicatum*, which can use bicarbonate, gives a much higher photosynthetic rate with free $CO_2$ at its optimum concentration than with $HCO_3^-$ at its optimum concentration.[468] Steemann Nielsen (1975) suggests that one of the reasons for the less effective use of $HCO_3^-$ is that energy must be used for its active transport into the plant, whereas $CO_2$ diffuses in freely. The data of Allen & Spence (1981) for fresh-water macrophytes in Table 11.1 show that in all cases, including the known bicarbonate users *Elodea canadensis* and *Potamogeton crispus*, the apparent $K_m$ for $HCO_3^-$ is 50- to 100-fold higher (indicating a much lower affinity) than the $K_m$ for $CO_2$. Furthermore, to give the same photosynthetic rate as a certain concentration of $CO_2$ (at pH 5.5), $HCO_3^-$ (pH 8.8) at 52 to 132 times the concentration was needed. In the case of the blue-green alga *Anabaena cylindrica*, on the other hand occurring typically in alkaline eutrophic waters, the $K_m$ values for $CO_2$ and $HCO_3^-$ are about the same, and the concentrations of $CO_2$ and $HCO_3^-$ which produce a certain photosynthetic rate are also about the same.

On the basis of their studies, Allen & Spence conclude that despite the ability of most fresh-water macrophytes to utilize bicarbonate, they do not in fact obtain much of their carbon from $HCO_3^-$ until the pH of the water exceeds 9.0, and at these high pH values their photosynthetic rates are greatly reduced anyway. Allen & Spence suggest that the natural rates of photosynthesis of fresh-water macrophytes and some planktonic algae are functions mainly of $CO_2$ (as opposed to $HCO_3^-$) concentration. This conclusion may not, however, apply to the more effective bicarbonate users amongst the macrophytes, such as *Myriophyllum spicatum*. Adams, Guilizzoni & Adams (1978) found that in a series of rather alkaline (pH values mainly 7.5–8.8) Italian lakes, all well supplied with phosphorus, the rate of photosynthesis of *M. spicatum* varied with the total dissolved inorganic carbon concentration in accordance with a Michaelis–Menten type of relation (Fig. 11.4). From calculations of the amounts of the different forms of inorganic carbon present, it appeared that photosynthetic rate was related primarily to the concentration of $HCO_3^-$ rather than to the concentration of $CO_2$.

Algae such as *Anabaena cylindrica*, which photosynthesize optimally at pH values in the region of 9.0, make very efficient use of bicarbonate. Two *Anabaena* species and *Chlamydomonas reinhardtii* have in fact been shown to possess an active transport system for $HCO_3^-$ which enables them to achieve intracellular levels of inorganic carbon far in excess of those outside the cells.[23] For these planktonic species and the more effective bicarbonate users amongst the macrophytes, the natural rate of photosynthesis (as far as carbon supply is concerned) can be regarded as a function of the concentration of $HCO_3^-$ plus $CO_2$. The ability of blue-green algae such as *Anabaena*, *Microcystis* and *Spirulina* to continue photosynthesizing in water of high pH, and effectively zero free $CO_2$ concentration could be one of the main factors contributing to their frequent domination of eutrophic lakes in late summer and their invariable domination of highly alkaline waters such as the African soda lakes.

The free $CO_2$ concentrations of 12–14 $\mu M$ referred to earlier apply only to waters in equilibrium with the atmosphere. Such concentrations might normally be found in the surface layer of the oceans, or of unproductive inland waters, or of most inland waters in the winter. However, any biologically productive water body is likely not to be in equilibrium with the atmosphere so far as $CO_2$ is concerned. The free $CO_2$ concentration at any time is likely to differ from the equilibrium value and will vary with time and depth in accordance with where and how fast the processes of consumption (photosynthesis) and production (respiration, decomposition) are taking place; free $CO_2$ can be even more affected by changes in pH caused by photosynthesis. Talling (1976, 1979) presents data for the changes in free $CO_2$ concentration in the eutrophic English lake Esthwaite

Fig. 11.4. Variation of photosynthetic rate of *Myriophyllum spicatum* with total dissolved inorganic carbon concentration in a series of Italian lakes (plotted from data of Adams *et al.*, 1978). Light intensities were at or near saturation in most cases. Using non-linear regression to the Michaelis–Menten equation, Adams *et al.* calculated a $K_m$ (half-saturation constant) value of 1.06 mM total dissolved inorganic carbon (46.5 mg $CO_2$ equivalents $l^{-1}$) and a $P_m$ value of 7.24 mg C $g^{-1}$ dry mass $h^{-1}$.

Water from April to July, 1971. On April 19th, before thermal stratification, the $CO_2$ concentration was the same from the surface down to 10 m depth. By May 3rd when the phytoplankton population had begun to increase, and there was some warming of the surface layer, $CO_2$ concentration had fallen by about 33% in the upper 4 m but had increased greatly between 6 and 10 m. During July, thermal stratification became established, the phytoplankton population above the thermocline increased substantially and free $CO_2$ levels in the epilimnion were reduced virtually to zero, while rising to even higher values than before in the 6–10 m layer below the thermocline. The amount of inorganic carbon removed by photosynthesis represented about 50% of the total inorganic carbon in the surface layer. The remainder of the *ca* 1000-fold decrease in free $CO_2$ was due to the rise in pH from $\sim 7.0$ to above 9.0, resulting from the photosynthetic activity of the dense phytoplankton population. Similar changes are likely to take place in any productive water body. For one of the components of the phytoplankton, the diatom *Asterionella formosa* (not a bicarbonate user), Talling (1979) calculated that on July 12th its photosynthetic rate was virtually zero in the surface layer due to lack of $CO_2$, that it rose between 2 and 3 m due to an increase in $CO_2$ accompanied by reasonably high irradiance values, and that it declined again below this depth because of the fall in irradiance. Thus with increasing depth, its photosynthesis was limited first by a $CO_2$ supply, then by $CO_2$ and light simultaneously, and then mainly by light availability alone. These changes with depth did not apply to total phytoplankton photosynthesis, however, since when the pH rose and the $CO_2$ concentration fell, *Asterionella* was largely replaced by the bicarbonate users *Ceratium* and *Microcystis*.

The sinking rate of diatom cells increases when $CO_2$ uptake becomes limited: the $CO_2$ depletion associated with summer stratification might lead to an increased sinking rate of diatoms, and this may contribute to the paucity of diatoms frequently observed in productive lakes in the summer.[205]

Immediately next to the surface of any aquatic plant, there is an unstirred layer of water that $CO_2$ and/or $HCO_3^-$ ions must diffuse across before they can enter the cells and be used for photosynthesis. The thickness of this layer diminishes if the plant is subjected to turbulence or rapid stirring but the layer never disappears completely. Under well-stirred conditions, the unstirred layer might be only about 5 $\mu$m thick around a small cell such as *Chlorella* but 30–150 $\mu$m thick around the surface of a macrophyte such as *Chara*:[395, 436] in still or slowly moving water the unstirred layer will often be much more than 150 $\mu$m thick. Raven (1970) and Smith & Walker (1980), who have reviewed this topic, conclude that diffusion of $CO_2$ (and presumably $HCO_3^-$) across this unstirred layer can be an important rate-limiting step in photosynthesis by aquatic macrophytes. An increase in photosynthetic rate of river plants associated with increased flow velocities in natural waters has in fact been demonstrated by Westlake (1967). Wheeler (1980*b*) found that the photosynthetic rate of the blade of the great kelp, *Macrocystis pyrifera*, in saturating light, increased about 4-fold when the current speed increased from 0 to 5 cm s$^{-1}$: in low light, however, the increase was only about 50%. Any morphological adaptation which increases the surface-to-volume ratio will help to overcome the diffusion problem and many aquatic macrophyte species have achieved this by evolving highly dissected leaves.[196] The more dissected the leaf form, the less the stimulation of photosynthesis brought about by increased turbulence.[143]

Whether the existence of an unstirred layer around each cell has any significant limiting effect on photosynthesis by phytoplankton remains uncertain.

### Temperature

In considering the effect of temperature on aquatic photosynthesis, it is necessary to distinguish between the effects immediately following a change in temperature and the effects obtained if the plants are allowed to adjust to the new temperature for one to several days. Considering the immediate effects of temperature change first, if the photosynthetic rate of a phytoplankton population or a marine or fresh-water macrophyte is measured under saturating light, at a series of temperatures covering the range from just above freezing to temperatures becoming unfavourable for life, say 5–40° C, it will be found that photosynthetic rate per unit biomass at first increases with temperature, then levels off, and finally begins to decrease again, i.e. there is a temperature optimum. The temperature optimum determined in the laboratory varies in accordance with the temperature of the normal habitat of the aquatic species. For example, benthic marine algae from tidal pools exposed to high temperatures in the summer have somewhat higher optimum temperatures for photosynthesis than algae, sometimes even of the same species, obtained from the sub-tidal zone.[318] Fig. 11.5 compares the dependence of photosynthetic rate on temperature for specimens of the green alga *Ulva pertusa* isolated from these two habitats: the temperature optimum for the subtidal sample is about 5° C lower than that for the pool sample. Yokohama (1973) found that for several species of green, brown and red benthic algae collected around the shore of the Izu peninsula, Japan (34°40′N), the temperature optimum was 5–10° C higher for material collected in summer than for samples obtained in winter. The reasons for the decrease in photosynthetic rate at temperatures above the optimum are not well understood: denaturation of enzymes, a runaway increase in respiration and other forms of thermal damage are likely to be involved.

If photosynthesis as a function of irradiance is measured at a series of temperatures (not exceeding the optimum), then it is observed that the maximum photosynthetic rate at light saturation increases with temperature: this is shown for the marine diatom *Skeletonema costatum* and the lower-littoral-zone red alga *Gigartina stellata* in Fig. 11.6. It is, however, generally found in experiments of this type (i.e. in which the plants are not given time to adapt to each temperature) that at low light intensities, in the linear part of the *P versus* $E_d$ curve, variation in temperature has little effect on the rate. We may reasonably attribute the fall-off in the light-saturated rate with decreasing temperature to the enzymic reactions of the dark carboxylation system working progressively more slowly as temperature decreases. The irradiance value, $E_k$, corresponding to onset of light saturation also decreases together with the temperature. When, due to low temperature, the photosynthetic dark reactions are working slowly, less light is required to produce $NADPH_2$ and ATP fast enough to saturate the dark reaction system. The comparative insensitivity of the

Fig. 11.5. Photosynthetic rate as a function of temperature in samples of the marine benthic alga *Ulva pertusa* (Chlorophyceae) collected from the subtidal zone and from a tidal pool (after Mizusawa *et al.*, 1978).

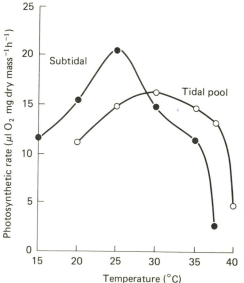

Fig. 11.6. Photosynthetic rate as a function of irradiance at different temperatures. (*a*) In the marine diatom *Skeletonema costatum* (after Steemann Nielsen & Jørgensen, 1968, and Steemann Nielsen, 1975). The cells were grown under an irradiance of $75 \times 10^{18}$ quanta m$^{-2}$ s$^{-1}$, and exposed to the experimental temperature for 30 min before measurements were carried out. (*b*) In the multicellular red alga *Gigartina stellata* (plotted from data of Mathieson & Burns, 1971). Plants were collected from the lower littoral zone of the coast of New Hampshire, USA. The original irradiance values (in foot-candles) have been converted to quantum units.

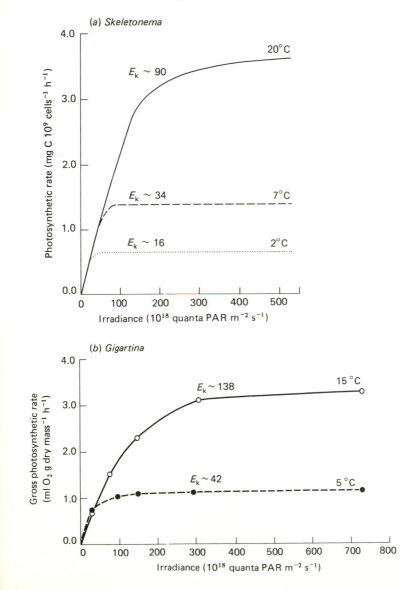

photosynthetic rate at low light intensity to temperature is commonly attributed to the fact that photochemical processes are not very sensitive to temperature, and it is mainly the light reactions which determine rate in the light-limited part of the curve. Respiration, however, does increase markedly with temperature. The failure of the light-limited photosynthetic rate to respond means that as temperature increases the light intensity required for photosynthesis to equal respiration, i.e. the light compensation point, also increases.

Although the photosynthetic rate of any individual aquatic plant or phytoplankton population sample is positively correlated with temperature (up to the optimum) in short-term experiments, in the case of the whole aquatic ecosystem undergoing slow seasonal temperature changes, there is time both for physiological adaptation by any given species, and for changes in the dominant species present. Thus the relation between primary production and temperature may not be so simple at the ecosystem level. Considering adaptation first, if samples of a phytoplankton population are exposed to each new temperature for some days before the $P$ versus $E_d$ curve is measured at that temperature, then in the case of some species the cells can adapt and the differences between the light-saturated rates of photosynthesis at the different temperatures is much smaller. For example, Steemann-Nielsen & Jørgensen (1968) found that in cells of the diatom *Skeletonema costatum*, grown at 20° C, an immediate transfer to 8° C reduced the light-saturated photosynthetic rate by about two-thirds: in contrast, cells grown and measured at 8° C had a light-saturated photosynthetic rate only about 10% lower than that of cells grown and measured at 20° C.

Given that we are interpreting the reduction in light-saturated photo-synthetic rate brought about by an immediate temperature lowering as being due to a lowered specific activity of the dark reaction enzyme system, then we might expect the adaptation to lower temperature to consist of an increase in the cellular content of this enzyme system.[468] Jørgensen (1968) found that the protein content of *S. costatum* per cell was twice as high in cells grown at 7° C as in cells grown at 20° C. In *Dunaliella*, Morris & Farrell (1971) observed that the level of photosynthetic enzymes increased as the growth temperature was lowered.

*Skeletonema costatum* is a ubiquitous diatom which in nature grows well over a wide range of temperature. It is therefore not surprising that marked adaptive abilities can be demonstrated in the laboratory in this species. Yentsch (1974), however, doubts if most phytoplankton species have this ability. Oceanic measurements from various parts of the world indicate that the photosynthetic rate per unit phytoplankton chlorophyll does not vary in any systematic manner with the water temperature:[553] in all temperature regions, areas with high and low rates of total primary production are found.[468] This could be due to adaptation but more recent studies suggest that there is little temperature adaptation of most oceanic phytoplankton species in nature, and that the lack of a clear dependence of photosynthetic rate on temperature in the oceans, is due to the dominance of different (presumably genetically adapted) species in different temperature regimes.[553]

Notwithstanding the apparent lack of a systematic effect of temperature on photosynthesis in the oceans, it is generally found for shallow, productive coastal waters and for inland waters, that both the *in situ* photosynthetic capacity (the maximum specific photosynthetic rate per unit phytoplankton or macrophyte) and the integral (areal) rate of photosynthesis, are positively correlated with the prevailing temperature. For example, in Nova Scotia (Canada) coastal waters, Platt & Jassby (1976) found that in the period July 1973 to March 1975 (temperature range 0–15° C) the photosynthetic capacity of the phytoplankton was linearly dependent on ambient temperature, with a regression slope of 0.53 mg C mg chl $a^{-1}$ $h^{-1}$ °C$^{-1}$. For the waters of the Hudson estuary and the New York Bight, through an 11-month period, Malone (1977) found that for both netplankton and nanoplankton, the photosynthetic capacity was an exponential function of ambient temperature from 8 to 20° C, and 8 to 24° C, respectively. The $Q_{10}$ values (proportionate increase in rate per 10° C rise in temperature) were 2.0–2.6 for the nanoplankton and 4.0 for the netplankton. Below 8° C the rates for both fractions were higher than expected on the basis of the exponential dependence on temperature. Above 20° C the photosynthetic capacity of the netplankton declined with temperature. In a number of temperate-zone eutrophic lakes, dominated by blue-green algae, the photosynthetic capacity of the phytoplankton has been found to vary throughout the year in an exponential manner with the prevailing temperature,[40, 141, 217] the $Q_{10}$ being about 2 over the approximate range 4 to 20° C. It should be noted that the changes in coastal and inland waters we have just been considering are seasonal changes which must include changes in the character (species composition, physiological adaptation) of the phytoplankton as well as in the temperature itself, and this may account for the very high seasonal $Q_{10}$ of 4.0 observed by Malone (1977) in the New York Bight. If a given phytoplankton population is exposed to a given series of temperatures (i.e. allowing no time for adaptation or species succession), photosynthetic capacity varies exponentially with temperature with a $Q_{10}$ of about 2.3.[482]

We noted earlier that when a given phytoplankton population or macrophyte sample is exposed to different temperatures, $E_k$, the irradiance value corresponding to onset of saturation, increases as temperature increases (Fig. 11.6). This also appears to be true on a seasonal basis (i.e. including any changes in the phytoplankton population). In Lough Neagh, N. Ireland, Jewson (1976) found the *in situ* $E_k$ to be correlated with prevailing temperature throughout a 2-year period, rising 2.5- to 4.5-fold between the winter (temperature 3–4° C) and late summer (18.5° C). This rise in $E_k$ is best seen as being part of the rise in photosynthetic capacity with temperature, rather than as a separate phenomenon. Given the increase in maximum specific photosynthetic rate, due to the increased rate of operation of the carboxylation system, it simply means that more light is required to produce $NADPH_2$ and ATP fast enough to saturate the dark reaction system.

In summary, there is no doubt that in coastal and inland waters temperature, operating mainly through its effect on photosynthetic capacity, is a major limiting factor for photosynthesis. In simple terms we may say that in any optically deep water body at a temperature below the photosynthetic optimum, photosynthesis per unit volume is likely to be limited by light in the lower part of the euphotic zone, by temperature in the upper part of the euphotic zone and by $CO_2$ everywhere. In water bodies so shallow that adequate light penetrates to the bottom, temperature (apart from $CO_2$) can become the dominant factor controlling photosynthesis in the whole system. In the Beaufort Channel, North Carolina, USA – a shallow estuary of average depth 1 m – Williams & Murdoch (1966) observed a pronounced seasonal cycle in phytoplankton primary production which appeared to follow the water temperature cycle but not to be related to insolation. They point out that such temperature-driven annual cycles in the productivity of marine phytoplankton are likely to be characteristic of shallow embayments in temperate regions.

### Indirect factors

There are environmental factors other than light, $CO_2$ and temperature which can have a major influence on the total amount of photosynthesis that takes place (i.e. on primary production in the ecosystem) by their effects on the amount of plant biomass present. Since our concern here is mainly with factors which influence photosynthesis directly we shall touch on these indirect factors only briefly.

Inorganic nutrition – particularly the concentrations of the key elements phosphorus and nitrogen – is the most important indirect factor. If the nutrient concentration is low, then although the existing phytoplankton may be photosynthesizing at a higher rate per unit biomass, they cannot increase their biomass and so the rate of photosynthesis per unit volume or area remains low. When nutrient levels increase, due to agricultural runoff or sewage input in inland waters, or an increase in river outflow in coastal waters due to increased rainfall, or the commencement of a seasonal upwelling off the west coast of a continental landmass, or a breakdown of stratification in the sea or a lake in the autumn bringing up fresh nutrients from below the thermocline, then the phytoplankton population and therefore total photosynthesis, both increase as well. While nutrition undoubtedly influences photosynthesis primarily through its effect on total biomass, it is possible that there may be some more direct effects on photosynthetic rate. It is well known, for example, that in the case of terrestrial plants and in algae such as *Chlorella*, nitrogen deficiency is associated with lowered chlorophyll content and photosynthetic activity per unit biomass. Thus when nitrogen is the limiting nutrient, as is believed commonly to be the case in marine waters, it is possible that the photosynthetic apparatus as a proportion of total cell biomass is lower than would be required to achieve the optimal photosynthetic rate by that amount of biomass. On the other hand the oceanic flora has had many millions of years to adapt to a low-nitrogen environment and so we may reasonably harbour the suspicion, on evolutionary grounds, that the particular species which grow in such waters distribute their internal resources in the optimum fashion, and that it would not be to their advantage to allot more of their cellular nitrogen to photosynthetic pigments and enzymes. During the course of a season, it may indeed happen that as nutrient levels fall, a particular species will find itself no longer well adapted to the nutrient composition of the water: when this occurs it seems likely that such a species would simply be replaced by other species better suited to the changed environment.

The other major environmental factor which limits total phytoplankton photosynthesis by limiting biomass is grazing by the aquatic fauna. The spring bloom of phytoplankton is typically followed by a bloom of zooplankton which graze upon the phytoplankton cells. It is undoubtedly true that zooplankton grazing can have a major impact on phytoplankton populations and in some cases can be the cause of the decline in phytoplankton numbers following the bloom: the relations, however, are complex and fluctuations in phytoplankton populations cannot always be interpreted easily in terms of zooplankton grazing. The matter is discussed in more detail for inland waters by Hutchinson (1967) and for the oceans by Raymont (1980) and Frost (1980). Large colonial phytoplankton forms, such as occur in the Peruvian upwelling, can be grazed directly by herbivorous fish such as the anchovy.[409]

The impact of zooplankton grazing on phytoplankton populations can be greatly affected by the extent to which the zooplankton themselves are subject to predation by larger animals. The addition of planktivorous fish to a pond in Minnesota was found to lead to an order-of-magnitude increase in total phytoplankton biomass.[297] In addition to the increased levels of those algal species already present, several new species appeared.

Phytoplankton growth can in turn limit primary production by the benthic flora. In eutrophic waters interception of light by dense phyto-plankton populations can prevent the growth of benthic macrophytes.[235] From their studies on the Norfolk Broads (England), Phillips, Eminson & Moss (1978) consider that during progressive eutrophication of water bodies, overgrowth of the macrophyte leaves by epiphytes and mats of filamentous algae, with consequent reduction in light availability, initiates the macrophyte decline even before dense phytoplankton populations have developed.

The density and therefore total photosynthetic rate of the benthic flora is also affected by grazing by invertebrate and vertebrate animals. In the case of macrophytes, snails and other gastropods feed on the photosynthetic tissue in both fresh and salt waters: in the sea, limpets and sea urchins are important invertebrate grazers, while in fresh waters insect larvae can be significant consumers. Some specialized species of herbivorous fish such as the grass carp (China) in fresh waters, and the parrot fish which feeds on seagrasses, eat macrophytes. Higher aquatic grazing animals such as sea turtles and dugongs, as well as ducks and geese, also eat macrophytes. In inland waters, grazing by invertebrates, fish and birds is a significant factor limiting biomass and photosynthesis in some macrophyte communities:[540] protection of *Potamogeton filiformis* in Loch Leven (Scotland) from grazing by waterfowl substantially increased its biomass throughout the growing season.[235] In shallow coastal sea waters, grazing by sea urchins can be a major factor limiting the growth of kelps and other seaweeds, and grazing by sea turtles can have a major impact on seagrass beds. The biotic relations are complex: whether or not a particular region of the sublittoral zone is colonized by brown algae, and the density at which they grow, are likely to be determined by the abundance of animals such as crabs, lobsters and sea otters (which prey on sea urchins), and the numbers of these animals may in turn be influenced by the activities of man.

The lower limit of *Laminaria hyperborea* at one site in the Isle of Man (UK) was found to be controlled by the sea urchin *Echinus esculentus*, which was present at a density of five animals m$^{-2}$.[237] In a study area in Nova Scotia (Canada), Breen & Mann (1976) found that the lobster *Homarus americanus*, a major predator of sea urchins, decreased by nearly

50% in 14 years: in the latter 6 years of this period the sea urchin *Strongylocentrotus* destroyed 70% of the *Laminaria* beds. Areas of bare rock protected from sea urchin grazing were rapidly recolonized by *Laminaria* plants. Similarly, off California, the mass mortality by disease of a localized population of sea urchins on the seaward side of a kelp forest was followed by the rapid seaward expansion of four species of brown algae.[357] On the basis of archaeological evidence from the Aleutian Islands, Simenstad, Estes & Kenyon (1978) concluded that when the Aleut people colonized this area and commenced hunting the sea otter, the diminution in otter numbers led to a population explosion of sea urchins, which was in turn followed by the decimation of the abundant kelp beds in the area: since fur hunting in the Aleutians ceased in 1911, both the sea otter and abundant kelp beds have become re-established. In Fiji, extinction of the dugongs and a reduction in sea turtle numbers has caused the seagrasses to increase to the point of becoming a nuisance.[311]

### 11.4 Temporal variation in photosynthesis

The short answer to the question 'When does aquatic photosynthesis take place?' is that it takes place when and to the extent that the various limiting constraints we have already discussed permit it to take place. Thus, to understand temporal variation in photosynthesis we need to know the manner in which these limiting factors vary with time.

Considering diurnal variation first, there is of course no photosynthesis at night. Photosynthesis begins at dawn and ends at dusk. The total photosynthesis of the whole water column roughly follows the approximately sinusoidal variation of solar irradiance during the day.[176, 231, 486] Under still conditions there may be a reduction of photosynthetic rate per unit volume near the surface due to photoinhibition. Also, where dinoflagellates are the dominant organisms, their downward migration to lower light intensities can lower photosynthetic rate per unit volume near the surface in the middle part of the day.[503] There have been many reports of variation in the photosynthetic capacity (i.e. photosynthesis in samples taken from the water body and exposed to saturating light) of phytoplankton populations and benthic algae during the day (see, for example, the reviews by Sournia (1974), Raymont (1980) and Harris (1980)). This is discussed later (§12.6).

In temperate or Arctic/Antarctic latitudes, there is marked seasonal variation in aquatic photosynthesis. This approximately follows, and is also a major contributing factor to, the seasonal variation in plant biomass. For the marine phytoplankton a detailed account has been given by Raymont (1980). In the winter, phytoplankton biomass and total photosynthesis are both very low, partly because of low solar irradiance[408] but even more because, as a consequence of lowered temperature and winter storms, thermal stratification is lost and the mixed layer through which the phytoplankton are circulated comes to exceed the critical depth for net column photosynthesis (see §11.1). In the spring, with increased heating of the surface water resulting from increased daily insolation, thermal stratification sets in, circulation below the critical depth is prevented, and this, combined with the increase in incident PAR and the availability of nutrients in the water, leads to a massive increase in phytoplankton biomass, typically of the order of 20-fold, measured as chlorophyll. Riley (1942) found that for a number of stations over the Georges Bank (Gulf of Maine) during the spring bloom the rate of phytoplankton increase (late March to mid-April) was approximately inversely proportional to the depth of the mixed zone. He concluded that the balance between the effects of vertical turbulence and the increasing vernal radiation determines the beginning of the spring bloom (see also §11.1).

Typically, the spring bloom is followed by a decline to a lower level of biomass and integral photosynthesis in the summer, caused by zooplankton grazing and possibly also by a loss of nutrients from the mixed layer resulting from the sedimentation of zooplankton faecal pellets down below the thermocline. In the autumn, the fall in temperature and increased wind strength lead to intermittent disruption of thermal stratification with consequent transport of nutrients up from below the thermocline. Thus, in oceanic and continental shelf waters in intermediate and high latitudes,

Fig. 11.7. Variation of photosynthetic rate per unit volume throughout the year in North Pacific coastal water (after an average curve derived by Koblents-Mishke, 1965, from literature data).

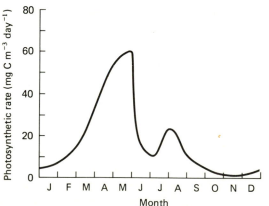

there is typically a second phytoplankton bloom and associated increase in areal photosynthesis in the autumn: this is terminated when disruption of the thermocline becomes so severe that stratification disappears, circulation below the critical depth sets in and productivity declines to the low winter values. Fig. 11.7 shows the variation in photosynthetic rate throughout the year in North Pacific coastal water illustrating the large spring, and small autumn, peaks.[268] Broadly speaking, as latitude increases, the growing season becomes progressively shorter and the spring and autumn phytoplankton peaks tend to merge into one.

The pattern of temporal variation described above is by no means universal; it can be greatly modified by local conditions. In the Kattegat (Baltic), lowered salinity (and therefore lower density) in the upper layer maintains stratification throughout the year, and so productivity is substantial at all times of the year, diminishing somewhat in the winter due to reduced light availability.[468] In coastal waters so shallow that the mixed depth rarely or never exceeds the critical depth, production also tends to remain high throughout the year. Seasonal variation in production may follow the temperature cycle if the water is so shallow that light is non-limiting.[548]

In tropical oceanic waters where there is no seasonal variation in the hydrological regime, there is usually no pronounced variation in phyto-plankton biomass and photosynthetic production. In regions where there is a seasonal upwelling of nutrient-rich deep water, such as parts of the west coast of Africa, there is an accompanying massive increase in phytoplankton and photosynthetic production. Seasonal variations in river outflow can also cause seasonal variations in primary production in tropical coastal waters.

Inland waters in intermediate and high latitudes show marked seasonal variation in biomass and total photosynthesis. Both are low in the winter due to a decrease in irradiance (sometimes accentuated by ice or snow cover) and temperature, and, in the deeper waters, circulation of the phytoplankton through a depth greater than the critical depth. Increased irradiance and temperature in the spring, together with the availability of nutrients and the onset of thermal stabilization, lead to a phytoplankton bloom and associated increase in areal photosynthesis. The behaviour during the rest of the seasons tends to be both complex and highly variable. Production may remain consistently high, or there may be wide fluctuations in populations and production caused by zooplankton grazing, fluctuations in nutrient content due to influxes of nutrients from the hypolimnion associated with temporary wind-induced disruption of the shallow (by marine standards) thermocline, or major changes in species composition resulting from the seasonal changes in water quality (e.g. higher pH values in late summer tend to favour blue-green algae rather than diatoms).

In certain clear-water lakes in northwestern Ontario, Canada, Fee (1976) observed that there was an increase in phytoplankton in the surface layer in the autumn, but that this appeared to be due not to increased algal growth, but to mixing of the algae from the deep chlorophyll layer (§11.1) up into the surface waters following the breakdown of thermal stratification. Primary production in the lake was in fact at its lowest at this time, because the phytoplankton recirculated from greater depth appeared to be inhibited by the high light intensity near the surface.

In some tropical lakes such as Lake George, Uganda, there may be no seasonal variation and primary production and phytoplankton biomass remain high throughout the year.[137] On the other hand, where, as in the case of Lake Chad (Chad, Africa), there are alternating dry and rainy seasons, there can be seasonal variation in temperature, insolation and water level, with associated changes in areal photosynthesis.[283]

Benthic macrophytes, being fixed in position, are not, unlike the phytoplankton, directly affected by the seasonal onset or disappearance of thermal stratification. They do, however, show seasonal variation in photosynthetic production in response to the yearly cycles of irradiance and temperature. Fig. 11.8 shows the variation in net photosynthetic rate (per unit area of thallus) of the sublittoral brown alga *Laminaria longicruris*

Fig. 11.8. Seasonal variation of the photosynthetic rate of the kelp *Laminaria longicruris* and of total daily irradiance (PAR) and water temperature at 10 m depth in a Nova Scotia, Canada, coastal water (St Margaret's Bay), from data of Hatcher *et al.*, 1977). Photosynthesis measurements were made *in situ* at the prevailing water temperature. The graph of total daily irradiance was obtained by drawing a smoothed curve through the data points of Hatcher *et al.*

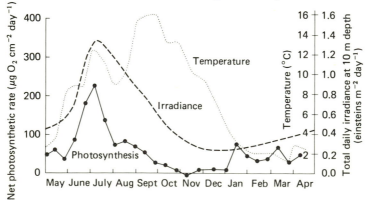

(a kelp) in Nova Scotia, Canada, coastal waters through a 12-month period, together with values of the total amount of PAR received per day at the sea bottom (10 m depth) on which the algae were growing, and the prevailing water temperature.[179] Photosynthetic rate was highest in summer, declined through the autumn to approximately zero in November–December, became significant once more in late winter and spring, and then rose steeply in early summer. The temporal variation in photosynthesis corresponded roughly to the variation in irradiance. By multiple regression analysis, Hatcher, Chapman & Mann (1977) found that 61% of the variance in daily photosynthesis could be accounted for by irradiance. Variation in temperature (from 1.5 to 13° C in these cold waters) did not account for any significant part of the variation in daily photosynthesis. Temperature did account for 56% of the observed variation in light-saturated photosynthetic rate (measured at ambient water temperature, back in the laboratory), but since for most of the time the plants *in situ* would be photosynthesizing at subsaturating irradiance values, the effect of temperature on the saturated rate is irrelevant.

Kirkman & Reid (1979) studied the seagrass, *Posidonia australis*, growing in shallow water – from low-water mark to a depth of 3 m below

Fig. 11.9. Seasonal variation in relative growth rate of the seagrass *Posidonia australis*, and in surface water temperature, at Port Hacking, NSW, Australia (by permission, from Kirkman & Reid (1979), *Aquatic Botany*, **7**, 173–83).

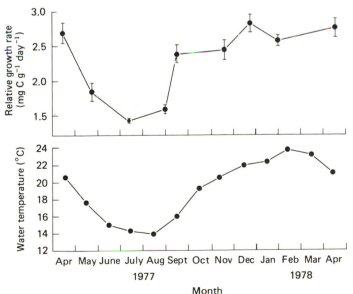

this – at Port Hacking, Australia (34° S). Fig. 11.9 shows the variation in relative growth rate (mg C g$^{-1}$ day$^{-1}$, which should be closely related to photosynthetic rate), and water temperature during a 12-month period. The growth rate was found to be strongly correlated ($r = 0.79$) with water temperature. It is possible that in this shallow, well-illuminated environment, light intensities were at or near saturation for much of the time, so that temperature became a limiting factor.

Hanisak (1979) carried out a combined field and laboratory study of the growth pattern of the siphonous green alga *Codium fragile*, growing in shallow water (average depth 1.5–2.3 m below mean low-water mark) on the northeastern (41°N) coast of the USA. The main period of growth (as dry matter accumulation, therefore closely related to photosynthesis) was from late spring (April–May) through to early autumn, the maximum rate being in July. It appeared that the total daily irradiance levels were above saturation for growth from April to August/September so that temperature was the major limiting factor, but that in the autumn, when temperatures were still high, the reduced light levels were limiting for growth. In the winter it seemed that both temperature and light were limiting.

In some coastal regions the diminution in underwater light availability in the winter due to the lowered solar elevation and shorter days is further aggravated by a marked increase in the vertical attenuation of light resulting from increased water turbidity caused by the stirring up of sediments in winter storms. At Banyuls-sur-Mer (western Mediterranean), $K_d$ rose from a minimum value of about 0.075 m$^{-1}$ in the summer to a maximum of about 0.19 m$^{-1}$ in the winter,[536] and off Helgoland (North Sea) also, water transparency was much lower from October to March than in the rest of the year.[296]

These admittedly sparse data suggest the tentative generalization that primary production of marine macrophytes is limited in deep water by light availability all the year, but in shallow water is usually limited by temperature except that in higher latitudes light becomes limiting in the winter, and that the temporal variation in primary production is in large part a function of the seasonal variation in these two physical parameters. It is likely that macrophyte photosynthesis is, in addition, always and everywhere limited by $CO_2$ availability, but this does not vary with the seasons. Growth, and therefore photosynthesis, are also determined by the availability of nutrients in the water, particularly nitrate, and in higher latitudes this varies on a seasonal basis being high in the winter but decreasing greatly during the spring phytoplankton bloom. This is less of a limiting factor for many macrophytes than it is for phytoplankton however. Seagrasses, for example, being rooted plants, can obtain nutrients from the sediments in which they are growing. In the summer, when nitrate concentration in the water is low, the kelps (Phaeophyta) continue to photosynthesize and accumulate carbohydrate reserves for later use, in the winter.[179] In the tropics we might *a priori* expect relatively little seasonal variation in macrophyte photosynthesis except where there is variation in the hydrological regime, associated for example with seasonal variation in river outflow. Experimental studies would be desirable.

The extent to which seasonal variation in marine macrophyte photosynthesis is influenced by internal, as well as external, changes is unclear. King & Schramm (1976) in their study of many green, brown and red algal species in the Baltic, found, for example, that light compensation points were lower in the winter than at other times of year, and that the light-saturated photosynthetic rates (per g dry mass) varied markedly within any species (but not in the same way in all species) according to the season. However, their measurements were in all cases carried out at the prevailing seasonal water temperature (which, of course, varied during the year) and so they were unable to conclude whether the seasonal adaptation was simply a direct response to temperature change or whether other factors were involved. An example of a seasonal change in photosynthesis caused by an internal physiological change is, however, provided by the seagrass *Posidonia oceanica*. Drew (1978, 1979) found that in plants of this species growing off Malta the photosynthetic rate (per unit area of leaf) declined markedly in the summer below the spring value, whereas in another seagrass species, *Cymodocea nodosa*, growing nearby, the photosynthetic rate remained high. The decline in *P. oceanica* photosynthesis

Fig. 11.10. Seasonal variation of the primary production rate of a mixed macrophyte community (predominantly *Chara vulgaris*, *Potamogeton Richardsonii* and *Myriophyllum alterniflorum*) in West Blue Lake, Manit., Canada (after Love & Robinson, 1977).

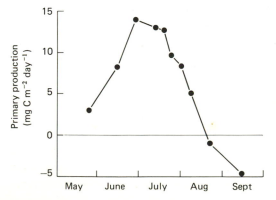

appeared to be due to senescence (leaf chlorophyll content declined markedly in the summer), possibly triggered by daylength changes.

In fresh waters, the typical pattern of benthic macrophyte primary production in temperate regions is that biomass accumulation does not occur in the winter, that it begins in the spring as a consequence of the increase in solar irradiance and temperature, and rises to its maximum rate (g C or dry matter per m² of bottom per day) in the early summer. The relative importance of light and temperature in determining the seasonal pattern of fresh-water macrophyte photosynthesis is not well understood. We may reasonably suspect that it is largely a matter of depth, temperature being the more important variable in shallow, and light in deeper, water. In the late summer and in the autumn, the rate of accumulation decreases for a number of reasons including a general deterioration in the vigour of the plant due to disease, grazing damage and excessive temperatures.[540] The rise in pH late in the summer, due to high phytoplankton photosynthesis, can also be inimical to some macrophytes. Since macrophytes can derive nutrients from the sediments in which they are growing, their productivity is not likely to be limited by the seasonal variations in levels of phosphorus and mineral nitrogen in the water. Fig. 11.10 shows the seasonal changes in primary production rate in a mixed macrophyte community in a Canadian lake.[294] Internal as well as external changes may influence the seasonal variation in photosynthesis: the photosynthetic capacity of fresh-water macrophytes is typically low in the winter and high in the spring.[539] Variations in photosynthetic capacity could influence the rate of photosynthesis in plants growing at saturating intensities in shallow water but would be of little consequence for plants growing at subsaturating intensities in deep water.

## 11.5    Photosynthetic yield per unit area

In our consideration of where and when aquatic photosynthesis takes place we have noted that some ecosystems are much more productive per unit area than are others. We shall now look more closely at the basis of this variation in productivity, particularly with respect to the role of light, and shall also examine some of the published data on the primary production rates actually achieved by aquatic ecosystems.

### Integral photosynthetic rate

Comparisons of phytoplankton productivity in different waters are best carried out in terms of the total phytoplankton production per unit area, i.e. the integral, or areal, photosynthetic rate for the whole water column beneath 1 m² of surface. We have already considered at some length the direct influences of light, temperature, $CO_2$ etc. and the inherent photosynthetic characteristics of the algal cells on photosynthesis at the level of individual samples of the phytoplankton population. We now consider how photosynthesis by the whole water column is influenced by these factors. To do this we need an expression for integral photosynthesis in terms of other parameters of the system. Such a relation was derived by Talling (1957*b*), and although it is non-rigorous it has proved to be very useful for calculating integral photosynthesis values, while at the same time providing theoretical insights. Its derivation and application are briefly considered here.

We saw earlier (§ 10.1) that the variation in phytoplankton photosynthetic rate with light intensity, if photoinhibition is ignored, conforms reasonably closely to the equation of Smith (1936) (eqn 10.1). We also know that irradiance of PAR decreases in an approximately exponential manner with depth. Combining these two relations, and assuming that the phytoplankton

Fig. 11.11. Idealized representation of the variation of photosynthetic rate per unit volume ($P_v$) with depth, assuming no photoinhibition, and that $P$ *versus* $E_d$ conforms with eqn 10.1 (after Talling, 1957*b*). The upper, vertical, part of the curve corresponds to the region of light saturation.

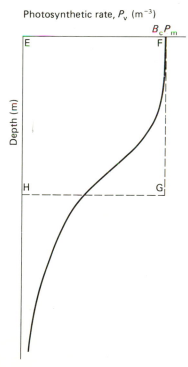

Photosynthetic rate, $P_v$ (m$^{-3}$)

concentration is uniform with depth, it follows that the photosynthetic rate $(P_v)$ per unit volume per hour changes with depth in the manner indicated in Fig. 11.11. The total (integral) photosynthesis $(P_A)$ underneath 1 m² of surface is equal to the area enclosed by the curve

$$P_A = \int_0^\infty P_v \, dz$$

A rectangle, *EFGH*, can be drawn with area equal to the area under the curve, where *EF* is the photosynthetic rate per unit volume at the surface and is equal to $B_c P_m$ (the product of the phytoplankton concentration, $B_c$ mg chl $a$ m⁻³, and the maximum photosynthetic rate per unit phytoplankton concentration per hour) and *FG* is a depth. Talling found by planimetry that the depth *FG* is approximately equal to

$$\frac{1}{K_d} \ln\left(\frac{E_d(0)}{0.5 \, E_k}\right)$$

where $E_d(0)$ is the downward irradiance of PAR just below the surface, $E_k$ is the irradiance value defining the onset of saturation (§ 10.1), and $K_d$ is the vertical attenuation coefficient for downwelling PAR. The integral photosynthetic rate, being equal to the area of the rectangle ($EF \times FG$) is therefore given by

$$P_A = \frac{B_c P_m}{K_d} \ln\left(\frac{E_d(0)}{0.5 \, E_k}\right) \tag{11.3}$$

The precise character of the curve of $P_v$ against depth varies in accordance with the relative size of $E_d(0)$ and $E_k$. For example, the higher the incident irradiance for a given $E_k$, then the longer will be the initial, vertical, part of the curve (i.e. the greater the depth required to reduce $E_d$ below saturation). Talling found, nevertheless, that eqn 11.3 gave reasonably accurate estimates of $P_A$ over a wide range of values of $E_d(0)/E_k$. Significant discrepancies did arise for low surface light intensities $(E_d(0) < E_k)$.

The major deficiency of the Talling expression is that it assumes there is no surface photoinhibition and more complex treatments have been devised by Fee (1969) and Vollenweider (1970) to take account of this. In practice, however, the error introduced by ignoring photoinhibition is quite small. Using data for Lough Neagh, N. Ireland, Jones (1978) found that the extent to which $P_A$ was underestimated when photoinhibition was ignored was only about 4–5% when surface photoinhibition was 50% (reduction of photosynthetic rate, relative to $P_m$) and only about 8% when surface photoinhibition was 75%. Given the unavoidable inaccuracies of field measurements of photosynthesis and considering also the fact that, because of wind-induced circulation, surface photoinhibition is probably in reality not a common occurrence, it seems that for most situations there is no significant advantage to be gained from using the more complex treatments.

Thus Talling's equation (of which eqn 11.3 is a slightly modified version) makes it possible to calculate with reasonable accuracy the areal photosynthesis from the measured maximum value of photosynthetic rate per unit volume ($B_c P_m$) achieved in the water column, the vertical attenuation coefficient for downward irradiance of PAR, the subsurface value of incident irradiance, and the saturation onset irradiance value, assuming this to have been determined for the phytoplankton population in question. Use of this equation makes it possible to estimate integral photosynthesis without always obtaining a complete depth series of *in situ* photosynthesis values. To obtain the daily total of photosynthesis per unit area, areal photosynthesis must be integrated against time for the daylight hours.

$$\text{areal photosynthesis per day} = \int_{t_1}^{t_2} P_A \, dt = \frac{B_c P_m}{K_d} \int_{t_1}^{t_2} \ln\left(\frac{E_d(0)}{0.5 \, E_k}\right) dt$$

where $t_1$ and $t_2$ are the times of sunrise and sunset, respectively. We now obtain

$$\text{areal photosynthesis per day} = \frac{B_c P_m}{K_d} \overline{\left[\ln\left(\frac{E_d(0)}{0.5 \, E_k}\right)\right]} (t_2 - t_1)$$

$$(11.4)$$

where $\overline{[\ln(E_d(0)/0.5 \, E_k)]}$ is the mean value of this function throughout the daylight hours. This can be replaced with the approximate, but useful, relation

$$\text{areal photosynthesis per day} \simeq \frac{B_c P_m}{K_d} \left[\ln\left(\frac{\overline{E_d}(0)}{0.5 \, E_k}\right)\right] N \qquad (11.5)$$

where $\overline{E_d}(0)$ is the mean value of subsurface downward irradiance during daylight hours and $N$ is daylength in h: the error in eqn 11.5, introduced by moving from the mean of a log to the log of a mean, can be largely overcome if the right-hand side is multiplied by the empirical correction factor of 0.9.[483]

If the ratio of the respiration rate of the phytoplankton population to the maximum photosynthetic rate is $\rho$ then the total amount of respiration in a 1 m² water column of depth $z$ m in a 24-h period is $24 B_c \rho P_m z$. When the depth through which the phytoplankton is circulating is the critical depth, $z_c$ (see §11.1) the total amount of $O_2$ liberated in the water column by photosynthesis during daylight (i.e. the areal photosynthesis per day) is just equal to the amount consumed by respiration in the whole 24-h period. From eqn 11.5, we may write

$$\frac{B_c P_m}{K_d} \left[ \ln \left( \frac{\overline{E_d}(0)}{0.5 \ E_k} \right) \right] N = 24 \ B_c \rho P_m z_c$$

and therefore

$$z_c = \frac{N}{24 \ K_d \rho} \ln \left( \frac{\overline{E_d}(0)}{0.5 \ E_k} \right)$$

which is eqn 11.1. Thus the Talling model leads to a useful approximate expression for critical depth.

In addition to being useful for calculating hourly or daily values of areal photosynthesis, the Talling model can assist in understanding the contribution of the different determinants of phytoplankton productivity in different waters, or in a given water at different times. From eqn 11.5 it follows that, as we might expect, daily production is approximately proportional to phytoplankton concentration and to daylength, and inversely proportional to the vertical attenuation coefficient for irradiance. It increases as average daily incident irradiance at the surface increases but in proportion to the logarithm, rather than linearly. Thus, the higher the incident irradiance, the smaller the increase in areal photosynthesis brought about by a given proportional increase in irradiance. An increase in the value of $\overline{E_d}(0)/0.5 \ E_k$ from 3 to 6 increases $P_A$ (daily) by 63%, but an increase from 12 to 24 increases $P_A$ (daily) by only 28%.

Areal photosynthesis is also affected by the inherent photosynthetic properties of the phytoplankton but not in a very simple way. On the face of it, eqns 11.3 and 11.5 seem to predict that productivity is directly proportional to the maximum specific rate of photosynthesis of the phytoplankton, $P_m$, but diminishes with the logarithm of the saturation onset parameter, $E_k$. This is correct to the extent that areal photosynthesis will indeed increase as the photosynthetic capacity of the cells increases and will diminish if they require too much light to achieve their maximum rate. In reality, these two parameters do not vary independently, however, and in fact $P_m = \alpha E_k$ (§10.1). Thus if, for example, a given increase in $P_m$ is entirely due to an increase in $\alpha$, perhaps because one phytoplankton species has been replaced by another with a higher quantum yield or one which captures light more effectively (see §10.2), then $P_A$ will increase proportionately. If, on the other hand, the increase in $P_m$ is caused by an increase in temperature which, in the given phytoplankton population increases both $E_k$ and $P_m$ together without affecting $\alpha$ (see §11.3), then the increase in $E_k$ (in the denominator) counteracts the effect of the increase in $P_m$, but only partially since it is within the logarithm. Thus, when an increase in $P_m$ is accompanied by an increase in $E_k$, as will often be the case, the increase in areal photosynthesis will be less than proportionate to the increase in $P_m$. The actual change obtained depends on the initial ratio of $\overline{E_d}(0)$ to $E_k$. If this is 6 and falls to 3 when $E_k$ is doubled, then $\ln(\overline{E_d}(0)/0.5\ E_k)$ falls by 38% and $P_A$ increases by only 23%. If, however, the ratio is 24 and falls to 12 when $E_k$ is doubled, $\ln(\overline{E_d}(0)/0.5\ E_k)$ falls by only 22% and $P_A$ increases by 56%. In short, the higher the incident light intensity the greater the increase in areal photosynthesis brought about by an increase in photosynthetic capacity.

### Geographical variation of photosynthetic yield

Eqn 11.5 predicts the amount of phytoplankton photosynthesis per unit area of the ecosystem per day for a specified set of values of the crucial parameters of the system: concentration ($B_c$) and photosynthetic characteristics ($P_m$, $E_k$) of the phytoplankton, penetration of light into the water ($K_d$), incident radiation ($\overline{E_d}(0)$, $N$). All these parameters vary not only throughout the course of the year, but also, taken over the whole year, with geographical location – from one part of the world's oceans to

Table 11.2. *Annual phytoplankton primary production in the oceans. Data from literature surveys by* (a) *Platt & Subba Rao (1975)*, (b) *Raymont (1980)*, (c) *Koblents-Mishke (1965)*, (d) *Walsh (1981)*

| | Primary production ($g\ C\ m^{-2}\ yr^{-1}$) | | |
| --- | --- | --- | --- |
| | Continental shelf | Deep ocean | Reference |
| *Ocean* | | | |
| Indian | 259 | 84 | *a* |
| Atlantic | 150 | 102 | *a* |
| Pacific | 190 | 55 | *a* |
| Antarctic | 325 | – | *a* |
| Arctic | 20 (inshore) | $\sim 1$ | *b* |
| *Regions within Pacific Ocean* | | | |
| Tropical deep ocean | 28 | | *c* |
| Tropical/temperate transition zone | 49 | | *c* |
| Temperate deep ocean | 91 | | *c* |
| Continental shelf, temperate | 102 | | *c* |
| Inshore coastal, temperate and tropical | 237 | | *c* |
| Peruvian upwelling | 1350–1570 | | *d* |

another, from one inland water body to another – because of differences in the average values of the controlling physicochemical parameters of the system such as depth, insolation, nutrient concentration, temperature, optical properties of the water, stability of the water column, etc. The consequent geographical variation in annual photosynthetic yield can conveniently be characterized in terms of the total amount of carbon fixed by photosynthesis per $m^2$ of aquatic ecosystem per year. Data for phytoplankton primary production in various parts of the sea are listed in Table 11.2. A detailed discussion of productivity in various parts of the ocean may be found in Raymont (1980).

The least productive waters are the deep oceans in tropical latitudes. Such waters have a permanent thermocline which greatly impedes transport of nutrients up from the depths where they are regenerated by mineralization of sedimenting phytoplankton and zooplankton faecal pellets. In the temperate oceans, nutrient levels in the upper layer are restored each winter when thermal stabilization breaks down. The highest rates of production are achieved when the high insolation and temperatures of tropical regions are combined with a plentiful nutrient supply in the form of upwelling water from the depths: the Peruvian upwelling is probably the most productive region of the oceans. Phytoplankton productivity is higher in the shallower waters off the continental shelf, and higher still close inshore, than in the deep oceans, even in tropical latitudes. Nutrient supply from the land is thought to be an important factor. Close inshore, recycling of nutrients from the bottom and the impossibility of deep circulation of the phytoplankton (below the critical depth), are also likely to contribute to the enhanced productivity. The Antarctic Ocean is vastly more productive than the Arctic Ocean because, associated with the presence of a land mass, there is extensive upwelling of nutrient-rich water.

In the case of inland waters, oligotrophic (nutrient-poor) lakes have phytoplankton primary productivities usually in the 4–25 g C m$^{-2}$ yr$^{-1}$ range whereas eutrophic (nutrient-rich) lakes typically fix 75–700 g C m$^{-2}$ yr$^{-1}$,[168, 403] although, for the saline Red Rock Tarn in Australia, an annual production of 2200 g C m$^{-2}$ has been reported.[168] Hammer presents an extensive compilation of productivity data for inland waters in the 1980 IBP report. Brylinsky's survey of a large number of lakes all over the world for the IBP report indicated that phytoplankton productivity is negatively correlated with latitude. This can reasonably be explained in terms of the diminution of annual solar radiation input with increasing latitude.

# 12

..........................................................................................

## *Ecological strategies*

Of the factors which limit the rate of primary production in aquatic ecosystems – light, nutrients, carbon dioxide, temperature – that which shows the most extreme variation within the aquatic medium is light. As we have seen (Chapter 6) the irradiance decreases with depth from intensities which are so high as to be damaging down to levels which cannot support photosynthesis, and the spectral distribution of the light also changes markedly. We have also seen that at any given depth the intensity and spectral quality of the light vary greatly in accordance with the optical properties of the water. Furthermore, to a much greater extent than the other limiting factors, light availability varies with time: both within the day – from darkness to the full noon Sun, and as clouds pass across the Sun – and with the seasons during the course of the year.

In this chapter we shall consider the ways in which the aquatic flora is adapted to this variability of the light climate.

### 12.1    Aquatic plant distribution in relation to light quality

As we saw in Chapters 8 and 9, there are major differences between the main taxonomic groups of aquatic plants with respect to the kinds of photosynthetic pigment present and, as a consequence, major differences in the absorption spectra. Given the variation in intensity and spectral quality of the light field in the aquatic environment, we may reasonably suppose that for any given location within a water body there will be certain species which are well equipped to exploit the particular prevailing light field and others which are not. It thus seems likely that photosynthetic pigment composition could be a major factor determining which species of aquatic plant grow where.

In the case of the benthic algae of marine coastal waters it has been observed that the different major algal groups are not mixed at random.

In some parts of the benthic environment browns predominate while in other the reds or the greens predominate, although domination by any type is not usually complete. Furthermore, progressive changes in the proportions of the different algal groups along the depth gradient can often be discerned. It has been a commonly held belief in marine biology since the last century that the most important factor determining algal zonation is the variation of the light field with depth. This theory has taken two forms which have, perhaps needlessly, been seen as opposed rather than complementary. According to the *chromatic adaptation* theory of Engelmann (1883) it is the varying colour (spectral distribution) of the light with depth which determines the algal distribution, i.e. as the predominant colour changes due to selective absorption, those algae which have absorption bands corresponding best to the spectral distribution of the surviving light can photosynthesize most effectively and so predominate. Berthold (1882) and Oltmanns (1892), on the other hand, proposed that it is the varying intensity of the light with depth which determines the distribution of the different algal types.

In fact both the colour and the intensity of the light field change simultaneously with depth and the plants must adapt to both. For example, any plant growing near the bottom of the euphotic zone must be able to make use not only of a restricted spectral distribution, but also of a very low total irradiance. One possible form of adaptation to low irradiance is a lowering of the respiration rate; another would be an increase in the concentration of all the photosynthetic pigments present, without alteration of the ratios. Harder (1923) suggested that both chromatic and intensity adaptation are involved: it is not in fact easy to disentangle the two. Nevertheless there is persuasive circumstantial evidence that chromatic adaptation plays an important role in determining the distribution of the different plant groups.

Before we examine this evidence, we must make clear the important distinction between *phylogenetic* and *ontogenetic*, chromatic adaptation.[388] Phylogenetic adaptation is adaptation that has taken place during phylogeny, i.e. during the evolution of the species. In the present context it refers to the genetically determined differences in pigment composition between the different taxonomic groups of aquatic plants. Within a given species, while the nature of the pigments formed is fixed, the proportion of the different pigments can alter, with significant effects on absorption properties, in accordance with the environmental conditions prevailing during growth and development, i.e. during ontogeny. This is ontogenetic chromatic adaptation. We shall consider both kinds of chromatic adaptation.

*Evidence for phylogenetic chromatic adaptation of benthic plants*

Changes in the pigment category of aquatic plants with depth are striking in sea water but hard to identify in fresh water, and so we shall here consider marine ecosystems. We shall begin by noting that in all marine waters there is ample light in all wavebands near the surface, and so the theory of chromatic adaptation, which is specifically concerned with light-limited situations, has nothing to tell us about the relative success of the different algal groups in this region: resistance to wave action, for example, or, in the case of upper sublittoral forms, to occasional exposure to the atmosphere, are likely to be more important factors.

While in all waters total irradiance diminishes exponentially with depth, the nature of the change of spectral distribution with depth varies in accordance with the absorption properties of the water. In all waters irradiance in the red waveband diminishes quite rapidly due to absorption by water itself. In very clear colourless waters attenuation is least in the blue region, and with increasing depth the underwater light becomes first blue-green and then predominantly blue in colour (Fig. 6.4a). In waters with significant amounts of yellow substances there is rapid attenuation in the blue region as well as the red, so that as depth increases the light becomes increasingly confined to the green waveband (Fig. 6.4b). On the basis of chromatic adaptation we might therefore expect somewhat different patterns of algal zonation in the different types of water, and this in fact turns out to be the case.

The depth distribution of the three major eukaryotic algal groups can be expressed in terms of either the biomass, or the number of species, in each group, at a series of depths. Both kinds of information are interesting, the first perhaps more so from our present point of view since it is more directly related to the competitive success through primary production of the different groups: species number, in contrast, can be regarded as a measure of the number of different solutions that the green, brown or red algae have found to the problems of growing at a given depth, or alternatively as a measure of the amount of green, brown or red algal genetic information to be found at that depth. Quantitative distribution data of either type seem to be rather scarce and unfortunately this is especially true in the case of biomass.

We shall first consider waters of the northern hemisphere temperate zone. For the west coast of Sweden (at the entrance to the Baltic Sea), Levring (1959) reported (although quantitative data were not given) that most members of the Chlorophyta occur in the littoral and upper sublittoral zones, the Phaeophyta occur mainly from the littoral zone down to the middle of the sublittoral zone ($\sim$ 15 m), and the Rhodophyta occur

Fig. 12.1. Variation of taxonomic composition of benthic algal flora with depth in northern hemisphere temperate waters. (*a*) West of Scotland (Argyll and Ayrshire, 56–57°N). Derived from data of McAllister, Norton & Conway (1967). (*b*) New England, USA (Maine and New Hampshire, 42–43°N). Plotted from data of Mathieson (1979). (*c*) Isle of Colonsay, Inner Hebrides, Scotland (56°N). Derived from data of Norton *et al.* (1969). (*d*) Isles of Scilly, England (50°N). Derived from data of Norton (1968).

The curves show the number of red (—●—), brown (– – ○ – –) and green (. . . △ . . .) algal species found at each depth. Curves (*a*), (*c*) and (*d*) were derived from the published depth distribution data for each algal species.

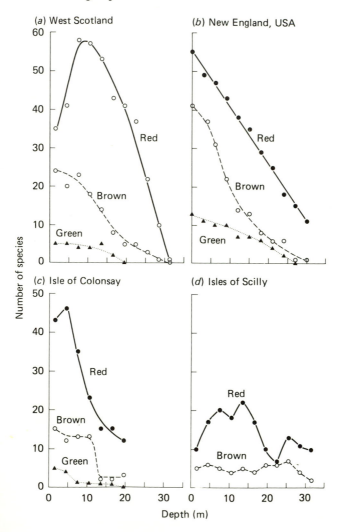

throughout the euphotic zone but are the predominant algal group in the lower sublittoral zone (15–30 m). Distributions described in other northern hemisphere temperate waters are broadly similar to this but clear-cut zonation appears to be absent. Fig. 12.1 shows the variation of species number in the three algal groups with depth at three sites in the British Isles and one from northeastern North America. The numbers of green and brown species decreased with depth. The number of red species at some places increased at first with depth but then began to decrease. At all sites green species were much less common than the other two, and brown species were less numerous than red. At the three sites where green algae were found (Chlorophyta apparently being absent or insignificant at the Scilly Isles site) they penetrated well down into the sublittoral – to about the middle or a bit further – but the brown algae penetrated deeper, and the red algae deeper still. In all cases the red algae dominated the lowest region of the sublittoral. Norton, Hiscock & Kitching (1977) measured the depth distribution of biomass of the more abundant seaweeds on a headland in southwest Ireland, and their data are plotted in Fig. 12.2. Green algae were apparently of no quantitative significance at this site. It can be seen that the brown algae constituted nearly all the underwater biomass throughout most of the sublittoral zone. Brown algal biomass

Fig. 12.2. Variation of biomass per unit area of brown and red algae with depth in a northern hemisphere temperate water (Carrigathorna, Lough Ine, southwest Ireland, 51°N). Plotted from data of Norton, Hiscock & Kitching (1977).

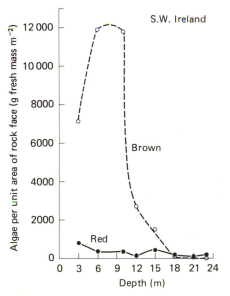

consisted almost entirely of the large kelp *Laminaria hyperborea*: it reached its peak between about 6 and 10 m depth but decreased sharply below this, falling virtually to zero biomass at 18 m. Below 18 m the vegetation consisted mainly of a comparatively sparse cover of red algal species.

These northern European waters contain sufficiently high levels of gilvin to ensure that the underwater light becomes predominantly green in colour,[167, 214] and the same is likely to be true of the northeastern American coastal waters. In such waters the rate of assimilation achieved by an alga with increasing depth is going to depend on, amongst other things, how much absorption it has in the green (500–600 nm) band. The green algae, which show relatively low absorption in this spectral region, are the most disadvantaged, which would explain their small contribution – both in species number and biomass – to the algal community, and the fact that they penetrate least deeply. The brown algae, which show, due to the presence of fucoxanthin, substantial absorption in the 500–560 nm region, hold their own through most of the sublittoral euphotic zone. At the lower fringe of the euphotic zone, where the spectral distribution becomes quite narrow, the red algae, whose biliprotein pigments have their absorption peaks within the green region, can compete better for the available light and so come to dominate.

We shall now consider the very different kinds of algal distribution that are observed in coastal waters very low in colour. Taylor (1959) found in the Caribbean that while red, brown and green species were all well represented in shallow water there was a progressive decrease in species number in all three groups with depth (Fig. 12.3). Proportionately, this

Fig. 12.3. Variation of taxonomic composition of benthic algal flora with depth in a tropical water. Plotted from data of Taylor (1959).

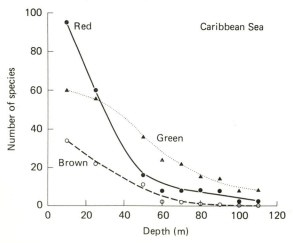

decrease was least for the green algae and in the lower 75% of the sublittoral zone there were more green species than either red or brown. Similar distributions have been observed in the Pacific. In Hawaiian coastal waters, green algal species, although overall less numerous, penetrated as deeply as red algal species, and more deeply than the brown.[94] In Eniwetok atoll lagoon, Gilmartin (1960) found that although the numbers of green and red algal species were comparable at depths down to 65 m (both greatly exceeding the number of brown species), the greens appeared (on the basis of visual observation) to be dominant in terms of biomass at all stations down to this depth, at which irradiance was 2–4% of that at the surface.

Algal distributions in the Mediterranean show some similarities. On a Corsican headland, Molinier (1960) found that green algae penetrated as deeply as 80 m, being replaced below this by red algae. A particularly valuable quantitative study, of the depth distribution of algal biomass on vertical rock faces off Malta, was carried out by Larkum, Drew & Crossett (1967). They measured the dry mass of algal biomass within the three taxonomic groups per unit area of the cliff face, as a function of depth:

Fig. 12.4. Variation of biomass per unit area of cliff face of green, brown and red algae with depth on vertical rock faces in the Mediterranean Sea (Malta, 36°N). (Plotted from data of Crossett, Drew & Larkum, 1965.)

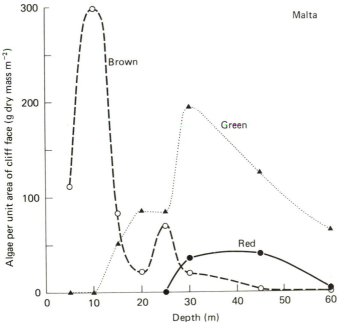

the results are shown in Fig. 12.4. Down to about 10 m depth brown algae were dominant but their contribution to total biomass decreased sharply below this. Green algae became significant at about 15 m and were the major component of the community from 20 to 60 m (the lowest depth studied). Red algae became significant components only at 30 m and decreased again below 45 m, proportionately as well as in absolute terms.

It is clear from these various studies that the most significant difference between the colourless and the slightly yellow coastal waters, so far as algal distribution is concerned, is the very much greater success of green algae in the former. This is just what would be expected on the basis of chromatic adaptation. As we have seen, the underwater light field in such waters becomes particularly rich in that blue waveband in which the green algae carry out most of their light harvesting. Except at the greatest depths there is a high proportion of green as well as blue light. Some species of green algae, in virtue of their containing the carotenoid siphonaxanthin (which when bound to protein *in vivo* absorbs in the 500–550 nm region), have an enhanced ability to absorb green light. Yokohama, Kageyama, Ikawa & Shimura (1977) in a survey of green algal species off eastern Japan found that species isolated from deep water contained siphonaxanthin, species from shallow water did not: a plausible example of chromatic adaptation. Amongst 14 species of the marine Chaetophoraceae (Chlorophyta), O'Kelly (1982) found five with lutein (the most abundant higher plant/green algal xanthophyll) and no siphonaxanthin, four with siphonaxanthin and no lutein, and five with both. The species having just lutein were found only in the mid- to upper-intertidal habitats, those possessing only siphonaxanthin were confined to the subtidal region, while those containing both pigments occupied intermediate, wide-ranging habitats.

Direct experimental evidence supporting the chromatic adaptation theory has been obtained by Levring (1966, 1968). He measured the photosynthetic rates of samples of green, brown and red algae suspended in bottles at a series of depths in turbid nearshore water (highest transmittance in the green) off the Swedish and North Carolina coasts, and in clear oceanic water (transmittance highest in the blue) in the Gulf Stream. To compare the ways in which photosynthesis and irradiance varied with depth he used a parameter, $q$, which may be regarded as the ratio of the vertical attenuation coefficient for photosynthetic rate to the vertical attenuation coefficient for irradiance. If a particular type of alga becomes increasingly ill-adapted to the spectral distribution of the light with increasing depth, then $q$ will be greater than 1 (i.e. photosynthesis will decrease faster than irradiance); if the alga is better adapted to the spectral distribution found at greater depth, then $q$ will be less than 1. Below 10 m, for green algae, $q$ was 1.2–1.3 in the turbid/coloured water and $\sim 0.8$ in the colourless water, indicating that they were better adapted for photosynthesis at depth in the less-coloured water. In the case of red algae, $q$ was $\sim 0.8$ in the turbid/coloured water, indicating improved adaptation with depth, and $\sim 1.0$ in the colourless water, indicating little change in adaptation with depth. For the brown alga, *Fucus*, $q$ was $\sim 1.0$ in the coloured/turbid water, but variable (above and below 1.0) in locations with colourless water.

### Evidence against chromatic adaptation

On the basis of the relation between water optical type and algal distribution outlined above, we would predict that in any clear, colourless coastal water, with maximum penetration in the blue waveband, green algae should be certainly a major, and probably a dominant, component of the algal biomass throughout much of the middle and lower sublittoral. The water bathing the coast of South Australia, in the region of the Great Australian Bight and the Gulf of St. Vincent, is of a clear, colourless oceanic type: there is little river runoff in this dry region. We might thus expect the depth distribution of benthic algae to be similar to that described above for Caribbean, central Pacific or Mediterranean locations. In fact, a series of thorough studies by Shepherd & Womersley (1970, 1971, 1976) and Shepherd & Sprigg (1976) has shown this not to be so: the distribution is in fact much more similar to that found on northern European coasts, with the mid-sublittoral zone dominated by large brown algal species, giving way in the lower sublittoral to a dense cover of red algae. In calmer waters the brown algae dominate the upper sublittoral too, whereas on rough-water coasts they are replaced in this zone by a short turf of the (presumably) surge-resistant coralline red alga *Corallina*.[549] In terms of biomass, green algae are generally a minor component of the algal community at all depths. At one site (Pearson Island, Great Australian Bight), however, where the water was particularly clear and colourless (oceanic water type IA), the Chlorophyta (mainly *Caulerpa* sp.) constituted a significant fraction of the community from 20 to 35 m (Fig. 12.5) but even here never attained the dominant position that green algae occupy in the underwater flora of Malta or Eniwetok atoll. The poor showing of the green algae in these waters cannot be explained on optical grounds. An alternative possibility is temperature. The South Australian waters are comparatively cool, attaining temperatures of 18–20° C at the surface in summer. The temperature of the surface water off Malta at the time of the study by Larkum *et al.* was about 27° C.[84] It may be that the siphonaxanthin-containing green algae, being perhaps mainly of tropical origin, require, as a group, higher temperatures for growth, than the brown and red algae.

Our consideration of chromatic adaptation has so far been couched in qualitative terms; in terms, for example, of whether a particular pigment absorption band is located within or outside the predominant waveband of light penetrating to a given depth. Dring (1981) has attempted to carry out a quantitative test of the theory by calculating how the amount of photosynthesis per unit irradiance might be expected to change with depth in different kinds of algae in waters of various optical types. The calculations for each alga were carried out using a measured photosynthetic action spectrum for the alga and a series of spectral distributions of irradiance at increasing depth computed for a given water type using the spectral transmission data of Jerlov (1976). In effect, what is being calculated is the extent to which the matching of the active pigment composition to the spectral distribution of available light, i.e. chromatic adaptation, should result in increased or diminished efficiency of utilization of available light with increasing depth, and therefore increased or

Fig. 12.5. Variation of biomass per unit area of rock face of green, brown and red algae with depth in a clear, colourless southern hemisphere temperate water (Pearson Island, Great Australian Bight, 34°S). Plotted from data of Shepherd & Womersley (1971).

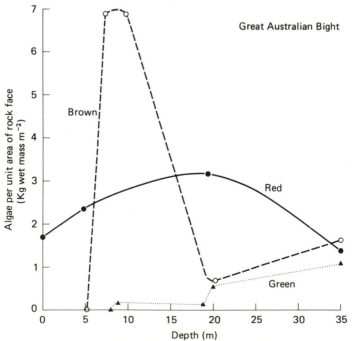

diminished competitive ability, of one algal type compared to another. The results of such calculations can then be compared with observed algal depth distribution data to see if the predictions of the theory correspond to observation.

The findings are somewhat mixed. Some of the predictions accord reasonably well with observation. For example, for coastal waters types 3 and 9 (both of which would have sufficient yellow colour to cause rapid attenuation of the blue waveband) calculations predict that with increasing depth there should be a decrease in photosynthetic effectiveness of green algae such as *Ulva* sp., little change for kelp (brown algae) such as *Laminaria saccharina*, and an increase for red algal species having phycoerythrin as the main biliprotein. This fits in quite well with the algal distributions in such waters on northern European coasts where the maximum depths to which the algae penetrate are generally found to be in the order red > brown > green.

On the other hand, these calculations and earlier ones by Larkum *et al.* (1967) indicate that in colourless oceanic waters, at the deepest limits of algal growth, green algae and thin-bladed brown algae should perform as well as or better than the phycoerythrin-containing red algae. We have in fact already noted that at least in warm oceanic waters green algae do indeed penetrate deeply and are sometimes dominant throughout most of the sublittoral zone. Nevertheless, at extreme depth in such waters they are eventually replaced by red algae. A problem with these calculations is that the validity of the results is highly dependent on the accuracy of the action spectrum attributed to the alga. Dring used action spectra data measured with just one wavelength at a time. As we saw in a previous chapter (§10.3), such action spectra can be misleading because some wavelengths of light when presented singly are incapable of exciting both photosystems to the same extent. The errors arising in this way are not great in the case of the green and brown algae but can be substantial in the case of the red algae in which photosystems I and II have rather different action spectra. If red-algal action spectra are measured with supplementary green light (to ensure that photosystem II is always functioning) then substantially higher relative activity in the blue region is obtained.[126] It may be that if the calculations were carried out again with a corrected red-algal action spectrum, the increased activity in the blue might be sufficient to give the red algae a significant advantage over the green algae in the blue-green light field at the bottom of the euphotic zone in oceanic water.

Another problem with these calculations is that the spectral distribution data were computed with the spectral transmission data used by Jerlov in his optical classification of marine waters. The original measurements on which these transmission values were based were, however, obtained round about 1950, before the development of narrow-band spectroradiometers, using rather broad-band filters. Comparisons with measurements obtained with accurate modern instruments indicate that some of these older transmission values deviate significantly from the true values, particularly at the blue end of the spectrum.[359] It is desirable that the calculations be repeated using more accurate spectral distribution data.

Another objection to the chromatic adaptation theory is based on the fact that if any particular pigment mixture, whether it be that of the green, the brown or the red algae, is raised to a high enough concentration per unit area of thallus, then there is eventually almost complete light absorption at all wavelengths and the algae become virtually black. Absorption by chlorophyll, for example, in the green waveband, although certainly low, is not infinitesimal and at high chlorophyll concentrations becomes substantial. A mature ivy leaf containing about 60 $\mu$g cm$^{-2}$ chlorophyll absorbs not only, as we might expect, $\sim 100\%$ of the red light incident upon it, but also $\sim 70\%$ of the green (550 nm) light it receives.[262] Ramus, Beale & Mauzerall (1976) argue that if a seaweed is optically thick as are, for example, *Codium fragile* (green) and *Chondrus crispus* (red), then it does not matter what colour it is, and they go on to conclude that the red algae are phylogenetically no better adapted to utilize the ambient light at great depth than are their green counterparts. That optically thick algae have roughly the same light-harvesting capacity (approaching 100% at all wavelengths) whatever the nature of their pigments is certainly true. What must be borne in mind, however, is that if an alga is faced with the problem of absorbing light from an ambient field which is rich in the green waveband, then it is much more efficient in terms of the biochemical economy of the cell to achieve this with a pigment such as phycoerythrin which has its absorption peaks in that waveband than to do it with chlorophyll which absorbs only weakly. The protein cost (remembering that chlorophyll has to be part of a pigment–protein) of achieving high absorption in the green with chlorophyll will be much greater than the protein cost of achieving it with phycoerythrin. It is true that within each of the three algal groups there are some species which have adopted the strategy of having a thick, deeply-coloured, thallus and thus absorbing most of the incident light at all wavelengths, and it would be reasonable to say that algae in this category do not make use of chromatic adaptation: as Ramus *et al.* point out, it does not matter which set of pigments they contain. In most marine algal species, however, absorption is far from complete in parts of, or throughout, the spectrum, and so the degree to which their spectral absorption matches the spectral quality of the field is of great significance for the efficient utilization of the incident radiation.

### Significance of phylogenetic chromatic adaptation

The evidence taken together seems to me to lead to the conclusion that chromatic adaptation is a major factor influencing the depth distribution of the three types of benthic marine algae, but that it is not the only one, and in some instances other factors prevail. The fact that all algae which grow at depths where the light is predominantly green or blue-green have pigments, whether specialized carotenoids such as fucoxanthin or siphonaxanthin, or biliproteins such as R-phycoerythrin, which enhance their absorption in that spectral region, cannot in my view be plausibly regarded merely as coincidence. The algae evolved in the sea; this means that they evolved in an environment throughout most of the illuminated part of which the light field had a predominantly blue-green character. Given the enormous importance of absorption spectra in determining the efficiency of collection of light energy, it is hard to believe that the types of pigment system which evolved would not be influenced by, and thus correspond to, the spectral character of the prevailing light. The argument that the specialized algal pigment systems are merely adaptations to low irradiance I do not find acceptable. That they are adaptations to limiting light levels seems unquestionable, but if the low-intensity light field has a markedly non-uniform distribution across the spectrum, then for efficient utilization of cellular resources it is better to make pigments whose absorption bands are well-placed to harvest the dim light. It is in principle possible to achieve any required level of light collection at any depth with, say, just the green pigment mixture found in surface-dwelling Chlorophyta provided the pigment concentration per unit area is made high enough: the fact that the algae have in the main not chosen that, biochemically expensive, form of adaptation to low irradiance, but have evolved pigments absorbing in the spectral regions where the light actually occurs, indicates that they are adapted not just to low light, but to low light of a particular spectral character.

Thus, the predominance of red algae in the deepest part of the euphotic zone can be plausibly attributed to their possession of phycoerythrin – the most efficient way (in terms of quanta collected per unit of protein invested) of harvesting the dim blue-green light prevailing at those depths. The failure of green algal species lacking siphonaxanthin to penetrate to great depths we may consider to be a consequence of their comparatively poor absorption capacity for the prevailing underwater light field. The greater depth distribution of the siphonaxanthin-containing green species, and the brown algae we may reasonably regard as being due to the enhanced capacity for absorption in the plentiful 500–550 nm waveband, conferred by their specialized carotenoids.

Chromatic adaptation is not, of course, the whole story. We noted early on that there is no reason why it should have anything to do with the relative success of the different algal groups near the surface. The presence of non-siphonaxanthin green species near the surface is not an example of chromatic adaptation, but their failure to penetrate the depths is such an example. The predominance of coralline reds near the surface in rough-water locations in South Australia has everything to do with their resistance to wave movement and nothing to do with their pigment composition. The presence of significant levels of green algae in the 20–35 m zone in the clear water of Pearson Island, South Australia (Fig. 12.5), we may regard as an example of chromatic adaptation, but their much poorer performance overall in South Australian than in the optically similar Mediterranean or central Pacific waters indicates that other factors such as, perhaps, temperature can prevail over chromatic adaptation in determining algal depth distribution. Within the deep-water red-algal zone, the transition at the lowest depths from fleshy macroscopic red algae to encrusting coralline red algae[419] has no obvious explanation in terms of pigment differences. In the very dim light available at those depths, production rates can only be very low, in which case the rates of loss by grazing and respiration become very important. It may be that the encrusting coralline reds are less susceptible to grazing than the fleshy species and/or have a lower respiration rate. The predominance of sea grasses on sandy bottoms down to great depths in clear colourless water, e.g. to 35 m in the St Vincent Gulf, South Australia[421] or 50 m in the Red Sea,[291] is not due to their being chromatically better adapted than algae (although they should be reasonably well adapted to the bluish light prevailing in such habitats) but to their ability, by means of their roots and rhizomes, to colonize these unstable substrates.

In short, phylogenetic chromatic adaptation is by no means the only factor responsible for the depth distribution of the different plant groups under water: we should, however, continue to regard it as one important factor in that set of interacting factors which finally determines that distribution.

## 12.2    Ontogenetic adaptation – intensity

Adaptation of the photosynthetic apparatus to changes in light quality (intensity and/or spectral distribution) within a given species can take the form of changes in the total amount of pigment per cell or in the ratio of different pigments, or both.

### Chlorophylls and photosynthetic units

It is generally found amongst the algae, both unicellular and multicellular, that as the light intensity during growth decreases (even without alteration of spectral composition), the content of their photosynthetic pigments increases:[263] two- to five-fold increases are commonly observed. There are exceptions. Jørgensen (1969) found that in four diatom and one green algal species, the chlorophyll *a* content per cell was the same whether the cells were grown at low or high light intensities. However, this is not a universal characteristic of the Bacillariophyceae: other diatom species have been found to substantially increase their chlorophyll *a* content as the light intensity during growth is decreased.[77]

In the higher plants and green algae, there is a general tendency for the level of chlorophyll *b* relative to chlorophyll *a* to increase as light intensity decreases. For example in the green flagellate *Dunaliella tertiolecta*, simultaneously with a 2.6-fold rise in chlorophyll *a* content, the *a/b* ratio fell from 5.6 to 2.3 when the growth irradiance was reduced from 400 to 20 $\mu$einsteins m$^{-2}$ s$^{-1}$.[119]

In green plants – the algae as well as the angiosperms – it appears that the increase in chlorophyll content resulting from growth at lower light intensity is largely due to an increase in the number of photosynthetic units per cell, or per unit mass, where by 'photosynthetic unit' we mean a photosystem I and a photosystem II reaction centre, each with its assemblage of light-harvesting pigment molecules. This has been shown for higher plants;[47] for the unicellular chlorophytes *Scenedesmus obliquus*[124] and *Dunaliella tertiolecta*;[119] and for the multicellular green species *Ulva lactuca*.[317] There appears generally to be little change in the number of chlorophyll molecules per photosynthetic unit. In *Chlorella pyrenoidosa*, Myers & Graham (1971) found that the number of chlorophyll molecules per photosynthetic unit of low-light-grown cells were 50% higher than in high-light-grown cells: however, the total chlorophyll concentration per unit mass in the low-light-grown cells was five times that in the high-light-grown cells, so the conclusion that the increase in chlorophyll is largely due to an increased number of photosynthetic units is true for this species also.

In the blue-green alga *Anacystis nidulans*, it appears that the shade-induced increase in chlorophyll content is, as in the green algae, due to an increase in the number, rather than the size, of the photosynthetic units.[522] In certain other algal groups, however, there is evidence that the increase in chlorophyll is largely due to an increase in the size of the photosynthetic unit. This has been shown for a number of diatom species:[119, 362] for example, Falkowski & Owens (1980) found that the (relatively modest) increase of 58% in chlorophyll *a* content of *Skeletonema costatum*, caused by a decrease in growth irradiance from 130 to 20 $\mu$einsteins m$^{-2}$ s$^{-1}$, was accompanied by a 26% decrease in the number of photosystem I reaction centres per cell and a doubling of the number of chlorophyll *a* molecules per photosynthetic unit. The number of chlorophylls in the photosynthetic units of the zooxanthellae (symbiotic dinoflagellates) of the coral *Stylophora pistillata* was found to be four times as large in corals growing in shaded sites (e.g. under ledges) where irradiance was 0.3–1.0% of the subsurface value, as in corals growing in exposed sites (irradiance 50–90% of subsurface value).[117] The number of photosystem I reaction centres per dinoflagellate cell was the same in both kinds of coral, and the number of algal cells per cm$^2$ of coral was also about the same. The amount of chlorophyll *a* per cm$^2$ of coral was about four times as high in the shaded as in the Sun-exposed forms; this was accounted for by the increase in chlorophyll content per cell which was in turn due to the increase in size of the photosynthetic unit. In the chrysophyte *Isochrysis galbana*, shade adaptation (growth at 4 rather than at 300 $\mu$einsteins m$^{-2}$ s$^{-1}$) was accompanied by a 62% increase in the number of chlorophyll molecules per photosynthetic unit.[362] In the multicellular red alga *Porphyra umbilicalis*, the number of chlorophyll molecules per photosynthetic unit varied from about 400 to 1300, roughly in proportion to the tissue chlorophyll content.[317]

### Other photosynthetic pigments

Light/shade adaptation affects other light-harvesting pigments as well as chlorophyll. The extent to which changes in these pigments are linked to those in chlorophyll varies from one species to another. In *Anacystis nidulans* the link is close but not rigid. Cultures grown at 10 $\mu$einsteins m$^{-2}$ s$^{-1}$ contained just over twice as much chlorophyll per cell as cultures grown at 100 $\mu$einsteins m$^{-3}$ s$^{-1}$, and the number of photosynthetic units per cell increased to the same extent in the low-light cultures.[522] The level of the biliprotein phycocyanin, however, increased about three-fold so that the number of phycocyanin molecules per photosynthetic unit rose but only by about 30%. In the cryptophyte *Chroomonas*, cultures grown at 1 W m$^{-2}$ contained about 20% more chlorophyll *a* but about 50% more phycocyanin (per cell) than cultures grown at 27 W m$^{-2}$.[121]

Amongst the red algae, more substantial changes in biliprotein/chlorophyll ratio are common. Plants of *Griffithsia pacifica* grown at a light intensity of 50 foot-candles had a phycoerythrin/chlorophyll ratio 2.5 times that in plants grown at 1000 foot-candles: the number of phycobilisomes per unit area of thylakoid was also correspondingly lower in the high-light-grown cells.[528] In plants of *Chondrus crispus* growing in sunlit sites at 3–4 m depth, the ratio of phycoerythrin to chlorophyll remained high during the winter but underwent a 60% fall in late spring/early summer: in plants in shaded sites at the same depth the ratio remained high during the summer.[398] When the sunlit plants became shaded due to a dense growth of an epiphytic diatom in August they regained much of their phycoerythrin.

Cultures of the unicellular red alga *Porphyridium cruentum* grown at 0.68 W m$^{-2}$ contained five times as much chlorophyll but 7.3 times as much phycoerythrin (per unit cell volume) as cultures grown at 11.4 W m$^{-2}$.[62] In the related species *P. aerugineum*, cells grown at 0.5 W m$^{-2}$ contained about 3 times as much total pigment (% dry mass) as cells grown at 50 W m$^{-2}$, and the phycocyanin/chlorophyll ratio was also increased about three-fold in the low-light-grown cells.[92]

Although in none of these cases except *Anacystis* were measurements of photosynthetic unit numbers carried out, given the evidence that the chlorophyll/photosynthetic unit ratio remains the same or increases during shade adaptation, then it seems likely that when there is a substantial increase in the biliprotein/chlorophyll ratio, there is indeed an at least commensurate increase in the average number of biliprotein molecules associated with each photosynthetic unit. In some instances the changes in biliprotein/chlorophyll ratio are so great that there must be changes in the number of biliprotein molecules per photosynthetic unit. For example, Calabrese & Felicini (1973) observed that in the sublittoral red alga *Gracilaria compressa* in the Adriatic Sea, the distal portions of the fronds which received direct illumination were yellow-green in colour and contained 0.065% (dry mass) chlorophyll and only traces of phycoerythrin. The proximal portions which were shaded were purplish-red, and contained 0.085% chlorophyll and 0.82% phycoerythrin.

In the marine dinoflagellate *Glenodinium*, Prézelin (1976) found that as the light intensity during growth was lowered over the range 30 to 2.5 W m$^{-2}$, the chlorophyll *a* content per cell rose progressively by, ultimately, about 80%. Over the same range of intensity, however, the cellular concentration of the light-harvesting peridinin/chlorophyll *a*–protein rose seven-fold. Although no data on accompanying changes in photosynthetic unit number were obtained, it does seem likely that there were many more peridinin/chlorophyll *a*–protein molecules associated with each photosynthetic unit in the low-light-grown than in the high-light-grown cells.

In the red alga *Callithamnion roseum*, Yu, Glazer, Spencer & West (1981) found that the proportion of the two different phycobilin chromophores – phycoerythrobilin (PEB) and phycourobilin (PUB) – present in the biliprotein phycoerythrin changed in accordance with the light intensity during growth. The PEB/PUB ratio in phycoerythrin I increased progressively from 2.17 to 3.06 as the irradiance during growth was increased from 6 to 37 $\mu$einsteins m$^{-2}$ s$^{-1}$. It seems likely that the total number of chromophore groups per molecule was unaltered, but since PEB absorbs less strongly than PUB the total absorption by the biliprotein presumably decreased at higher growth irradiance.

### Changes in electron carriers and carboxylase

In higher plant species which are able to adapt to low light conditions, the increase in pigment during shade adaptation is not accompanied by an increase in the content of the photosynthetic electron transfer components – the cytochromes, ferredoxin and plastoquinone. The levels of these per unit mass may remain about the same or may decrease somewhat.[47, 547] The level of ribulose diphosphate carboxylase per unit mass of leaf is markedly lowered in shade-adapted plants.[312] These changes are of adaptive value since at low light intensities the plant will not be able to carry out high rates of electron transfer and carboxylation anyway and so can achieve biosynthetic economies by refraining from increasing, or preferably decreasing, its content of electron transfer components and carboxylase at the same time as it increases its pigment content and photosynthetic unit number. Thus, in shade-adapted plants there is an increase in the ratio of pigment assemblies to the pools of electron carriers, and an even bigger increase in the ratio of pigment assemblies to carboxylase.

A study by Senger & Fleischhacker[124, 420] indicates that the same sorts of changes accompany shade adaptation in the unicellular chlorophyte *Scenedesmus obliquus*. While the chlorophyll content per g fresh mass increased by 64% (in the low-light-grown, relative to the high-light-grown cells) the cytochrome *f* content decreased by about 33%. The carboxylase activity (per mg protein) of cell extracts was also lower in the low-light cultures. In the marine diatom *Phaeodactylum tricornutum*, grown in continuous culture, while the chlorophyll and carotenoid content rose progressively by about 100% as light intensity was lowered from 12 to 0.5 klux, the ribulose diphosphate carboxylase activity per cell fell to less than 25% of the value present in high-light-grown cells.[33] Thus, although more biochemical information on the mechanism of light/shade adaptation in algae is needed, we may reasonably suspect that the preferred strategy for coping with low light levels is an increase in the number or size (or both) of the pigment assemblies (photosynthetic units) with no increase, or an actual decrease, in the synthesis of carboxylase and electron carriers.

### Photosynthetic consequences of light/shade adaptation

The physiological consequences of these biochemical changes are manifested as changes in the dependence of photosynthetic performance on light intensity. If photosynthetic rate per unit chlorophyll is measured as a function of irradiance then it is in some cases found that rates exhibited by low- and high-light-adapted cells or tissues are much the same at low irradiance, and initially show the same rate of increase with increasing irradiance, but the low-light-adapted plants level off and reach light saturation first. The high-light-adapted cells require a higher light intensity to reach saturation and achieve a higher light-saturated photosynthetic rate, as is shown for *S. obliquus* in Fig. 12.6: the $E_k$ values were about 40 and 110 W m$^{-2}$ for the low- and high-light-adapted cells, respectively, in this green alga. In the dinoflagellate *Glenodinium*, the $E_k$ value increased from 14 to 28 W m$^{-2}$ as the light intensity used for growth increased 10-fold.[378]

In the case of *S. obliquus*, the high-light-adapted cells also achieve a higher light-saturated photosynthetic rate per unit of cellular biomass (packed cell volume), presumably reflecting the higher cellular content of electron transfer components and carboxylase. This is not observed in all

Fig. 12.6. Photosynthetic characteristics (*P versus* $E_d$) of cells of *Scenedesmus obliquus* (Chlorophyceae) grown under high (○) or low (●) light intensity. The specific photosynthetic rate is expressed either per mg chlorophyll (——) or per unit cellular biomass (packed cell volume, -----). The cells were grown in continuous culture under high (28 W m$^{-2}$) or low (5 W m$^{-2}$) irradiance, and contained 7.8 and 12.8 mg chlorophyll ml$^{-1}$ packed cell volume, respectively. Plotted from data of Senger & Fleischhacker (1978).

species. In some, such as the dinoflagellate *Glenodinium*,[378] the high-light-adapted forms achieve a higher light-saturated rate per mg chlorophyll but only about the same light-saturated rate per unit cellular biomass, as the low-light-adapted forms. It seems likely that in such species the carboxylase content does not increase during high-light-adaptation.

We have noted that both high- and low-light-adapted forms of *S. obliquus* and certain other algae have about the same photosynthetic rate per mg chlorophyll at low light intensities: the *P* versus $E_d$ curve has about the same slope for both types of cells in the light-limited region. This is because at such intensities, electron transfer and carboxylation capacity are present in excess. Photosynthetic rate is determined entirely by the rate of photon capture, and so is determined by the amount of pigment present. Strictly speaking, it is the rate of photosynthesis per unit absorptance rather than per unit chlorophyll which we would expect to be the same for both types of cell in low light, and the ratio of absorptance to chlorophyll can change. The slope, $\alpha$, of the *P* versus $E_d$ curve in the light-limited region is equal to $k_c \phi_m$ where $k_c$ is the specific vertical attenuation coefficient (per mg chlorophyll), and $\phi_m$ is the maximum quantum yield, for the algal cells (§10.2). We do not expect $\phi_m$ to change during shade adaptation: however, we saw earlier (§9.5) that $k_c$ is a function of the intracellular pigment concentration as well as of cell size and shape. The more concentrated the pigments within the cell, the less efficiently they collect light, and so the lower the value of $k_c$. Thus the increase in pigment content that takes place during shade adaptation is likely (assuming no change in the ratio of pigments present) to lead to some reduction in $k_c$, and therefore in $\alpha$: moderate decreases (7–34%) in $\alpha$ have in fact been observed in certain diatom species during shade adaptation.[362] In an alga in which there was a substantial increase in the ratio of some other light-harvesting pigment to chlorophyll during shade adaptation, and therefore a diminution in the proportionate contribution of chlorophyll to light absorption, we might expect the shade-adapted cells to show a higher rate of photosynthesis per unit chlorophyll at low irradiance than cells grown at high light intensity. Such an effect could be looked for in, for example, those red-algal species which greatly increase their phyco-erythrin/chlorophyll ratio during shade adaptation.

In the marine diatom *Phaeodactylum tricornutum*, it has in fact been observed that cells grown at low light intensity show a higher photosynthetic rate per mg chlorophyll *a* at low irradiance than cells grown at high intensity.[33] In this alga, however, there is evidence that in the cells grown at high light intensity the physicochemical state of fucoxanthin is modified in such a way that excitation energy is inefficiently transferred to chlorophyll *a*.[426] Thus the increase in photosynthetic rate per mg chlorophyll at low irradiance in the low-light-grown cells is likely to be due to more efficient utilization of the absorbed light. There is no reason to suppose that the light absorption capacity per mg chlorophyll *a* increases in the low-light-grown cells of this diatom; in fact the fucoxanthin/chlorophyll *a* ratio decreases slightly.[426]

When shade-adapted cells achieve about the same photosynthetic rate per unit pigment at low irradiance as high-light-grown cells, but contain more pigment per unit cellular biomass, they consequently have a higher photosynthetic rate per unit biomass than the high-light-grown cells under these conditions. This can be seen for *Scenedesmus obliquus* in Fig. 12.6. The difference becomes more pronounced in accordance with the extent to which pigment content increases and is particularly marked for example in the case of *Chlorella vulgaris*.[468] It is at these low light intensities, when the supply of excitation energy is the limiting factor, that the higher concentrations of light-harvesting pigment within the low-light-grown cells gives them the advantage over the high-light-grown cells.

It is sometimes assumed that the light-saturated photosynthetic rate per cell is determined by the number of photosynthetic units per cell, and it has been argued that if shade adaptation were accompanied (as we here suppose sometimes to be the case) by an increase in the number of photosynthetic units per cell, then the light-saturated photosynthetic rate per cell or other unit of biomass would increase during shade adaptation.[378] The light-saturated photosynthetic rate per cell would only be determined by the number of photosynthetic units per cell if there was some mechanism which ensured that the number of electron transfer and carboxylase molecules always increased in parallel with the number of photosynthetic units. There is no evidence for the existence of such a mechanism and there are well-established examples where the contrary happens. In shade-adapting higher plants, for instance, it seems that photosynthetic units (pigment assemblies) increase in number while carboxylase decreases.[47] In *Glenodinium*, the fact that $P_{max}$ per cell does not increase in parallel with pigment content during low-light adaptation can most plausibly be attributed to a failure of carboxylase content to increase rather than to a lack of increase in the number of photosynthetic units (which need not affect $P_m$ anyway).

As mentioned earlier, there are some algal species whose chlorophyll *a* content does not vary with the light intensity at which they are grown. This does not, however, mean that they do not undergo light/shade adaptation. Jørgensen (1969) found that the diatom species and a green alga (*Scenedesmus quadricauda*), whose chlorophyll contents were unaffected by light intensity during growth, all showed a substantially (31–95%) higher light-saturated photosynthetic rate per cell when grown at 30 klux than when grown at 3 klux. Measurements of photosynthesis as a function of light intensity in one of the diatoms, *Nitzschia palea*, showed that the rates per $10^6$ cells (or per unit chlorophyll) were identical and increased at the same rate for both kinds of cells over the lower range of light intensity, but that whereas photosynthesis by the low-light-grown cells ceased to increase beyond 10 klux ($E_k = 8$ klux), photosynthesis in the high-light-grown cells continued to increase, levelling off only above 20 klux ($E_k = 13$ klux) at a light-saturated rate 68% greater than that achieved by the low-light-grown cells (Fig. 12.7). It is a plausible supposition, which could be tested by direct measurement, that the high-light-grown cells contain substantially more carboxylase and thus are able to take advantage of high irradiance: in the low-light-grown cells the lower amounts of carboxylase reach their maximum rate of working at comparatively low light levels. Jørgensen only measured chlorophyll *a* content in *N. palea*, so changes in other pigments such as chlorophyll *c* and/or fucoxanthin may have occurred. However, the fact that the low-light-grown cells do not achieve a higher rate of photosynthesis per unit cellular biomass than the high-light-grown cells at low light intensity (Fig. 12.7) confirms that in this alga, unlike those in the other groups, shade adaptation does not involve increasing its ability to harvest light from low-intensity fields and

Fig. 12.7. Photosynthetic characteristics (*P versus* $E_d$) of cells of the freshwater diatom *Nitzschia palea* grown under high (○) or low (●) light intensity. The cells were grown at irradiance values of 30 klux and 3 klux, respectively. (After Jørgensen, 1969). For the light source used, 30 klux is approximately equivalent to $500 \times 10^{18}$ quanta m$^{-2}$ s$^{-1}$.[468]

indeed may consist simply of achieving economies by making only as much carboxylase and perhaps other non-pigment components as can be effectively used at the prevailing light intensity. These considerations suggest that algae of this type are not as well fitted for living at low irradiances as are other algae.[468]

### Shade adaptation in aquatic angiosperms

All the adaptive changes we have considered so far are in the biochemical composition and consequent functioning of the photosynthetic apparatus. There are also other forms of ontogenetic adaptation, particularly in multicellular species, which can enable plants to cope with variation in the intensity of the light field. In higher plants, both aquatic and terrestrial, adapting to shade, at the same time as chlorophyll content (% of dry mass) increases, the leaves increase in area so as to intercept more light and also become thinner. Because of the increase in specific leaf area (area per unit dry mass of leaf), the chlorophyll content per unit area increases less than content per unit mass during shade adaptation, or may remain the same, or may decrease.

Spence & Chrystal (1970) studied light-intensity adaptation in certain species of the fresh-water angiosperm genus *Potamogeton* from Scottish lochs. In *P. polygonifolius*, a shallow-water species (occurring from the waters edge to 0.6 m depth), shade plants (grown at 6% of full sunlight) had a specific leaf area three times that of plants grown in full sunlight: the shaded leaves were 0.04 mm and the sunlit leaves 0.12 mm thick. The chlorophyll content in the shaded leaves was about one third higher than that in the sunlit leaves on a dry mass basis. Because of the great increase in area, however, the shade leaves contained only about half as much chlorophyll per unit area as the sunlit leaves. The dark-respiration rate per unit area was 27% lower in the shaded leaves – a greater reduction might have been expected given that leaf thickness was reduced by two-thirds. Presumably as a consequence of the diminution in respiration rate, the light compensation point in the shaded leaves was lowered by about the same amount.

*P. obtusifolius*, a deeper-water species (depth range 0.5–3.0 m) appeared to lack the ability to change its specific leaf area in accordance with light intensity, there being little difference in this respect between plants grown at 100% and 6% of full sunlight. There was also little change in the chlorophyll content per unit mass or leaf area. However, the specific leaf area of *P. obtusifolius*, ($\sim 2$ cm$^2$ mg$^{-1}$) is already greater than that of *P. polygonifolius* (sunlit leaves, 0.48 cm$^2$ mg$^{-1}$; shaded leaves, 1.43 cm$^2$ mg$^{-1}$), and so we may suppose that *P. obtusifolius* has already taken this step during evolution, as part of its phylogenetic adaptation to the deeper water, and it may be that further increases in specific leaf area would not be possible. The respiration rate of *P. obtusifolius* per unit area of leaf was, even in the sunlit leaves, as low as one third, and the light compensation point about half that of the shaded leaves of *P. polygonifolius*, further evidence for the superior phylogenetic adaptation of *P. obtusifolius* to low light. This species still retained some capacity for ontogenetic adaptation, however, since its shaded leaves had a very much lower respiration rate than its sunlit leaves, so much so that its light compensation point was reduced by $\sim 90\%$.

## 12.3    Ontogenetic adaptation – spectral quality

Within the underwater environment the spectral composition of the light varies greatly, both with depth within a given water body and from one water body to another, in accordance with the absorption spectrum of the water. As we saw in the previous section, laboratory studies show that variation in intensity alone can bring about marked changes in pigment composition, photosynthetic characteristics and morphology of aquatic plants. Laboratory studies have also shown that intensity is not the whole story and that the spectral quality of the incident light has, for some plants, a specific role to play.

### *Chromatic adaptation within the blue-green algae*

The most clear-cut examples are to be found amongst the blue-green algae (Cyanobacteria), and it was indeed in a blue-green algal species that ontogenetic chromatic adaptation was first described. Gaidukov (1902) observed that *Oscillatoria rubescens* was red in colour when grown in green light, and blue-green when grown in orange light: he attributed these colour changes to the synthesis of different kinds of pigment. Boresch (1921) showed that the colour changes are due to shifts in the types of biliproteins synthesized by the algae: the red cells contain predominantly phycoerythrin, the blue-green cells mainly phycocyanin. This phenomenon was considered by Engelmann & Gaidukov (1902) to be an example of *complementary chromatic adaptation*, the pigment induced by a specific waveband being one which absorbs that waveband (phycoerythrin and phycocyanin have their absorption peaks in the green ($\sim 565$ nm) and the red ($\sim 620$ nm), respectively): 'complementary' because the pigment has the complementary colour to that of the light which induces it.

Not all blue-green algal species show chromatic adaptation, and amongst those that do there are variations in the form that adaptation takes. Tandeau de Marsac (1977) separated cyanobacteria into three groups. In group I strains the biliprotein composition is not affected by the spectral quality of the light in which they are grown, i.e. they lack chromatic adaptation. In group II strains, only the phycoerythrin content of the cells is affected by light quality, being very low in red light and high in green light; phycocyanin content remains high in cells grown under either kind of light. In group III strains the synthesis of both the major biliproteins is affected by the spectral composition of the light: like the previous group their phycoerythrin content is high in green light and low in red, but their phycocyanin content, while still substantial in green light, is 1.6- to 3.7-fold higher in red light. Of 69 strains examined, 25 were in group I, 13 in group II and 31 in group III.[63, 494]

The phycocyanin formed in group I and group II strains has two polypeptide subunits, $\alpha$ and $\beta$. Bryant (1981) found that in 24 of the 31 group III strains, the phycocyanin formed in green light also contained just two subunits, but the phycocyanin from cells grown in red light contained four subunits, $\alpha_1$, $\alpha_2$, $\beta_1$, and $\beta_2$. The particular subunits present in phycocyanin in the green-light-grown cells were $\alpha_2$ and $\beta_1$. These, being always present, Bryant referred to as constitutive: the other two, $\alpha_1$ and $\beta_2$, formed in red light, he referred to as inducible. Thus, in these 24 strains it appeared that the increased phycocyanin formation induced in red light consisted of synthesis of different kinds of phycocyanin subunit. It was not possible to determine from the data whether the new phycocyanin subunits associated to form a distinct phycocyanin species, $(\alpha_1\beta_2)_n$, or whether they aggregated with the two constitutive subunits, $\alpha_2$ and $\beta_1$, to form a hybrid phycocyanin. In the remaining seven group III strains, no $\alpha_1$ and $\beta_2$ phycocyanin subunits were found in red-light-grown cells: the extra phycocyanin in these cases may have consisted just of additional $\alpha_2$ and $\beta_1$ subunits.

The specific nature of the light treatments required to bring about changes in biliprotein composition have been studied in one of the group III strains, *Calothrix* 7101 (formerly *Tolypothrix tenuis*), by Fujita & Hattori (1960, 1962*a*, *b*) and Diakoff & Scheibe (1973). It has been found convenient to study the controlling influence of light in *Calothrix* 7101 by first depleting the cells of biliproteins by exposure to intense light for 24 h in the absence of nitrogen, then giving a brief light treatment of the desired spectral quality, and then placing the cells in the dark with a nitrogen source, and following biliprotein synthesis. After a saturating dose of red light the cells synthesize phycocyanin but not phycoerythrin during the subsequent dark incubation: after a saturating dose of green light they continue to make phycocyanin but now also make phycoerythrin. If red and green light treatments are given alternately, the subsequent pattern of biliprotein synthesis is determined by the colour of the last light treatment, i.e. green or red light will each reverse the effect of the other.[134]

The action spectrum for the promotion of phycoerythrin synthesis has a peak in the green at about 550 nm with an additional small peak in the UV at about 350 nm (Fig. 12.8): the action spectrum for inhibition of phycoerythrin synthesis has a peak in the red at about 660 nm, and also has a subsidiary peak in the UV at about 350 nm.[91, 133] The shape of these action spectra are consistent with the photoreceptive pigment system being itself biliprotein in nature. Bogorad (1975) has suggested the term 'adaptochrome' for any photoreceptive pigment involved in the regulation of biliprotein synthesis.

Scheibe (1972) has proposed that this controlling pigment is analogous to phytochrome and has two interconvertible forms: one ($P_G$) would have its absorption maximum in the green and on irradiation with green light undergo photoconversion to another form ($P_R$) with an absorption maximum in the red region. On irradiation with red light $P_R$ would be photoconverted to $P_G$

$$P_G \underset{\text{red } h\nu}{\overset{\text{green } h\nu}{\rightleftharpoons}} P_R$$

Scheibe points out that both forms of the pigment could be biologically active, $P_G$ promoting phycocyanin and $P_R$ promoting phycoerythrin

Fig. 12.8. Action spectra for the promotion and inhibition of phycoerythrin synthesis in the blue-green alga *Calothrix* 7101 (*Tolypothrix tenuis*) (after Diakoff & Scheibe, 1973). The reciprocal of the amount of radiant energy (J cm$^{-2}$) required to produce a 25% increase (promotion) or decrease (inhibition) in the amount of phycoerythrin (as a proportion of total biliprotein) synthesized in the dark after the light treatment is (after normalization to 500 or 680 nm) plotted against wavelength.

formation, or else just one form could be active, its presence or absence determining which of two possible differentiative pathways is to be followed.

There have been various reports of the isolation from blue-green algae of pigments which show a small degree of green–red reversibility *in vitro*.[43, 415] It appears that such pigments are in fact allophycocyanin:[348] there is no evidence yet that allophycocyanin has any role in the photo-control of biliprotein synthesis.

Although in light-grown cells the spectral quality of the light influences the pattern of synthesis of biliproteins, light is not essential for the synthesis of these pigments to occur: facultatively heterotrophic cyanobac-teria continue to make biliproteins when grown on carbohydrate medium in the dark. When such strains in group III are grown in the dark their biliprotein composition is similar to that of cells grown in red light: high in phycocyanin and low in phycoerythrin.[494] Bryant (1981) found that the phycocyanin synthesized by *Calothrix* strains 7101 and 7601 in the dark included the inducible subunits, $\alpha_1$ and $\beta_2$, found in red-light-grown but not green-light-grown cells. This was the case whether the inocula used were derived from cultures grown in red or green light. Bryant concluded that it is not exposure to red light, but the absence of green light which turns on the synthesis of the inducible phycocyanin subunits. If there is indeed a photoreversible pigment involved in the control of biliprotein synthesis, then it is presumably the $P_G$ form which is synthesized by cells growing in the dark.

### Chromatic adaptation in red algae

Red algae do show changes in their pigment composition in response to changes in the spectral quality of the light field in which they are grown. The direction of the pigment change, however, quite apart from its quantitative extent, seems to depend on the intensity of the light. Brody & Emerson (1959) determined the ratio of phycoerythrin to chlorophyll in the unicellular red alga *Porphyridium cruentum* grown in green light (546 nm – absorbed mainly by phycoerythrin) or blue light (436 nm – absorbed mainly by chlorophyll), at low ($\sim 0.1$ W m$^{-2}$) or high (25–62 W m$^{-2}$) intensity. At low irradiance the cells grown in green light had a phycoerythrin/chlorophyll ratio more than twice that of the cells grown in blue light. This we may regard as complementary chromatic adaptation: the cells increased the proportion of that pigment which best absorbed the light to which they were exposed.

In the cells grown at high monochromatic irradiance, however, the position was reversed: the phycoerythrin/chlorophyll ratio of cells grown in green light was less than 50% of that in cells grown in blue light. Similarly, Ley & Butler (1980) found that cells of *P. cruentum* grown in high-intensity red (112 $\mu$einsteins m$^{-2}$ s$^{-1}$) or blue (50 $\mu$einsteins m$^{-2}$ s$^{-1}$) light has about twice the phycoerythrin/chlorophyll ratio of cells grown in high-intensity green (99 $\mu$einsteins m$^{-2}$ s$^{-1}$) light (Fig. 12.9). In the multicellular red alga *Porphyra*, Yocum & Blinks (1958) found that plants exposed to high-intensity blue light (436 nm, 24 W m$^{-2}$) for 10 days contained more phycoerythrin and less chlorophyll than plants exposed for the same time to high-intensity green light (546 nm, 17.5 W m$^{-2}$).

These changes in pigment composition induced by high-intensity mono-chromatic illumination are accompanied by changes in the photosynthetic characteristics of the plants. Yocum & Blinks (1958) found that marine red algae which had just been collected or had been kept under green light for 10 days showed low photosynthetic efficiency in the region of the chlorophyll red absorption band, the action spectrum falling well below the absorption spectrum between 650 and 700 nm. Plants which had been kept for 10 days under blue light, however, were highly efficient in the red region (as well as showing some increase in the blue region) the action and absorption spectra now approximately coinciding between 650 and

Fig. 12.9. Effect of spectral quality of light field during growth on the photosynthetic pigment composition of the unicellular red alga *Porphyridium cruentum* (by permission, from Ley & Butler (1980), *Plant Physiology*, **65**, 714–22). The spectra are normalized to the same absorbance value at 676 nm. The letter labelling each spectrum indicates the light field in which the cells were grown: R = red, B = blue, L = low-intensity white, H = high-intensity white, G = green.

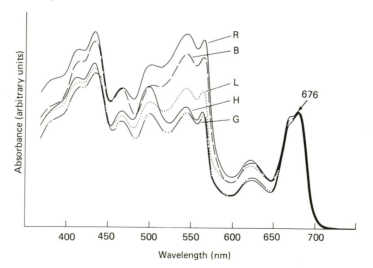

700 nm. Clearly, what had resulted from the exposure to blue light (itself absorbed by chlorophyll) was an increase in the efficiency of utilization of light absorbed by chlorophyll.

The nature of adaptive changes that take place within the photosynthetic system under these various moderate-to-high intensity, spectrally selective light regimes has been clarified by the detailed study of Ley & Butler (1980) on *Porphyridium cruentum*. By careful analysis of the absorption spectra and the fluorescence behaviour they have been able to arrive at conclusions concerning the absorption characteristics of, and energy transfer between, the two photosystems in cells grown in light of various spectral compositions. In red algae, chlorophyll *a* is the main light-harvesting pigment for photosystem I and phycoerythrin is the main pigment in photosystem II. Cells growing in red or blue light will be receiving much more excitation energy in chlorophyll than in phycoerythrin, which, given a photosynthetic system initially adapted to a broad spread of wavelengths, will lead to a much greater input of energy into photosystem I than photosystem II. To ensure a balanced operation of the photosynthetic system, the cells must increase the absorption cross-section of photosystem II relative to that of photosystem I. This they achieve partly by increasing the ratio of phycoerythrin to chlorophyll, but also by putting a much larger proportion of their chlorophyll into photosystem II. In addition, Ley & Butler found that these red- or blue-light-grown cells have a diminished probability of energy transfer from photosystem II to photosystem I, which helps to keep excitation energy in photosystem II. The increased effectiveness of red light in *Porphyra*, observed by Yocum & Blinks after prolonged exposure of the alga to blue light, we may now plausibly explain in terms of an increased proportion of chlorophyll in photosystem II, and therefore a more balanced functioning of the photosystems in red light. Cells growing in green light will, if they are initially adapted to a mixture of wavelengths, receive far more excitation energy in photosystem II than I. Their adaptive response is to lower the absorption cross-section of photosystem II relative to that of photosystem I by reducing the phycoerythrin/chlorophyll ratio, and including nearly all their chlorophyll in photosystem I: PSI chlorophyll/PSII chlorophyll is $\sim$ 20 in green-light-grown, compared to $\sim$ 1.5 in red- or blue-light-grown cells.[290] In addition, the green-light-grown cells show a greater probability of energy transfer from photosystem II to I than cells grown in red or blue light.

The red- or blue-light-grown cells have a higher phycoerythrin content than the green-light-grown cells on a per cell basis, as well as in proportion to chlorophyll. It might be argued that although it makes sense to put more chlorophyll into photosystem II, to make more phycoerythrin in red light is not a useful adaptive response since this biliprotein has little absorption above 600 nm and so will not collect the red photons. However, it may be that what the cell 'detects' is not specifically the spectral nature of the incident light, but simply the fact that photosystem I is receiving more energy than photosystem II, and so it responds with a general increase in the pigment complement of photosystem II. Red-dominated light fields do not occur naturally in the marine environment and so we should not expect red algae to show specific adaptation to them.

From the work of Ley & Butler it seems that the general principle on which chromatic adaptation to relatively high intensity, spectrally confined, light fields takes place is adjustment of the composition and properties of the photosystems in such a way as to lead to their being excited at about the same rate, and thus to efficient photosynthesis. The complementary chromatic adaptation observed by Brody & Emerson (1959) in *P. cruentum* exposed to low-intensity blue and green light, since it operates in a contrary fashion so far as pigment changes are concerned, must have a different basis. At low light intensity the total rate of supply of excitation energy rather than imbalance between the photosystems becomes the major constraint on photosynthesis. It may be that the best strategy for the algae growing in dim green or blue-green light is to make whatever pigments will best capture the available light, to incorporate these predominantly into photosystem II (biliproteins are in this photosystem, anyway) and transfer a proportion of the absorbed energy to photosystem I.

The fact that the chromatic adaptation in cyanobacteria, involving changes in the ratio of phycoerythrin to phycocyanin, is entirely of the complementary type is not surprising. Both the pigments involved are biliproteins and both feed their excitation energy to photosystem II, so the problem of imbalance between the photosystems does not arise. Growth of the blue-green alga *Anacystis nidulans* in red ($\lambda > 650$ nm) light does lead to a 75% fall in chlorophyll content with little change in phycocyanin content.[229] We may now interpret this as an attempt by the cells to reduce excitation of photosystem I to the point at which it is in balance with photosystem II.

### The blue light effect

The underwater light field in oceanic waters becomes, as we have seen earlier, predominantly blue-green and eventually blue with increasing depth. There is evidence for the existence of specific effects of blue light on the development of the photosynthetic apparatus of certain phytoplankton species. Wallen & Geen (1971) found that cells of the marine diatom *Cyclotella nana* grown in blue light (0.8 W m$^{-2}$) contained 20% more chlorophyll *a* and had a 70% higher light-saturated photosynthetic rate per cell than cells grown in white light at the same intensity. Jeffrey & Vesk (1977, 1978) observed that in cells of the marine diatom *Stephanopyxis turris* grown in blue-green light (4 W m$^{-2}$), the cellular content of all the chloroplast pigments (chlorophylls *a* and *c*, fucoxanthin and other carotenoids) was about twice that in cells grown under the same intensity of white light (Fig. 12.10). Both the number of chloroplasts per cell and the number of three-thylakoid compound lamellae within each chloroplast were higher in the cells grown in blue-green light (Fig. 12.11). Furthermore, the blue/green-light-grown cells were better able to utilize low intensity (the same irradiance as during growth) blue-green light for photosynthesis than the white-light-grown cells, achieving a 42% higher rate. Vesk & Jeffrey (1977) examined a number of other marine phytoplankton species. Substantial increases in pigment content (55–146%) in blue-green-light-grown cells compared to white-light-grown (both at 4 W m$^{-2}$) cells were observed in five diatom, one dinoflagellate and one cryptomonad species. Minor increases (17–39%) were found in two diatom, two dinoflagellate, one prymnesiophyte, one chrysophyte and one chlorophyte species. There was no chlorophyll increase in two diatom and one prymnesiophyte species. Thus, the ability to increase pigment content in a blue-green light field is common, but not universal amongst marine phytoplankton. There was little alteration in the proportions of the different photosynthetic

Fig. 12.10. Absorbance spectra of cells of the marine diatom *Stephanopyxis turris* grown in white (——) or blue-green (------) light (4 W m$^{-2}$, in each case). (By permission, from Vesk & Jeffrey (1977), *Journal of Phycology*, **13**, 280–8.) The cells are suspended at a concentration of 200 000 cells ml$^{-1}$.

Fig. 12.11. Chloroplast number and structure in cells of the marine diatom *Stephanopyxis turris* grown in white or blue-green light (by permission, from Jeffrey & Vesk (1977) *Journal of Phycology*, **13**, 271–9). *Upper*. Electron micrograph of chloroplast in cell grown in white (4 W m$^{-2}$) light. Inset is light micrograph of whole cells. *Lower*. Electron micrograph of chloroplasts in cell grown in blue-green (4 W m$^{-2}$) light showing increased number of three-thylakoid compound lamellae. Inset is light micrograph of whole cells showing increased number of chloroplasts.

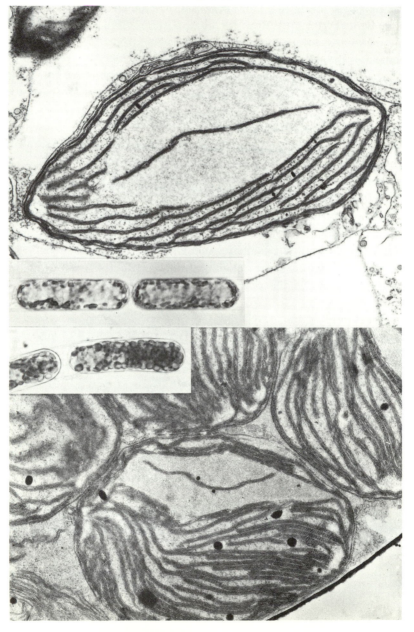

pigments, and so these blue-light-induced changes are not examples of complementary chromatic adaptation. Evidence that the increases in pigment content were accompanied by increases in the number of thylakoids per chloroplast was obtained in a dinoflagellate and a cryptomonad, as well as certain diatom species.

## 12.4    Ontogenetic adaptation – depth

We have seen that aquatic plants undergo adaptive changes in response to changes in the intensity and spectral quality of the ambient light. It is not surprising, therefore, that in species which in nature occur over a range of depths, adaptive changes with depth can often be observed. In this section we shall examine some of the changes that have been observed in aquatic plants accompanying increases in depth, and consider to what extent the lowered intensity or the altered spectral composition are responsible.

### Depth variation of pigment composition in multicellular plants

Ramus *et al.* (1976) suspended samples of two green- (*Ulva lactuca* and *Codium fragile*) and two red-algal species (*Chondrus crispus* and *Porphyra umbilicalis*) for 7-day periods at depths of 1 and 10 m in harbour water at Woods Hole, Massachusetts, USA. Pigment compositions were determined. The samples were then reversed in position – the 1-m samples being lowered to 10 m, and *vice versa* – and after a further 7-day period, further pigment analyses were carried out. In all four species the level of photosynthetic pigments increased in the plants held at 10 m, relative to those held at 1 m, the increase for chlorophyll *a* being about 1.4-fold in the red algae and 2.3 to 3.4-fold in the green. In *U. lactuca*, chlorophyll *b* increased about 5-fold, so that the *b*/*a* ratio rose by 50%, but in *C. fragile* the *b*/*a* ratio remained the same. In the red-algal species the increase in phycoerythrin in the 10 m relative to the 1 m samples was greater than the increase in chlorophyll *a*, so that the phycoerythrin/chlorophyll ratio rose by 50–60% with increased depth.

The light field at 10 m depth at this site, being predominantly yellow-green, differed greatly from that at 1 m depth in spectral composition as well as intensity. In an attempt to determine the effects of diminished intensity alone, Ramus *et al.* analysed the pigments from algae taken from sunny and shaded sites in the intertidal region. Shade increased pigment levels to about the same extent as submersion to 10 m depth. In *U. lactuca* and *P. umbilicalis*, the chlorophyll *b*/*a* and the phycoerythrin/chlorophyll ratios, respectively, did increase in the shade but to a lesser extent than that induced by increased depth. Ramus *et al.* concluded that the greater shifts in pigment ratio associated with increased depth were indeed a response to the changed spectral composition of the light, not just to the lowered intensity. Confirmation with plants kept at a known ratio of intensities would be desirable.

Wheeler (1980a) carried out transplantation experiments with the sporophytes of the giant kelp *Macrocystis pyrifera* (Phaeophyta). When plants growing at 12 m depth were moved up to 1 m depth, then over a 10-day period the pigment content per unit area of thallus diminished markedly, the decreases being 52% for chlorophyll *a*, 35% for chlorophyll *c* and 68% for fucoxanthin. When the plants were replaced at 12 m depth pigment levels increased again to their original values in 18 days. Re-adaptation to increased depth was accompanied by an increase in the molar ratio of fucoxanthin to chlorophyll *a* from 0.50 to 0.77 but by a decrease in the chlorophyll *c*/*a* ratio from 0.50 to 0.31. When samples of the intertidal brown algae *Ascophyllum nodosum* and *Fucus vesiculosus* were suspended at 4 m depth for 7 days, concentrations of fucoxanthin and chlorophylls *a* and *c* per g fresh mass rose 1.5- to three-fold.[393] However, in these species fucoxanthin increased somewhat less than chlorophyll *a* so that the fucoxanthin/chlorophyll *a* ratio diminished by about 20%: the chlorophyll *c*/*a* ratio remained the same during depth adaptation.

Interesting though transplantation experiments are, the alternative approach of studying plants of a given species growing naturally at different depths is of greater ecological relevance. Plants of the siphonous green alga *Halimeda tuna* growing in the Adriatic Sea at 5 m had a total chlorophyll content per g dry mass which was 50% greater than that in plants growing at 2 m depth:[450] the chlorophyll *a*/*b* ratio fell only slightly with increased depth, from 2.03 to 1.86. The chlorophyll content of the green benthic alga *Udotea petiolata* growing in Spanish coastal water increased with depth by 50% over the range 5 to 20 m, most of the increase taking place between 5 and 10 m:[360] the *a*/*b* ratio did not change with depth.

Wiginton & McMillan (1979) studied a number of seagrass species growing at various depths off the Virgin Islands and the coast of Texas, USA. Three of the species showed no significant increase in chlorophyll content with increasing depth down to their maximum depths of 12 to 18 m. In a fourth species, *Halophila decipiens*, however, which penetrated to 42 m, the chlorophyll content changed only slightly from 7 to 18 m, but approximately doubled between 18 and 42 m. The chlorophyll *a*/*b* ratio fell from 1.72 at 18 m to 1.49 at 42 m. Plants of this species from a shallow but shaded site had the same chlorophyll content and *a*/*b* ratio as plants from 42 m depth: this was taken to indicate that the pigment changes associated with increased depth were due to the decrease in total intensity rather than to the change in spectral quality. In the fresh-water macrophyte *Myriophyllum spicatum* in a lake in northeastern America, no significant difference in chlorophyll content was found between plants growing at 1.5, 2.5 and 3.5 m.[302] The differences in light intensity were not, however, very great, the irradiance values being 55, 44 and 32%, respectively, of the subsurface value. The chlorophyll $a/b$ ratio did decrease with depth, being 2.27, 2.03 and 1.79, respectively, at the three depths.

In the brown alga *Dictyota dichotoma* growing in Spanish coastal water, Perez-Bermudez, Garcia-Carrascosa, Cornejo & Segura (1981) found that the chlorophyll $a$ and $c$ contents (on a mass basis) increased by 17 and 53%, respectively, between 0 and 20 m depth, the $a/c$ ratio decreasing from 2.35 to 1.80: at 10 m depth the values were intermediate. The fucoxanthin content, on the other hand, decreased by 42% between 0 and 10 m depth, and then increased again to its original value between 10 and 20 m depth: the possible functional significance of this is discussed in a later section (§12.5). Plants growing in a shaded site at the surface contained 35% more chlorophyll $a$, 63% more chlorophyll $c$ and 20% more fucoxanthin than plants growing in a sunny site. In a study on the red alga *Chondrus crispus* growing from 0 to 20 m depth off Massachusetts, Rhee & Briggs (1977) found that the phycoerythrin/chlorophyll ratio in midsummer remained approximately constant down to about 10 m depth, but approximately doubled between 10 and 13 m, remaining at this value at greater depth.

On the basis of the small number of studies that have so far been carried out, we may conclude that some, but not all, benthic aquatic species can increase the content and/or alter the proportions of their photosynthetic pigments with increasing depth. More field studies of this type are required.

### Depth variation of pigment composition in phytoplankton

The great majority of phytoplankton cells are subjected to continually varying irradiance as they circulate through the mixed layer and so do not have the time to change their pigment composition in accordance with the prevailing light field. When as in some of the clearer, less-coloured waters, light penetrates so well that the euphotic zone is deeper than the mixed layer, however, then in the stable water below the thermocline the phytoplankton can remain at approximately the same depth for long periods and there is evidence that they do indeed adapt to the light conditions prevailing at such depths. We noted earlier (§11.1) that phytoplankton isolated from the deep chlorophyll maximum in the oceans have about twice as much chlorophyll per unit biomass as phytoplankton from the surface layer.[246] This is not necessarily adaptation within a species: whether members of the same species are found both in a highly pigmented form in the oceanic deep chlorophyll maximum and in a less-pigmented form near the surface, or whether the taxonomic composition of the phytoplankton population is entirely different at the two depths, is not known.

In inland waters with their usually more rapid attenuation of light with depth, it is usually the case that the depth of the mixed layer is greater than or equal to that of the euphotic zone, so there is little opportunity for depth adaptation by the phytoplankton. There are, however, some lakes such as Crater Lake, Oregon, USA, and Lake Tahoe, California–Nevada, USA, with water so clear and colourless that the euphotic zone is deeper than the mixed layer and a deep chlorophyll maximum develops at 75–100 m depth in the growing season.[279, 504] Even in lakes in which light does not penetrate deeply, if there is for some reason particularly stable thermal stratification so that circulation is restricted to a very shallow layer, then the euphotic depth can exceed the mixed depth. This appears to be the case in Lake Lovojärvi in Finland: in the summer this was found to have an epilimnion about 2.5 m deep but with photosynthesis extending down to about 3.5 m.[231] A chlorophyll maximum, consisting largely of cells of the blue-green alga *Lyngbya limnetica* with a particularly high chlorophyll content, developed at about 3.5 m.

In Lake Tahoe, Tilzer & Goldman (1978) showed that in June and September (particularly the latter) the chlorophyll content of the phytoplankton (mainly diatoms) per unit biomass increased with depth. In September, for example, the chlorophyll *a* content ($\mu$g per mg wet mass) was 0.61, 1.8 and 3.06 at 0, 50 and 105 m depth, respectively. Since one diatom species, *Fragilaria vaucheriae*, constituted about 58% of the total biomass at all three depths, it seems likely that the increase in chlorophyll content of the total biomass represented true adaptation within this, and perhaps other, species. In June also, when three other species constituted most of the biomass, a doubling of chlorophyll *a* content between 20 and 50 m was accompanied only by comparatively small changes in the population composition, suggesting that chlorophyll content had increased within species. In Lake Lovojärvi, in contrast, there was a marked change in population composition between the mixed layer and the chlorophyll maximum in the hypolimnion, with *Lyngbya* being unimportant in the former and dominant in the latter: the chlorophyll maximum in this lake we may regard as being due to phylogenetic rather than ontogenetic adaptation. Similarly, in clear-water lakes in northwestern Ontario, Canada,[123] the deep chlorophyll maximum (4–10 m depth) was in each case dominated by a single large, colonial chrysophycean flagellate species (the particular species varying from one lake to another), presumably adapted to this particular niche.

Given the demonstrated *in vitro* pigment adaptation of marine phytoplankton species to blue light, and taking into account also the data for Lake Tahoe (optically similar to ocean water), we may reasonably suspect that the deep chlorophyll maximum in oceanic waters is caused at least in part by increased pigment synthesis promoted by the low intensity and/or the predominantly blue character of the prevailing light, within species which occur over a range of depths. It seems likely that there are present, in addition, specialized, highly pigmented species for which the bottom of the euphotic zone is the preferred niche. Studies of the taxonomic composition of the phytoplankton population as a function of depth in stratified oceanic waters could provide information on the relative importance of phylogenetic and ontogenetic adaptation in the establishment of the deep chlorophyll maximum.

Not all unicellular algae are planktonic. Any species which normally occur attached to, or within, fixed structures are thereby freed from the continually varying character of the light field experienced by the plankton and so have the opportunity to adapt to the light field prevailing at the depths where they are located. Little seems to be known about pigment and photosynthetic adaptation within those unicellular species which occur on the surface of sediments or rocks, or epiphytically upon larger algae. Studies have, however, been carried out on the symbiotic dinoflagellates (zooxanthellae) which occur within corals. There is a general tendency, but to varying extents in different coral species, for the pigment content of the zooxanthellae to increase with depth. Leletkin, Zvalinskii & Titlyanov (1981) compared the zooxanthellae of the coral *Pocillopora verrucosa* growing at 20 and 45 m depth in the Timor Sea. Those from coral at 45 m contained about 1.5 times as much chlorophylls *a* and *c*, $\beta$-carotene and diadinoxanthin, and 2.4 times as much peridinin per cell as zooxanthellae from coral at 20 m. The number of chlorophyll molecules per photosynthetic unit was 42% higher in the cells from 45 m than in those from 20 m, but the number of photosynthetic units per cell was about the same. Thus the pigment adaptation to increased depth appears largely to consist of an increase in the pigment complement per photosynthetic unit.

In the coral *Montastrea annularis* on a West Indian reef, Dustan (1979) found a somewhat complex dependence of pigment content on depth. The content of chlorophylls *a* and *c*, and carotenoid per zooxanthella increased by about 100% from the surface down to 28 m depth. Between 28 and 35 m depth the values decreased to somewhat below the surface value and then increased slightly again at greater depths. Dustan suggested that in this coral there are two ecotypic races of zooxanthellae, one adapted to the high-intensity light conditions above 15–20 m and the other to the lower intensity light conditions below 20 m.

### Depth variation of photosynthetic characteristics in multicellular algae

Plants of the subtidal red alga *Ptilota serrata* collected from 24 m depth off northeastern America showed light saturation of photosynthesis at about 116 $\mu$einsteins m$^{-2}$ s$^{-1}$, whereas plants from 6 m saturated at about 182 $\mu$einsteins m$^{-2}$ s$^{-1}$.[307] At very low light intensities (7–14 $\mu$einsteins m$^{-2}$ s$^{-1}$) both kinds of plant achieved the same photosynthetic rate per g dry mass but at higher intensity the deep-water plants levelled off first and achieved a maximum photosynthetic rate only half that of the shallow-water plants (Fig. 12.12). It seems likely that the shallow-water plants contained a more active carboxylation system.

Transplantation experiments with seaweeds have given some puzzling results. Ramus *et al.* (1976) found, in their studies (referred to earlier) on green and red algae moved from one depth to another for 7 days, that in all four species the light-saturated photosynthesis rate per unit biomass (fresh mass) was about twice as high in the deep-water-adapted as in the shallow-water-adapted plants. The light-saturated photosynthetic rate per mg chlorophyll was approximately the same in the surface and deep-water-adapted plants in all except *Ulva lactuca* in which, as the result of a particularly great increase in chlorophyll content, the rate per mg chlorophyll was lower in the deep-water material. In the brown algae

Fig. 12.12. Photosynthesis as a function of irradiance in plants of the subtidal red alga *Ptilota serrata* collected from 6 m and 24 m depth (after Mathieson & Norall, 1975).

*Ascophyllum nodosum* and *Fucus vesiculosus* also, tissue held at increased depth for 7 days increased its light-saturated photosynthetic rate per g fresh mass as well as its pigment content.[393] The results seem to suggest that in these species subjected to increased depth, the carboxylation activity increases together with pigment content. This is the opposite of what has been found for higher plants and for phytoplankton species subjected to lower light intensity, which, as we saw in the previous section, do not increase and may even decrease their carboxylation capacity. It is also contrary to the observations (above) on the changes in photosynthetic properties of the red alga *Ptilota serrata* with depth. To increase carboxylation capacity with increased depth, when the diminished light intensity will prevent them using it, seems on the face of it to be a disadvantage rather than an advantage for these algae. Studies over a longer time period might reveal a different pattern of behaviour.

A different adaptive response has been observed by Glenn & Doty (1981) in the tropical red alga *Eucheuma striatum*. They attached thalli of this alga (grown in shallow water) at a series of points to a line extending from 1 to 9.5 m depth on a Hawaiian reef: irradiance diminished by about a factor of 10 over this range. After a month the thalli were recovered for photosynthetic measurements. A progressive decrease in photosynthetic capacity (light-saturated photosynthetic rate per g dry mass) with depth was observed, the value in thalli from 9.5 m being about half that in material from 1 m. Thus, *Eucheuma* seems to divest itself of surplus carboxylating capacity as depth increases: a response which should be of benefit to the alga.

In the case of the giant kelp *Macrocystis pyrifera*, a single frond can extend from the holdfast on the sea floor up through as many as 30 m of water column and thus through a very considerable light gradient. Wheeler (1980*a*) found that in immature fronds, 12–14 m long, growing in 12 m of water the pigment content rose by about 50% in the first 2–4 m above the holdfast and then remained approximately constant up to about 12 m. The light-saturated photosynthetic rate per unit area rose progressively with distance from the holdfast, 1.5- to three-fold, reaching a maximum at about 10–12 m. Thus, over this part of the frond the photosynthetic capacity per unit area rose with diminishing depth, which we may regard as an adaptive response since it is only at the lesser depths that there is enough light to support a high photosynthetic rate per unit area.

*Depth variation of photosynthetic characteristics in phytoplankton*

As we have noted earlier, depth adaptation of the photosynthetic system in phytoplankton is only to be expected in waters in which thermal stratification permits some of the cells to remain for long periods at depths where circulation is minimal but there is sufficient light for photosynthesis to take place. This situation exists throughout most of the ocean in the lower part of the euphotic zone. By what criteria can we recognize true depth adaptation of the phytoplankton photosynthetic system? The primary criterion is that cells from deep water should achieve a higher photosynthetic rate per unit cellular biomass than cells from the surface, in the light field prevailing in deep water. The deep-water cells would generally achieve this by investing an increased proportion of their cellular substance in light-harvesting pigments, possibly accompanied by lower respiration. A secondary criterion is that the deep-water cells should have a lower photosynthetic rate per unit cellular biomass than the surface cells at saturating light intensity: this would be a consequence of the deep-water cells, in the interests of biochemical economy, synthesizing only as much carboxylase as they can make use of at the low irradiance values existing at the greater depths.

On its own, a reduction in the light saturation parameter, $E_k$, is not an entirely satisfactory indication of adaptation to low irradiance.[554] Since $E_k = P_m/\alpha$ (see §10.1), any environmental factor which reduces the light-saturated photosynthetic rate per mg chlorophyll to a greater extent than it affects the slope of the initial part of the $P$ (per mg chlorophyll) *versus* $E_d$ curve, will reduce $E_k$. Yentsch & Lee (1966) showed, by analysing a large amount of published data for both natural and cultured phytoplankton populations, that $E_k$ tends to be linearly related to $P_m$ and they emphasize the desirability of presenting $P_m$ as well as $E_k$ values in studies on light adaptation by phytoplankton. Nevertheless if, when measured under standardized conditions, one phytoplankton population is found to have a lower $E_k$ value (or light saturation value) than another, then while we cannot on this ground alone conclude that its light-harvesting capability has increased, we may reasonably suppose that its ratio of light-harvesting capacity to carboxylation rate has increased, i.e. it has either increased its pigment content or divested itself of some of its surplus carboxylation capacity, or both, all of which can be regarded as forms of shade adaptation. It is generally found in the oceans that once thermal stratification has been established, phytoplankton isolated from the lower part of the euphotic zone have a lower $E_k$ value than phytoplankton from near the surface.[468] Phytoplankton from the surface, 50 m and 100 m depth (100%, 10% and 1% of subsurface irradiance) in the Sargasso Sea in

October, had $E_k$ values of about 600, 300 and 60 $\mu$einsteins m$^{-2}$ s$^{-1}$ (Fig. 12.13a).[410] In the winter when the thermocline breaks down, the algae are circulated rapidly enough to prevent their becoming adapted to the irradiance value at any depth, and it is found that the cells from all depths show the same degree of light adaptation and indeed are rather similar in their properties to summer surface phytoplankton (Fig. 12.13b).[410, 468] It can be seen from the curves in Fig. 12.13a that the phytoplankton from the lower part of the euphotic zone not only have a lower $E_k$, but are much more susceptible to photoinhibition by higher light intensity than are the surface phytoplankton.

Shimura & Ichimura (1973) found that phytoplankton from near the bottom of the euphotic zone in the ocean had a higher photosynthetic activity than surface phytoplankton when both were placed deep in the euphotic zone. In addition the phytoplankton from the deep layer showed a higher ratio of photosynthetic activity in green light to that in red light than surface layer phytoplankton. Thus the phytoplankton growing deep in the euphotic zone of the ocean appeared to be the better adapted to the dim, predominantly blue-green, light prevailing in that region.

In nature in thermally stratified waters, the cells below the thermocline will be at a lower temperature than those near the surface and as we saw in an earlier section (§ 11.3), lowering the temperature automatically lowers $E_k$ (by diminishing $P_m$ without affecting $\alpha$), quite apart from any alterations in the biochemical composition of the depth-adapted cells. That there are nevertheless inherent changes in the light relations of the cells is readily

Fig. 12.13. Depth adaptation by marine phytoplankton (after Ryther & Menzel, 1959). These *P versus $E_d$* curves for Sargasso Sea phytoplankton were determined by the $^{14}$C method at the temperature prevailing at the sea surface. Samples taken from 0 m (○), 50 m (◑) and 100 m (●) depth. (a) October; water stratified with thermocline at 25–50 m. (b) November; water isothermal down to > 150 m depth.

shown (as in the experiments in Fig. 12.13) by comparing the deep- and shallow-water phytoplankton at the same temperature. The *in situ* lowering of the $E_k$ value of the deep phytoplankton due to lowered temperature is in fact of little ecological significance since the light intensity at that depth will be well below saturation in any case.

To what extent the shade adaptation of the deep phytoplankton in the ocean represents true ontogenetic adaptation (i.e. physiological adaptation within species occurring throughout the illuminated water column) or is phylogenetic adaptation (the predominance in deeper water of particular species genetically adapted to low irradiance values) is unknown.

In the clear, colourless water of Lake Tahoe in which, as in the ocean, the illuminated region extends below the thermocline, Tilzer & Goldman (1978) found that the increase in chlorophyll content of the deep-water cells at the time of maximum thermal stratification (September) was accompanied by a diminution in the light saturation onset parameter: the $E_k$ values were about 80, 50 and 15 W (PAR) m$^{-2}$, for cells from 0, 50 and 105 m depth, respectively. The phytoplankton in this lake satisfied the primary criterion for true depth adaptation of the photosynthetic system.

Fig. 12.14. Variation with depth of specific photosynthetic rates of phytoplankton originating from different depths, in Lake Tahoe (California–Nevada, USA) (by permission, from Tilzer & Goldman (1978), *Ecology*, **59**, 810–21. Copyright © 1978, the Ecological Society of America). The phytoplankton were collected from the surface (. . . . .), 50 m (– – – –) and 105 m (——), and incubated with $^{14}CO_2$ at a series of depths between the surface and 105 m. (*a*) Specific photosynthetic rate per unit chlorophyll. (*b*) Specific rate per unit cellular biomass, expressed as carbon. The lower maximum photosynthetic rate per unit chlorophyll of the deeper-water cells is due to their higher pigment content.

In the light field prevailing in deep water, the deep-water cells achieved a substantially higher photosynthetic rate per unit biomass than cells from the surface (Fig. 12.14). Their light-saturated rate of photosynthesis per unit biomass was not, however, significantly lower than that of the shallow-water cells, suggesting that the deep-water cells in this lake had not adopted the additional shade adaptation strategy of reducing their carboxylase content.

### Morphogenetic adaptation to depth

We have noted earlier that the adaptation of angiosperms to increased shade includes a morphogenetic response: the leaves increase in area while simultaneously becoming thinner. Spence, Campbell & Chrystal (1973) studied five submerged species of *Potamogeton* in several Scottish lochs. They found that specific leaf area (SLA, cm² mg⁻¹ leaf dry mass) increased linearly with water depth. The rate of increase with depth appeared to be typical of the loch (and therefore, presumably, of the water

Fig. 12.15. Change in length and area of leaves of the seagrass *Halophila stipulacea* (Sinai, northern Red Sea) with depth (by permission, from Lipkin, 1979. *Aquatic Botany* 7, 119–28). ITZ = intertidal zone.

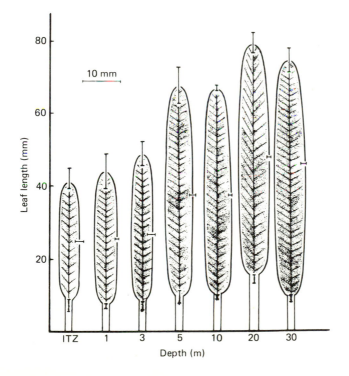

type) rather than of the species, the more rapid rates of increase being observed in the more highly coloured and therefore more attenuating waters. For example, the specific leaf area of *P. perfoliatus* increased from about 0.6 to 1.3 cm² mg⁻¹ from near the surface to 4 m depth in the brown water of L. Uanagan, but increased only from about 0.7 to 1.1 cm² mg⁻¹ from the surface down to 6 m depth in the comparatively colourless waters of L. Croispol. It seemed that a given specified leaf area could be achieved in light fields of markedly different spectral quality, and it was thought likely that the total irradiance (400–750 nm) was of more importance in determining the SLA than any aspect of the spectral quality such as the blue/red or red/far-red ratios.

Lipkin (1979) found that in communities of the seagrass *Halophila stipulacea* growing in the northern Red Sea, leaf area increased about 2.5-fold with depth from the surface down to 30 m (Fig. 12.15). Although this was at first considered to be an ontogenetic adaptation controlled by light intensity, culture experiments indicated that the differences were genetically controlled and Lipkin concluded that true ecotypes of this seagrass were present in the same locality within short distances, i.e. adaptation was apparently phylogenetic rather than ontogenetic.

The siphonous green alga *Halimeda tuna* growing in the Adriatic Sea shows depth-dependent changes in thallus morphology. Mariani Colombo, Orsenigo, Solazzi & Tolomio (1976) observed an increase in the total surface number of branches and number of segments with increasing depth over the range 7 to 16 m. The overall effect is to increase the photosynthetic surface exposed to light.

## 12.5 Significance of ontogenetic adaptation of the photosynthetic system

From our consideration of this topic so far, it is clear that the ability to adapt the characteristics of their photosynthetic systems to the prevailing light field is widespread amongst aquatic plants. To what extent is such adaptation of ecological significance?

In broad terms, the significance of ontogenetic adaptation is that it enables a species to exploit a wider range of habitats. Suitability of habitat for aquatic plant photosynthesis varies with depth: with water type, particularly with regard to optical character; with time, as the seasonal cycle progresses; and in the intensity of competition between plants. The significance of ontogenetic adaptation of the photosynthetic apparatus might therefore be assessed in terms of the extent to which it enables particular species to increase the range of depth and of water types that they can exploit, the proportion of the year in which they can grow, and the success with which they withstand competition from other species.

*Depth adaptation*

Given the demonstrated ability of so many phytoplankton species under laboratory conditions to increase their light-harvesting capability in dim light, especially in dim blue-green light, and given also the observed shade adaptation of natural phytoplankton populations taken from deep in the euphotic zone of stratified waters, we may reasonably conclude that ontogenetic adaptation of the phytoplankton photosynthetic apparatus to the low-intensity light field in these deep layers is a real phenomenon, although it may well coexist with phylogenetic adaptation, certain species being perhaps permanently adapted to the greater depths. The shade-adapted phytoplankton growing at the bottom of the euphotic zone in the ocean contribute only a small proportion of the integral photosynthesis of the water column. It may therefore be concluded that their shade adaptation, while highly relevant to our understanding of how that ecological niche comes to be filled, is not of much significance for the total primary production of the oceans. Primary production in the region of the deep chlorophyll maximum, although a small proportion of the total is not a trivial one (see, for example, Fig. 11.1). Furthermore, there is evidence for a significant enhancement of zooplankton biomass around the deep chlorophyll maximum; Ortner, Wiebe & Cox (1980) consider that this is a depth zone of particularly intense trophic activity. Thus, when all trophic levels are taken into account these shade-adapted phytoplankton may be of more significance for the ecology of the ocean than their contribution to primary production would indicate.

So far as total primary production is concerned, it is true that most of it takes place in the upper region of the euphotic zone and so, as Steemann Nielsen (1975) points out, the most relevant form of adaptation is that of the phytoplankton in the surface layer to the high irradiances prevailing there; for example, their greater ability to withstand photoinhibition. There is evidence that one of the roles of carotenoids, specifically those xanthophylls not involved in light harvesting, is to protect plants against damage by excessive light. There is also evidence that in red algae, for example, in which carotenoids contribute little to photosynthesis, the ratio of carotenoid to chlorophyll increases if the cells are grown at high light intensity.[92, 290] It is thus a plausible supposition that the acquisition of resistance to photoinhibition by phytoplankton involves increased synthesis of protective carotenoids.

In the case of the multicellular benthic algae and angiosperms, while there is evidence for an increase in light-harvesting capacity as growth irradiance is decreased, it is less well documented than in phytoplankton species, perhaps because of the greater difficulty in culturing these plants under controlled laboratory conditions. Although more studies are required, particularly in the field, we may reasonably conclude on the basis of the information already available that in some species which grow over a range of depth, adaptation in the deeper-growing individuals consists in part of an increase in pigment content per unit biomass. An alternative or additional strategy is a reduction in respiration rate; morphological changes which increase light interception by the plants may also occur.

Any adaptive ability which enables an aquatic plant species to cope with the variation in light regime over a range of depth should in principle also enable it to cope with the variation in light quality that will be found within a range of optical water types. The extent to which the ability of particular species to grow in a range of water types is due to ontogenetic adaptation of the photosynthetic apparatus has not been the subject of systematic study.

Given that a common ontogenetic response of aquatic plants to variations in light regime is to vary the light-harvesting capacity of the cells, how important is chromatic adaptation as a component of this? How important, that is, are changes in the shape of the absorption spectra (due to changes in the proportions of the pigments) of the cells, as opposed to changes in absolute absorption across the whole spectrum? In aquatic plants which do not contain biliproteins, i.e. higher plants and algae other than the Rhodophyta, Cryptophyta and Cyanophyta, changes in pigment ratios in response to changes in the ambient light are relatively small, where they occur at all. In certain green species (algae and higher plants) there is some increase in the ratio of chlorophyll *b* to *a*, during shade adaptation: this should somewhat increase the absorption in the 450–480 nm waveband relative to other parts of the spectrum. In the dinoflagellate *Glenodinium*, the increase in concentration of the peridinin/chlorophyll *a*–protein associated with growth at low irradiance causes absorption to rise more in the 500–550 nm region than elsewhere in the spectrum.[381] These comparatively subtle changes in the shape of the absorption spectrum will indeed have the effect of matching the absorption spectrum of the cells better to the spectral distribution of the blue-green light field which prevails at increased depth in many marine waters and so can legitimately be regarded as examples of ecologically useful chromatic adaptation. Nevertheless, the effects of such changes are quantitatively less important for total light capture than the increase in absorption throughout the spectrum resulting from the general increase in pigment concentration. For clear-cut instances of chromatic adaptation we must turn to the biliprotein-containing groups, the Cyanophyta and Rhodophyta.

The changes in the levels of phycoerythrin and phycocyanin in blue-green algae in response to growth in light of differing spectral quality are such as to lead to drastic alterations in the absorption spectra. As shown in Fig. 12.16, the phycocyanin-dominated cells absorb much more strongly in the 600–650 nm region than the phycoerythrin-dominated cells, which in turn absorb more strongly in the 500–575 nm region. There is no doubt that the former cells would be better equipped to collect light from an underwater light field of an orange/red character, while the latter cells would harvest more photons from a greenish light field. Underwater light fields with a predominantly orange/red character would almost never be found in marine systems, but do exist in inland waters strongly coloured by soluble and particulate humic materials (see Fig. 6.4*d*). Green light fields are, of course, common in natural waters. Thus, light climates to which these two kinds of cell would be well matched do occur in aquatic ecosystems. Unfortunately there appear to be no field data on the extent to which adaptation of this type actually happens in nature.

The ratio of biliprotein to chlorophyll can also, as we have seen, change in accordance with the prevailing light. As light intensity diminishes, even when the spectral composition is unaltered, the biliprotein/chlorophyll

Fig. 12.16. Absorption spectrum of blue-green algae as influenced by spectral quality of light field during growth (after Bennett & Bogorad, 1973). Cultures of *Calothrix* 7601 (formerly *Fremyella diplosiphon*) were grown in red light or in fluorescent light (enriched in the green-yellow waveband), and the absorbance spectra of the whole cells were determined on suspensions containing 0.68 mg dry mass ml$^{-1}$. The presence of high levels of phycoerythrin (PE) in the fluorescent-light-grown cells (——) and of phycocyanin (PC) in the red-light-grown cells (-----) is apparent from the spectra.

ratio increases to an extent varying from one blue-green- or red-algal species to another. Since this change can be brought about by variation in intensity alone, it could be argued that this is simply 'intensity adaptation' and is not true chromatic adaptation since it does not require a change in spectral composition to bring it about. On the other hand, diminution of light intensity in the underwater environment, as opposed to the laboratory, is most commonly associated with increase in depth in which case it is associated with a change in spectral quality, in particular with a removal of red light. In the course of evolution, therefore, the common association between low irradiance and the virtual absence of red light might well lead to the development of regulatory mechanisms which respond automatically to lowered intensity by increasing the synthesis of pigments, such as phycoerythrin, which absorb in spectral regions other than the red waveband. The plant does not necessarily have to possess a chromatically specific detection system in order to undergo chromatic adaptation. The increase in biliprotein/chlorophyll ratio as light intensity diminishes might reasonably be regarded as both intensity and chromatic adaptation: while it is triggered off by falling intensity, at the same time it makes the cells chromatically better adapted to the particular spectral character that dim underwater light fields usually have.

Little information is presently available on the extent to which this kind of adaptation is of ecological significance. Limited field studies indicating an increase in biliprotein/chlorophyll ratio with depth have been carried out on two red-algal species (§12.4).

The non-complementary kind of chromatic adaptation that occurs at relatively high light intensities in red algae, in which the adaptation seems to be directed towards achieving a balanced excitation of the two photosystems (§12.3) is accompanied by substantial changes in the absorption spectrum (Fig. 12.9). This kind of adaptation has not yet been described in the field. It is most likely to be found amongst algae growing at relatively shallow depths. On the basis of the laboratory studies and theoretical explanations[62, 290, 556] one might predict that with increasing depth and therefore increasing green/red ratio in the underwater light, the phycoerythrin/chlorophyll ratio would at first decrease to ensure equal excitation of the photosystems and then increase again as the need to capture all available quanta became paramount. Although this has not yet been described for a red alga, what could be an analogous phenomenon has been reported for a brown alga. As we noted earlier, in *Dictyota dichotoma* at a Spanish coastal site, the fucoxanthin/chlorophyll ratio decreased by 42% between 0 and 10 m depth, and then increased again to its original value between 10 and 20 m depth.[360] If, as is likely to be the case, fucoxanthin transfers its excitation energy primarily to photosystem II, then the increasingly blue-green character of the light with increasing depth might lead to greater excitation of photosystem II than photosystem I, and a decrease in the fucoxanthin/chlorophyll ratio could rectify this. At greater depths still, the need to harvest all available quanta becomes of dominating importance and so the fucoxanthin content is increased again.

In the more densely vegetated parts of the aquatic environment, the ability to compete with other plants for light and/or to tolerate shading by other plants is of great importance. On *a priori* grounds it seems likely that plants which find themselves in dim light, not because of great depth but because of overshadowing by other plants, will undergo similar kinds of shade adaptation to those associated with increased depth, but the topic has received little experimental attention in the case of aquatic ecosystems. An example of shade tolerance enabling a plant to withstand competition for light is the previously mentioned increase in phycoerythrin content in *Chondrus crispus* plants overgrown by epiphytic diatoms.[398]

### Seasonal adaptation – multicellular benthic algae

Away from the tropics the average irradiance during daylight hours changes cyclically with the seasons during the year and we may reasonably suppose that it would be of advantage to the aquatic flora to adapt their photosynthetic systems to the seasonally varying light climate. This should especially be the case for the perennial benthic algae which survive throughout the year and therefore must photosynthesize as best they can at all seasons. However, the seasonal variation in temperature which accompanies the variation in irradiance does itself impose certain changes on the photosynthetic behaviour of aquatic plants. Light compensation and saturation points increase with temperature since the rates of respiration and carboxylation both increase with temperature whereas the rate of the light reactions is relatively unaffected. King & Schramm (1976) in their studies on a large number of benthic algal species from the Baltic found that there was a marked lowering of the compensation point in the winter compared to the summer or autumn. Since their measurements were carried out at the water temperature prevailing at the time, it was impossible to say whether there was any adaptation or whether the lower winter temperatures accounted entirely for the lower compensation points. In most of the species there was no marked seasonal change in the light saturation point but in two *Laminaria* (Phaeophyta) species the irradiance required to achieve saturation was much lower in winter than in autumn. In one of these species, *L. saccharina*, the light-saturated photosynthetic rate per g dry mass was about the same in both seasons suggesting that the reduction in light saturation point in the winter was not merely due to a fall in carboxylation rate caused by the lower temperature.

In a number of red-algal species it has been shown that the ratio of phycoerythrin to chlorophyll is higher in the winter than in the summer.[73, 324, 398] It seems likely in such cases that the irradiance required to achieve saturation would, quite apart from temperature effects, be lower in the winter, but experimental confirmation is required.

In the intertidal red alga *Bostrychia binderi*, growing in a subtropical location (Florida, USA), Davis & Dawes (1981) found little seasonal variation in chlorophyll content, but the light-saturated photosynthetic rate was at its highest in the late summer and autumn and at its lowest in the winter, the summer/autumn values being two- to four-fold greater than the winter values. Since all the measurements were carried out at 30° C, this probably represents a true seasonal variation in carboxylation capacity.

### Seasonal adaptation – phytoplankton

Seasonal temperature changes make interpretation difficult in the case of the phytoplankton also. Steemann Nielsen & Hansen (1961) measured the light saturation onset parameter, $E_k$, of phytoplankton in the surface layer in Danish coastal waters at intervals throughout a year.

Fig. 12.17. Seasonal variation of saturation onset parameter ($E_k$) of phytoplankton, temperature and total daily irradiance at a Danish coastal station. Plotted from data of Steemann Nielsen & Hansen (1961). The temperature (– – –○– – –) and $E_k$ (—●—) values are for the surface layer and $E_k$ values were measured at the prevailing temperature. Irradiance values have been converted from lux to quanta m$^{-2}$ s$^{-1}$ using an approximate factor. The values for total daily irradiance (· · · · · ·) are for days of average cloudiness throughout the year.

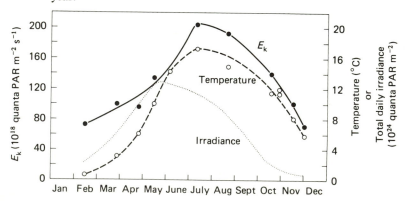

It varied about three-fold being highest in midsummer and lowest in midwinter. It did not, however, appear to be solely a function of the average irradiance in which the cells were growing: $E_k$ values were substantially higher throughout the midsummer to midwinter period than in times of equivalent irradiance in the midwinter to midsummer period. The $E_k$ values were all determined at the prevailing water temperature and this was higher in the post- than in the pre-midsummer period. A comparison of the variation in $E_k$, daily insolation and water temperature throughout the year (Fig. 12.17) suggests that most of the seasonal change in the light saturation parameter can be attributed to the change in temperature: however, the value of $E_k$ in the spring seems to be somewhat greater than would be expected on the basis of temperature alone and it may be that the rise in $E_k$ at this time of year is in part due to adaptation of the photosynthetic system to the higher irradiance.

The values of $\alpha$ and $P_m^B$ for the phytoplankton of a Nova Scotia coastal water were found by Platt & Jassby (1976) to vary about five-fold over a period of 2.75 yr. Measurements were carried out at the prevailing water temperature. Values tended to be highest in the summer and early autumn but subsidiary peaks at other times of the year were also sometimes observed. Statistical analysis indicated that photosynthetic capacity ($P_m^B$, light-saturated photosynthetic rate per mg chlorophyll) was strongly correlated with temperature but that $\alpha$ (the initial slope of the $P$ versus $E_d$ curve) was not. These findings are in accordance with the known insensitivity of the photochemical processes, but sensitivity of carboxylation, to temperature.

The $E_k$ value of the phytoplankton in a Japanese pond was found to vary about three-fold during the year, rising to a maximum in the summer and falling to a minimum in the winter:[10] however, these measurements were carried out at the prevailing water temperature.

Even when the effects of temperature are allowed for, there is the problem of determining whether any apparent seasonal change in the photosynthetic behaviour of the population is true ontogenetic adaptation within species, or merely represents changes in which species are present. A study by Durbin, Krawiec & Smayda (1975) has shown how the relative importance of different types of phytoplankton (separated in terms of size), both as components of total biomass and as contributors to primary production, varied in a northeast American coastal water throughout most of one year (Fig. 12.18). Larger cells (from 20 $\mu$m to > 100 $\mu$m), mainly diatoms, dominated the winter-spring bloom, while small cells (< 20 $\mu$m), mainly flagellates, dominated the summer population. Even within given size classes the species composition changed with time:

for example, different diatom species were dominant in each of the three abundance peaks during the February to May period.

The photosynthetic capacity of the phytoplankton population in a Canadian Pacific coastal inlet was found by Hobson (1981) to rise rapidly

Fig. 12.18. Seasonal variation in the contribution of different size fractions of the phytoplankton to total biomass and to primary production in a north temperate coastal water (by permission, from Durbin, Krawiec & Smayda (1975), *Marine Biology*, **32**, 271–87). The data are for Narragansett Bay, Rhode Island, USA. (*a*) Cumulative graph of chlorophyll *a* content of the different size fractions. (*b*) Upper: cumulative graph of photosynthetic carbon assimilation (per unit volume) of the different phytoplankton size fractions. Lower: photosynthetic carbon assimilation by the different size fractions expressed as a percentage of assimilation by the total population.

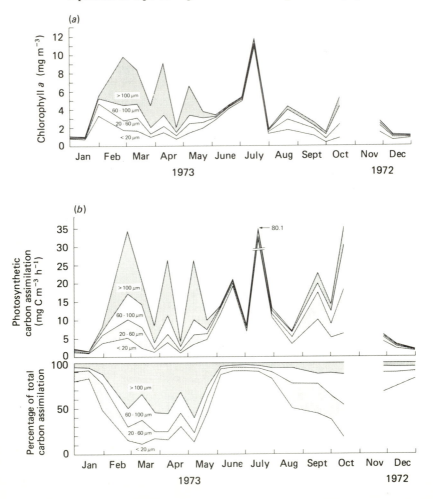

many-fold from a very low value in March–April up to a peak in the June to August period, and then to decrease rapidly in the mid-August to September period, down to a very low value again in October. Although the measurements were carried out at the prevailing water temperature, and although $P_m^B$ for any given species tends to increase with temperature (§11.3), it seemed that only a small part of the variation ($\sim 17\%$) was accounted for by the changes in temperature. The major changes in $P_m^B$ in early April and August/September coincided with changes in the taxonomic composition of the phytoplankton, this being dominated by unidentified nanoflagellates at times when photosynthetic capacity was low, and by diatoms (*Chaetoceros*, *Thalassiosira*) or dinoflagellates (*Gymnodinium*, *Peridinium*) when capacity was high. Hobson suggested that the nanoflagellates which dominated the phytoplankton in the winter were genetically adapted to conditions of short days and low irradiance.

Situations like these, with a continually varying species composition of the phytoplankton population, are probably the rule rather than the exception in fresh as well as marine waters and so any apparent seasonal adaptation of phytoplankton photosynthetic characteristics (after allowing for temperature effects) is in most cases attributable to changes in the algal species present, i.e. is phylogenetic rather than ontogenetic adaptation. It is possible, of course, that species such as *Skeletonema costatum*, which in many coastal waters are present in significant amounts throughout the year, do indeed adapt to the changing irradiance values, as they are known to do under laboratory conditions, but this is hard to demonstrate in the presence of all the other species.

In high-latitude and high-altitude lakes, the phytoplankton in the dim light under the ice in winter have a higher cellular chlorophyll content and a lower light-saturation intensity than the phytoplankton present in the summer.[232, 238, 506] In one such case, Lake Pääjärvi, Finland, different algal species were dominant at the different times of the year.[232] In an Austrian alpine lake, however, the species composition did not vary greatly through the year, being mainly dominated by the dinoflagellate species *Gymnodinium uberrimum*,[503] and Tilzer & Schwarz (1976) attributed the much higher (two- to four-fold) chlorophyll content and lower $E_k$ values, of cells under the ice to actual adaptation within the species.

Evidence for comparatively short-term photosynthetic adaptation of a phytoplankton population – over a period of a few days – has been obtained by Jones (1978) for Lough Neagh in N. Ireland. He found that the light saturation onset parameter, $E_k$, for the dominating blue-green-algal population was positively correlated with the average daily irradiance over the previous 5 days, i.e. the higher the average irradiance in the previous

5 days, the higher the light intensity required to achieve saturation. The changes in $E_k$ appeared to be due more to changes in $\alpha$ (the initial slope of the $P$ versus $E_d$ curve) rather than in the light-saturated rate ($P_m$) per mg chlorophyll ($\alpha = P_m/E_k$). Variation in the ratio of biliprotein to chlorophyll would account for the observed changes in photosynthetic behaviour. In view of the short time period during which these changes occurred, we may reasonably assume that they correspond to true ontogenetic adaptation within the species rather than to changes in the population structure.

Thus it may be that while there are substantial seasonal changes in the photosynthetic properties of the phytoplankton population as a whole, due to the succession of species, there are also, on a shorter time scale, adaptive changes in the photosynthetic properties of individual species, during the few weeks that they persist as significant members of the population.

## 12.6    Rapid adaptation of the photosynthetic system

Quite apart from the seasonal cycle in solar irradiance, there are of course extreme changes in the intensity of PAR every day: not only those due to the regular diurnal change in solar altitude, but also the rapid fluctuations in irradiance resulting from variations in cloud cover (Fig. 2.7).

### Dinoflagellate migration

Within the illuminated water column, there is at a given time for any particular species in the phytoplankton population a particular depth at which the light intensity is optimal for photosynthesis: high enough to give the maximum rate without causing photoinhibition. This optimal depth will vary with solar altitude during the day. The ability to move within the water column to whatever depth suits it best would clearly be of advantage to an alga; many groups have this capability.

The possession of flagella is common in all the algal divisions other than the Rhodophyta, Phaeophyta, Bacillariophyceae and Cyanophyta. Vertical migration by flagellated cells has been studied most amongst the dinoflagellates. In some cases the behaviour pattern is for the cells to migrate downwards away from the surface as the light becomes more intense, towards noon, and then to ascend again during the remainder of the day and the night.[45, 503] Migration appears to be in part phototactic, but can be positive or negative within any given species, the direction of migration being determined by the light intensity at the water surface. For example, Blasco (1978) found that off the coast of Baja California, USA, the marine dinoflagellate *Ceratium furca* migrated towards the sea surface at 07.00 h when the solar irradiance was 280 W m$^{-2}$, but away from the sea surface at 12.00 h when the irradiance was 900 W m$^{-2}$.

Not all dinoflagellate species show a regular up-and-down diurnal migration but instead appear to use their powers of locomotion to move themselves to and remain at a particular depth. Heaney & Talling (1980) found that in Esthwaite Water, England, in the spring and summer, *Ceratium hirundinella* showed surface avoidance but not vertical migration. The dinoflagellate cells were found in maximum density at the depth, 3–4 m, at which irradiance was reduced to about 10% of that penetrating the surface.

Patterns of dinoflagellate movement can be influenced by the availability of nitrate in the water. Cullen & Horrigan (1981) studied the movement of the marine species *Gymnodinium splendens* in a 2 m-deep laboratory tank under a 12 h dark/12 h light regime. When nitrate was present throughout the container, the cells spent most of the day near the surface and most of the night at greater depth: they started swimming upwards before the end of the dark period and downwards before the end of the light period. When the nitrate in the water became depleted, the light-saturated photosynthetic rate per unit biomass decreased, and the cells formed a layer at a depth corresponding to the light level at which photosynthesis was saturated. The behaviour pattern of this species was also observed in the sea off the California coast. During the night (01.10 h) the cells were at 18 m depth, in the region of the nitrocline (the layer in which nitrate concentration is rising from the low value in the upper waters to the high value of deep water). By 10.50 h in the morning the cells had moved up into nitrate-depleted water at 14 m, a depth at which the light intensity was sufficient to saturate photosynthesis. It therefore seems that *G. splendens* moves down at night to a depth where it can accumulate nitrogen and then up during the day to a depth where, although nitrate is lacking, it can photosynthesize at its maximum rate. Its behaviour may be chemotactic as well as phototactic.

### Vertical movement by blue-green algae

Blue-green algae have a completely different mechanism for moving up and down in the water column, involving the formation and collapse of gas-filled vacuoles within the cells. A comprehensive account of these structures and their likely ecological significance may be found in the monograph on the blue-green algae by Fogg, Stewart, Fay & Walsby (1973) and the reviews by Walsby (1975) and Reynolds & Walsby (1975).

The gas vacuoles (Fig. 12.19) are cylindrical structures about 0.07 $\mu$m in diameter and of variable length, 0.3–0.4 $\mu$m being common but up to 2 $\mu$m being recorded. There are many in each cell. The membrane by which they are bounded is composed of protein. The gas mixture within the vacuole is the same as that in the surrounding solution and at the same partial pressures. When the turgor pressure within the surrounding cytoplasm rises above a certain critical value, the gas vacuoles collapse. New vacuoles are formed by *de novo* synthesis rather than by re-inflation of collapsed vacuoles: small vesicles are formed first which then increase in length. Formation of gas vacuoles increases the buoyancy of the blue-green-algal cells, making them float towards the surface: collapse of the vacuoles makes the cells sink. It is the gas-vacuole volume/cell volume

Fig. 12.19. Freeze-fractured cell of *Anabaena flos-aquae* showing cylindrical gas vacuoles longitudinally and in cross-section (by courtesy of Professor D. Branton). The bar corresponds to 1.0 $\mu$m.

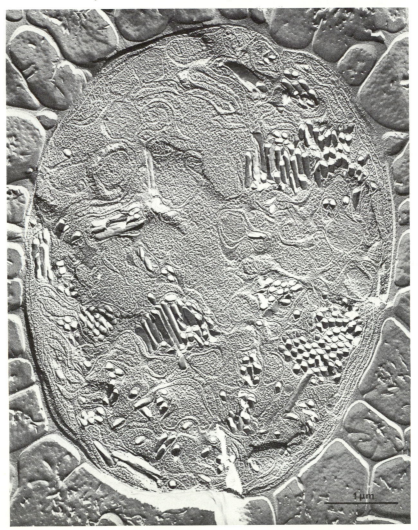

ratio that determines whether, and how fast, the cells rise or sink, or remain at the same depth.

Vacuole formation is favoured by dim light, by a lack of inorganic carbon ($CO_2$, $HCO_3^-$) for photosynthesis, and by a sufficiency of inorganic nitrogen.[125, 267, 355, 534] Under these circumstances photosynthesis proceeds rather slowly and the availability of nitrogen both facilitates the synthesis of vacuole membrane and diverts the products of photosynthesis towards the manufacture of cell materials rather than sugar accumulation. Having formed vacuoles the cells float up towards the surface. With increased light intensity and $CO_2$ availability (resulting from diffusion of atmospheric $CO_2$), the rate of photosynthesis increases. If, as will commonly be the case, mineral nitrogen concentration decreases towards the surface, photosynthesis leads increasingly to sugar accumulation rather than cell growth, turgor pressure within the cell increases and, eventually, the vacuoles begin to collapse. Depending on how far this process goes the cells will either simply slow down in their upward motion and eventually come to a halt at a particular depth, or will begin to actually sink down again towards the deeper layers where nitrogen concentrations are higher. Thus gas-vacuole buoyancy control can be used to enable the blue-green-algal population to find a particular depth that suits it and this is what seems to happen to forms such as *Oscillatoria*: these occur as small-diameter single filaments with high resistance to movement through the water, respond slowly to buoyancy changes, and so normally remain in a layer at a particular depth, usually several metres. Species with large colonies, on the other hand, can sink or rise rapidly through the water and since marked buoyancy changes can take place within time periods as short as an hour or so, can respond rapidly to changing conditions during the day. It is in the forms such as *Microcystis* and *Anabaena*, with large colonies, that the algae will frequently rise all the way up to the surface under still conditions, forming the surface scums so commonly seen on eutrophic lakes in the summer. They are redistributed again from the surface either by collapse of the gas vacuoles or by wind action.

### Benthic diatom movements

Planktonic diatoms are unable to move within the water, although they of course undergo passive movement determined by sinking and turbulence within the water. On surfaces, however, diatoms can move by a mechanism involving secretion of mucilage. Diatoms in sediments both in fresh and marine waters start to move up to the surface of the sediments before dawn.[75, 405] In the afternoon the cells begin to migrate down into the sediments again, this movement continuing during the first few hours of darkness. In the case of benthic diatoms in a fresh-water lake it was found that they would not leave the surface of the sediments for as long as these were illuminated at 750 lux, but migrated downwards again at light intensities of 75 lux or lower.[175] The advantage of moving up onto the illuminated surface during daylight is clear enough: what is less obvious is the benefit the cells derive from burrowing down into the sediments during the hours of darkness – the greater availability of mineral nutrients is a possible explanation.

In turbid estuarine waters, the arrival of the flood tide causes a marked reduction in light intensity. Diatoms in tidal mud flats where the water is turbid can show a tidal rhythm superimposed on the diurnal one. In the River Avon estuary, England, 1–2 h before tidal flooding of the sediment at any particular point, the diatoms burrow beneath the surface again: the disappearance of the green or brown film of algae on the surface of the mud moves as a wave along the banks in advance of the tide.[406] In contrast, in the clear water of the River Eden estuary, Scotland, the movement of the benthic diatoms was found to be strictly diurnal.[361]

Diurnal movements onto the surface during the day and down into the sediments during darkness occur with benthic euglena and blue-green-algal species as well as diatoms

### Chloroplast movements

The rate of light collection by chloroplasts depends not only on their pigment complement but also on their position and orientation within the cell. In some aquatic plants this can vary with light intensity. Amongst the higher plants[525] the general pattern of behaviour is that in low-intensity light the chloroplasts move to a position in the cell such that light absorption is maximized – they spread themselves out adjoining, and parallel to, those cell walls which face the incident light. In high-intensity light (saturating for photosynthesis), the chloroplasts move to a position such that light absorption is minimized, perhaps to reduce photoinhibition: they move away from that cell wall which is directly exposed to the light and align themselves adjoining, and parallel to, those cell walls which are most shaded, typically the side walls so that not only are they edge-on rather than face-on to the light, but also some are shaded by others. In the case of the aquatic angiosperm *Potamogeton crispus*, growing in a Scottish loch, leaves collected from 0.25 m depth had their chloroplasts around the side walls of the cells, whereas leaves from 2.5 m depth had their chloroplasts distributed over the surface facing the light.[453] An alternative way of reducing light collection in high-intensity light is for the chloroplasts to assemble in a cluster around the nucleus: this has been

Fig. 12.20. Light-induced movement of the chloroplasts in the brown alga *Laminaria saccharina* (by permission from Nultsch & Pfau (1979), *Marine Biology*, **51**, 77–82). Arrangement of the chloroplasts in (*a*) low intensity (1000 lux) and (*b*) high intensity (10000 lux) light. Magnification ~ 4000 × .

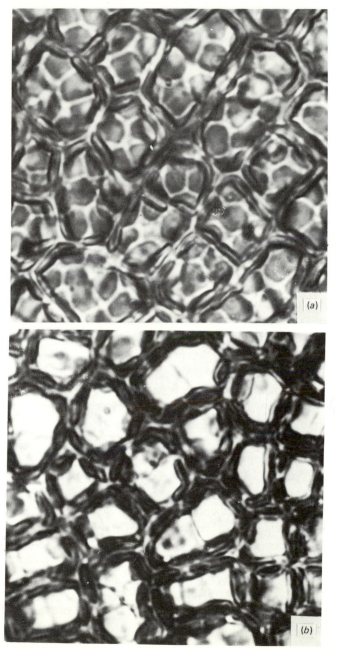

observed in leaves of the seagrass *Halophila stipulacea* growing at 0.5 m depth in the Mediterranean.[97]

The action spectrum for chloroplast movement in higher plants has a peak in the blue region at about 450 nm: the photoreceptor may be a flavin or a carotenoid.[180] The mechanism may be related to that of cytoplasmic streaming: the chloroplasts appear to move together with the cytoplasm rather than through it.

Nultsch & Pfau (1979) studied chloroplast movements in a large number of littoral and sublittoral marine algae. In most of the brown algal species, the chloroplasts moved to the cell walls facing the light at low irradiance and to the side walls parallel to the light direction at high irradiance (Fig. 12.20): the change from one position to the other took 1–2 h. No clear-cut light-induced movements of the chloroplasts were observed in the green- or red-algal species studied.

In the filamentous green algae *Mougeotia* and *Mesotaenium*, which have a single, flat centrally-located rectangular chloroplast in each cell, the characteristic movement of the chloroplast in response to a change in the light regime is to turn on its longitudinal axis, rather than to move around the wall. It turns face-on to moderate light, and edge-on to intense light. Haupt (1973) and co-workers have shown that the photoreceptor controlling this movement is phytochrome. In another filamentous, coenocytic alga, *Vaucheria sessilis* (Xanthophyceae), if the filament is illuminated at one point with low-intensity blue light, the chloroplasts and other organelles which are normally carried along by the streaming cytoplasm are caused to aggregate in the illuminated part of the cell.[46] The photosynthetic implications of this are not yet clear.

In the case of floating, planktonic algae, oriented at random with respect to the light, movement alone to one part of the cell or another is not likely – if the cell contains only one or a few chloroplasts – to make much difference to the rate of light collection. In some cases the chloroplast can reduce its absorption cross-section in bright light by shrinking: this has been observed in a dinoflagellate[479] and in diatoms.[178, 245] In the marine centric diatom *Lauderia borealis*, which has about 50 chloroplasts per cell, Kiefer (1973) observed that in the first 2 min of exposure to intense light (244 W m$^{-2}$) the chloroplasts contracted in size (Fig. 12.21a). During the following 30–60 min the chloroplasts, which in low light were distributed evenly around the periphery, moved to the valvar ends of the cell forming two aggregates of equal size (Fig. 12.21b). The changes in size and position of the chloroplasts were accompanied by a decrease of about 40% in the absorbance of the suspension at 440 nm (probably an underestimate since scattering always contributes some spurious absorbance): thus by a

Fig. 12.21. Chloroplast shrinkage and aggregation induced by high light intensity in the marine diatom *Lauderia borealis* (by permission, from Kiefer (1973), *Marine Biology*, **23**, 39–46). In each case the series of micrographs shows increasing light-induced change from left to right. (*a*) Contraction of chloroplasts. (*b*) Movement of chloroplasts to valvar ends of the cell, followed by aggregation.

(*a*)

(*b*)

combination of shrinkage and aggregation this diatom is able to substantially reduce its rate of energy collection in bright light.

### Rhythms in the photosynthetic system

Diurnal rhythms in photosynthetic capacity (photosynthetic rate per unit biomass at light saturation) have been described for phytoplankton[95, 173] and for multicellular benthic algae.[61, 149, 236] Photosynthetic capacity rises, often several-fold, during the early part of the day to a maximum which can be reached any time from about mid-morning to mid-afternoon and then declines during the rest of the day and remains low during the night. Fig. 12.22 shows the time course of variation in $P_m$ for the phytoplankton in the St. Laurence estuary during the course of a week.[89] In some species, both benthic and planktonic, it has been shown that this diurnal variation is not directly controlled by environmental parameters but is a true circadian rhythm which continues for several cycles in cells kept in the dark.

With cultures, a daily periodicity in $P_m$ has been shown to be present in some, but not all, marine diatom, dinoflagellate and chrysophyte species:[172] the species that showed no day/night differences in $P_m$ were small, rapidly dividing types, whereas larger, slower growing cells almost uniformly showed marked changes. Detailed studies of this rhythm in marine dinoflagellates have been carried out by Sweeney and co-workers. The photosynthetic rate per cell in cultured *Glenodinium* and *Ceratium* varies two- to six-fold during the cycle, when measured under light-limiting as well as light-saturating conditions.[382, 383] The fact that the rate-variation occurs under light-limited conditions led Prézelin & Sweeney (1977) to conclude that it is variation in the light reactions rather than the dark reactions of photosynthesis which are primarily responsible for the diurnal rhythm: it had previously been shown that carboxylase activity does not vary during the cycle.[71] There is no significant change in the pigment content or the absorption spectrum of the cells during the cycle, but the initial slope ($\alpha$) of the photosynthesis *versus* irradiance curve varies in parallel with the photosynthetic capacity. This implies that the quantum yield of photosynthesis – the efficiency with which the cells use the quanta that have absorbed – changes in a cyclical manner. For natural phytoplankton populations also, off the Southern California coast, Harding, Prézelin, Sweeney & Cox (1982a) found that $\alpha$ and $P_m$ varied (three- to nine-fold) in parallel during the day: this applied to both diatom- and dinoflagellate-dominated assemblages. Changes in chlorophyll content were much smaller and did not correlate with the changes in photosynthetic activity (which was, in any event, expressed per unit chlorophyll). Prézelin

& Sweeney suggest that there is a cyclical variation in the proportion of the total photosynthetic units present which are functional. In the multicellular brown alga *Spatoglossum pacificum* the mechanism may be different. There is evidence that the approximately two-fold daily variation in photosynthetic capacity in this alga is due to variation in the activity of the $CO_2$-fixing enzyme system.[550]

What, if any, advantage algae derive from cyclical variations in their photosynthetic capability remains unknown. Conceivably, after the main photosynthetic period is finished each day, certain proteins of the photo-synthetic apparatus might be broken down so that their amino acids can be used for other cellular purposes: the proteins would then be resynthesized each morning. These daily variations in photosynthetic properties may affect the amount of primary production that is actually carried out. The effect of variation in light-saturated photosynthetic rate may be, at least in part, offset by the fact that solar irradiance itself varies roughly in parallel. The rate of light-limited photosynthesis also shows diurnal variation, however, and this may reduce the amount of primary production that the cells achieve early and late in the day. Harding, Prézelin, Sweeney & Cox (1982*b*) calculated, for Californian coastal phytoplankton, that as a result of the daily periodicity in the parameters of the photosynthesis *versus* irradiance curves, the primary production achieved would be 19–39% less than if the values of $P_m$ and $\alpha$ remained constant all day at their maximum values.

In calculations of integral daily photosynthesis, although the diurnal variation of solar irradiance is taken into account, it is commonly assumed that the values of $\alpha$ and $P_m$ observed at mid-day, apply throughout the day. Making this assumption for the Californian phytoplankton yielded values for daily integral photosynthesis ranging from 15% greater to 20%

Fig. 12.22. Diurnal rhythm in the photosynthetic capacity ($P_m$) of phytoplankton in the St. Laurence estuary, Canada, over a 7-day period (by permission, from Demers & Legendre (1981), *Marine Biology*, **64**, 243–50).

less than values calculated taking account of periodicities in α and $P_m$. The variation in size and direction of the discrepancy is caused by variation in the timing of the photosynthetic activity maximum: when this is not at mid-day, the mid-day values of α and $P_m$ come closer to the mean values, and so the errors may not be so great.

Although, as we have seen, no diurnal fluctuation in pigment content was observed in marine dinoflagellates under laboratory conditions, Yentsch & Scagel (1958) have described such a fluctuation in a natural phytoplankton population dominated by diatoms. In a north American coastal (Pacific) water they found that the chlorophyll content per cell fell by 50–60% during late morning, remained approximately constant during the afternoon, and then rose again during the evening. Auclair *et al.* (1982) studied fluctuations in chlorophyll content of estuarine phytoplankton under semi-natural conditions. Water from the St Lawrence estuary was transferred to a 1200 l tank, exposed to sunlight and the chlorophyll content per litre of water was followed over the next 45 h. Cell numbers showed little change. Chlorophyll content fluctuated up and down in a cyclical manner, through an approximately two-fold range with maxima in chlorophyll content occurring about every 6 h. Auclair *et al.* were able to correlate the variation in pigment content with tidal movements. Assuming the rhythm in the tank was a continuation of that in the estuary, it seemed that the times of maximum chlorophyll content corresponded to the periods of maximum current speed and turbulence and therefore probably, as a result of enforced circulation, to periods of lower average irradiance on the cells. The times of minimum chlorophyll content corresponded to periods of slack water in which a stability gradient was established resulting in a higher irradiance on the cells. It makes sense that the cells should increase their light-harvesting capacity during periods of low light intensity: what is noteworthy about this system is that they do it so rapidly, doubling their chlorophyll content in 2–3 h. Regular cyclical changes in phytoplankton pigment content may be more common in nature than we realize. The patchiness of the horizontal distribution of phytoplankton and the additional complicating factor of water movement make it difficult to unequivocally demonstrate such changes in large water bodies.

## 12.7    Highly productive aquatic ecosystems

To achieve a high rate of primary production, an aquatic plant community must achieve a high rate of collection of light energy and an efficient utilization of this absorbed energy by its photosynthetic system, followed by conversion of photosynthate to new cell material. In this section we are concerned with the ways in which some natural and man-made ecosystems manage to do this.

The most effective strategy which aquatic plants can follow to minimize the extent to which their photosynthesis is limited by lack of light is to attach themselves to surfaces at depths sufficiently shallow to ensure that for most of the day sufficient light is available to permit active photosynthesis. By 'choosing' to be benthic rather than planktonic, plants avoid the problem of being carried by water circulation down to depths where the light is insufficient for growth. Furthermore, the attached benthic flora can, as a consequence of water movement (tides, currents, wind-induced circulation), exploit the nutrients in a much larger volume than that in its immediate vicinity at any moment. The non-motile members of the phytoplankton, by contrast, are carried around in the moving mass of water and so have less opportunity to find new nutrient supplies. The rooted members of the benthic flora – seagrasses in marine waters and the various kinds of aquatic angiosperms in fresh waters – have the additional advantage that they can derive nutrients from the sediments in which they are growing.

It is thus not surprising that the most productive natural aquatic plant communities are benthic. In marine waters the most productive systems are brown algal beds, seagrass beds and coral reefs. The kelps, brown algae of the genera *Laminaria* and *Macrocystis*, form dense forests in the sublittoral zone of rocky coasts in cool waters. According to Mann & Chapman (1975), these plants achieve annual net production rates in the range 1000–2000 g C m$^{-2}$. The intertidal brown seaweeds such as *Fucus* and *Ascophyllum* in temperate and subarctic latitudes have annual net production rates of 500–1000 g C m$^{-2}$. Tropical sublittoral seagrass beds dominated by the genus *Thalassia* achieve annual net production rates of 500–1500 g C m$^{-2}$ and in temperate waters beds dominated by genera such as *Zostera* fix in the range of 100 to 1500 g C m$^{-2}$.

In coral reefs primary production is carried out by multicellular algae and seagrasses, as well as by the symbiotic zooxanthellae living within the cells of the coral. Total primary production on coral reefs is typically in the range 300–5000 g C m$^{-2}$.[288]

If the immediate products of photosynthesis, which are carbohydrate in nature, are to be used for cell growth and multiplication, proteins, nucleic acids and other cell constituents must be synthesized, for which mineral nutrients are required. A problem faced by aquatic plants is that at the time of year when high solar altitude favours photosynthesis in any one species, it favours it in all so that the period of maximum potential primary production is also a period of low nutrient levels in the water. Certain productive brown algae in the genus *Laminaria* circumvent this problem by separating the period of maximum growth from that of maximum photosynthesis.[237, 295, 301] In the summer, the plants accumulate photosynthate in the form of mannitol and laminaran, but grow only slowly. In the winter, although the light intensity is low, the nutrient levels in the water are at their highest and this is when the kelps grow rapidly at the expense of their stored carbohydrate.

Taking the world ocean as a whole, most of the primary production is carried out by phytoplankton. Within coastal embayments and estuaries, however, it is commonly the case that the benthic macrophytes account for most of the primary production. For St Margaret's Bay, Nova Scotia, Canada, Mann (1972) found that seaweed (mainly brown algae) productivity averaged over the whole 138 km² of the bay was 603 g C m$^{-2}$ yr$^{-1}$, compared to 190 g C m$^{-2}$ yr$^{-1}$ for phytoplankton. In the Newport River estuary, North Carolina, USA, the contribution of seagrasses to annual primary production was estimated to be 2.5 times that of phytoplankton.[498]

Beds of submerged macrophytes in inland waters have productivities comparable to those of marine macrophytes. In the temperate zone, values range from below 10 to about 500 g C m$^{-2}$ yr$^{-1}$, whereas in tropical regions annual production may exceed 1000 g C m$^{-2}$.[539]

The major factor responsible for the lower productivity of the phytoplankton, compared to the benthic plants, is, as we have noted, the lower average irradiance they receive as a consequence of vertical circulation. One solution to this problem, adopted by the blue-green algae (§12.6), is the evolution of a flotation mechanism enabling the algae to move vertically within the water to a depth which suits them, without (as in the dinoflagellates) the need for continual expense of energy to drive flagella. This may in part account for the fact that the natural water bodies with the highest phytoplankton productivity are dominated by blue-green algae: annual net yields from such water bodies are commonly in the region of 300–1000 g C m$^{-2}$ but values in the region of 2000 g C m$^{-2}$ have been reported.[168]

The circulation problem is also in some measure solved if the water body is very shallow, so that the cells never get very far away from the light. This is the solution generally adopted in man-made high-yield aquatic ecosystems such as sewage oxidation ponds or algal mass culture systems[492] in which a high rate of phytoplankton primary production is the aim: depths of 10–90 cm are commonly used. It is generally arranged in such systems that mineral nutrients are available in excess, so that the rate of primary production is limited by the rate of supply of PAR, and sometimes also of $CO_2$, to the system. Bannister (1974a, b, 1979) has developed a theoretical treatment of phytoplankton growth under mineral nutrient- and $CO_2$-saturated conditions by means of which the steady-state growth rate can be expressed as a function of the incident irradiance and the photosynthetic and respiratory characteristics of the cells. Goldman (1979a, b) has comprehensively reviewed the topic of outdoor mass culture of algae. He concludes that yields in excess of 30–40 g dry matter $m^{-2}$ $day^{-1}$ (equivalent to $\sim$ 5000–7000 g C $m^{-2}$ $yr^{-1}$) are unlikely to be exceeded: to date, such rates have been achieved only for short periods (less than a month), and over the long term, yields of 10–20 g $m^{-2}$ $day^{-1}$ are more typical. A crucial limiting factor in such systems is light saturation of photosynthesis. Even in the very dense algal suspensions which develop, the cells near the surface are exposed to irradiances well above their saturation points. Of the three requirements for high productivity listed at the beginning of this section, such cells will have a high rate of collection of light energy, and (given the plentiful nutrient supply) rapid conversion of photosynthate to new cell material: what they will lack is efficient utilization of the absorbed energy in photosynthesis, since their carboxy- lation system will be unable to keep pace with the high rate of arrival of excitation energy. The inability of algae to increase their photosynthetic rate linearly with irradiance all the way up to full sunlight is likely to remain an insuperable obstacle in the path of increasing yields from outdoor mass culture. To develop, by mutation or genetic engineering, algal types with increased carboxylase and lowered pigment content would not be a solution, because such cells would be poorly adapted to the lower light intensities existing at greater depths. It is essential that the algal suspension should be sufficiently deep and dense to absorb nearly all the light incident upon it. This means that a gradient of irradiance, from full sunlight to near darkness, must exist within the suspension, and no one type of alga can be optimally adapted to all the intensities present. In principle, a laminar arrangement with a range of algal types stacked one above the other in thin layers separated by transparent boundaries, with the most Sun-adapted type at the top and shade-adapted type at the bottom, could increase the yields, but the technical problems and capital cost would be considerable.

# REFERENCES AND AUTHOR INDEX

*The numbers in italic following each item are pages where
the author's work is mentioned*

1. Abiodun, A. A. & Adeniji, H. A. (1978). Movement of water columns in Lake Kainji. *Remote Sens. Env.*, 7, 227–34. *146, 164*

2. Adams, M. S., Guilizzoni, P. & Adams, S. (1978). Relationship of dissolved inorganic carbon to macrophyte photosynthesis in some Italian lakes. *Limnol. Oceanogr.*, 23, 912–19. *266–7*

3. Agami, M., Beer, S. & Waisel, Y. (1980). Growth and photosynthesis of *Najas marina* L. as affected by light intensity. *Aquat. Bot.*, 9, 285–9. *256*

4. Alberte, R. S., Friedman, A. L., Gustafson, D. L., Rudnick, M. S. & Lyman, H. (1981). Light-harvesting systems of brown algae and diatoms. *Biochim. Biophys. Acta*, 635, 304–16. *186, 188*

5. Allen, E. D. & Spence, D. H. N. (1981). The differential ability of aquatic plants to utilize the inorganic carbon supply in fresh waters. *New Phytol.*, 87, 269–83. *263, 265–6*

6. Amos, C. L. & Alföldi, T. T. (1979). The determination of suspended sediment concentration in a macrotidal system using Landsat data. *J. Sedim. Petrol.*, 49, 159–74. *164*

7. Anderson, G. C. (1969). Subsurface chlorophyll maximum in the Northeast Pacific Ocean. *Limnol. Oceanogr.*, 14, 386–91. *258–9*

8. Antia, N. J. (1977). A critical appraisal of Lewin's Prochlorophyta. *Br. phycol. J.*, 12, 271–6. *178*

9. Arnold, K. E. & Murray, S. N. (1980). Relationships between irradiance and photosynthesis for marine benthic green algae (Chlorophyta) of differing morphologies. *J. exp. mar. Biol. Ecol.*, 43, 183–92. *227–8*

10. Aruga, Y. (1965). Ecological studies of photosynthesis and matter production of phytoplankton. I. Seasonal changes in photosynthesis of natural phytoplankton. *Bot. Mag. (Tokyo)*, 78, 280–8. *341*

11. Arvesen, J. C., Millard, J. P. & Weaver, E. C. (1973). Remote sensing of chlorophyll and temperature in marine and fresh waters. *Astronaut. Acta*, 18, 229–39. *137, 139, 164*

12. Ashley, L. E. & Cobb, C. M. (1958). *J. Opt. Soc. Amer.*, 48, 261–8. Quoted by Hodkinson & Greenleaves (1963). *73*

13. Atlas, D. & Bannister, T. T. (1980). Dependence of mean spectral extinction coefficient of phytoplankton on depth, water colour and species. *Limnol. Oceanogr.*, **25**, 157–9. *217, 240*

14. Auclair, J. C., Demers, S., Frechette, M., Legendre, L. & Trump, C. L. (1982). High frequency endogenous periodicities of chlorophyll synthesis in estuarine phytoplankton. *Limnol. Oceanogr.*, **27**, 348–52. *354*

15. Aughey, W. H. & Baum, F. J. (1954). Angular-dependence light scattering – a high resolution recording instrument for the angular range 0.05–140°. *J. Opt. Soc. Amer.*, **44**, 833–7. *79*

16. Austin, R. W. (1974*a*). The remote sensing of spectral radiance from below the ocean surface. In *Optical aspects of oceanography*, ed. N. G. Jerlov & E. Steemann Nielsen, pp. 317–44. London: Academic Press. *39–40, 135, 137, 143, 150, 235*

17. Austin, R. W. (1974*b*). Ocean colour analysis, part 2. San Diego: Scripps Inst. Oceanogr., Ref. 74–10, 28 pp. *117*

18. Austin, R. W. (1980). Gulf of Mexico, ocean-colour surface-truth measurements. *Boundary-layer Meteorol.*, **18**, 269–85. *143*

19. Austin, R. W. (1981). Remote sensing of the diffuse attenuation coefficient of ocean water. In *Special topics in optical propagation*, Conference Proceedings No. 300, pp. 18–1 to 18–9. Neuilly-sur-Seine: AGARD (NATO). *161*

20. Austin, R. W. & Petzold, T. J. (1977). Considerations in the design and evaluation of oceanographic transmissometers. In *Light in the sea*, ed. J. E. Tyler, pp. 104–20. Stroudsbury: Dowden, Hutchinson & Ross. *77*

21. Austin, R. W. & Petzold, T. J. (1981). The determination of the diffuse attenuation coefficient of sea water using the Coastal Zone Colour Scanner. In *Oceanography from space*, ed. J. F. R. Gower, pp. 239–56. New York: Plenum. *161–2*

22. Bader, H. (1970). The hyperbolic distribution of particle sizes. *J. geophys. Res.*, **75**, 2822–30. *72*

23. Badger, M. R., Kaplan, A. & Berry, J. A. (1978). A mechanism for concentrating $CO_2$ in *Chlamydomonas reinhardtii* and *Anabaena variabilis* and its role in photosynthetic $CO_2$ fixation. *Carnegie Inst. Wash. Yearbook*, **77**, 251–61. *266*

24. Baker, K. S. & Smith, R. C. (1979). Quasi-inherent characteristics of the diffuse attenuation coefficient for irradiance. *Soc. Photo-opt. Instrum. Eng.*, **208**, 60–3. *18, 20*

25. Bannister, T. T. (1974*a*). Production equation in terms of chlorophyll concentration, quantum yield, and upper limit to production. *Limnol. Oceanogr.*, **19**, 1–12. *357*

26. Bannister, T. T. (1974*b*). A general theory of steady state phytoplankton growth in a nutrient-saturated mixed layer. *Limnol. Oceanogr.*, **19**, 13–30. *357*

27. Bannister, T. T. (1979). Quantitative description of steady state, nutrient-saturated algal growth, including adaptation. *Limnol. Oceanogr.*, **24**, 76–96. *357*

28. Barrett, J. & Anderson, J. M. (1977). Thylakoid membrane fragments with different chlorophyll *a*, chlorophyll *c* and fucoxanthin compositions isolated from the brown seaweed *Ecklonia radiata*. *Plant Sci. Lett.*, **9**, 275–83. *188*

29. Barrett, J. & Anderson, J. M. (1980). The P-700-chlorophyll *a* protein complex and two major light-harvesting complexes of *Acrocarpia paniculata* and other brown seaweeds. *Biochim. Biophys. Acta*, **590**, 309–23. *188*

30. Bauer, D., Brun-Cottan, J. C. & Saliot, A. (1971). Principe d'une mésure directe dans l'eau de mer du coefficient d'absorption de la lumière. *Cah. Oceanogr.*, **23**, 841–58. *48*

31. Bauer, D. & Ivanoff, A. (1970). Spectroirradiance-metre. *Cah. Oceanogr.*, **22**, 477–82. *99*

32. Bauer, D. & Morel, A. (1967). Etude aux petits angles de l'indicatrice de diffusion de la lumière par les eaux de mer. *Ann. Geophys.*, **23**, 109–23. *79*

33. Beardall, J. & Morris, I. (1976). The concept of light intensity adaptation in marine phytoplankton: some experiments with *Phaeodactylum tricornutum*. *Mar. Biol.*, **37**, 3777–87. *308, 310*

34. Beardsley, G. F. (1968). Mueller scattering matrix of sea water. *J. Opt. Soc. Amer.*, **58**, 52–7. *79*

35. Beardsley, G. F. & Zaneveld, J. R. (1969). Theoretical dependence of the near-asymptotic apparent optical properties on the inherent optical properties of sea water. *J. Opt. Soc. Amer.*, **59**, 373–7. *102*

36. Belay, A. (1981). An experimental investigation of inhibition of phytoplankton photosynthesis at lake surfaces. *New Phytol.*, **89**, 61–74. *221, 223, 230–1*

37. Belay, A. & Fogg, G. E. (1978). Photoinhibition of photosynthesis in *Asterionella formosa* (Bacillariophyceae). *J. Phycol.*, **14**, 341–7. *231*

38. Bennett, A. & Bogorad, L. (1973). Complementary chromatic adaptation in a filamentous blue-green alga. *J. Cell Biol.*, **58**, 419–35. *337*

39. Berthold, G. (1882). Uber die Verteilung der Algen im Golf von Neapel nebst einem Verzeichnis der bisher daselbst beobachten Arten. *Mitt. Zool. Sta. Neopol.*, **3**, 393–536. *292*

40. Bindloss, M. E. (1974). Primary productivity of phytoplankton in Loch Leven, Kinross. *Proc. Roy. Soc. Edin. (B)*, **74**, 157–81. *216, 273*

41. Biospherical Instruments Inc. (San Diego, Calif.) MER-1000 Spectroradiometer. *99*

42. Biscaye, P. E. & Eittreim, S. L. (1973). Variations in benthic boundary layer phenomena: nepheloid layer in the North American basin. In *Suspended solids in water*, ed. R. J. Gibbs, pp. 227–60. New York: Plenum. *89*

43. Björn, L. O. & Björn, E. S. (1980). Photochromic pigments and photoregulation in blue-green algae. *Photochem. Photobiol.*, **32**, 849–52. *317*

44. Bjornland, T. & Aguilar-Martinez, M. (1976). Carotenoids in red algae. *Phytochem.*, **15**, 291–6. *186*

45. Blasco, D. (1978). Observations on the diel migration of marine dinoflagellates off the Baja California coast. *Mar. Biol.*, **46**, 41–7. *344*

46. Blatt, M. R. & Briggs, W. R. (1980). Blue-light-induced cortical fibre reticulation concomitant with chloroplast aggregation in the alga, *Vaucheria sessilis*. *Planta*, **147**, 355–62. *350*

47. Boardman, N. K. (1977). Comparative photosynthesis of sun and shade plants. *Ann. Rev. Plant Physiol.*, **28**, 355–77. *305, 308, 311*

48. Bochkov, B. F., Kopelevich, O. V. & Kriman, B. A. (1980). A spectrophotometer for investigating the light extinction of sea water in the visible and ultraviolet regions of the spectrum. *Oceanology*, **20**, 101–4. *77*

49. Boczar, B. A., Prézelin, B. B., Markwell, J. P. & Thornber, J. P. (1980). A chlorophyll *c*-containing pigment–protein complex from the marine dinoflagellate, *Glenodinium* sp. *FEBS Lett.*, **120**, 243–7. *189*

50. Bodkin, P. C., Posluszny, U. & Dale, H. M. (1980). Light and pressure in two freshwater lakes and their influence on the growth, morphology and depth limits of *Hippuris vulgaris*. *Freshwater Biol.*, **10**, 545–52. *256*

51. Bogorad, L. (1975) Phycobiliproteins and complementary chromatic adaptation. *Ann. Rev. Plant Physiol.*, **26**, 369–401. *190, 316*

52. Bold, H. C. & Wynne, M. J. (1978). *Introduction to the algae.* Englewood Cliffs, N.J.: Prentice-Hall. *172*

53. Booth, C. R. (1976). The design and evaluation of a measurement system for photosynthetically active quantum scalar irradiance. *Limnol. Oceanogr.*, **21**, 326–36. *98*

54. Boresch, K. (1921). Die komplementäre chromatische Adaptation. *Arch. Protistenkd.*, **44**, 1–70. *314*

55. Borowitzka, M. A. & Larkum, A. W. D. (1976). Calcification in the green alga *Halimeda*. *J. exp. Bot.*, **27**, 879–93. *263*

56. Borstad, A., Brown, R. M. & Gower, J. F. R. (1980). Airborne remote sensing of sea surface chlorophyll and temperature along the outer British Columbia coast. *Proc. 6th Canadian Symp. Rem. Sens.*, pp. 541–9. *139, 157, 165*

57. Braarud, T. & Klem, A. (1931). *Hvalradets Skrift.*, **1**, 1–88. Quoted by Steemann-Nielsen (1974). *255*

58. Breen, P. A. & Mann, K. H. (1976). Changing lobster abundance and the destruction of kelp beds by sea urchins. *Mar. Biol.*, **34**, 137–42. *276*

59. Bricaud, A., Morel, A. & Prieur, L. (1981). Absorption by dissolved organic matter of the sea (yellow substance) in the UV and visible domains. *Limnol. Oceanogr.*, **26**, 43–53. *55–7*

60. Bristow, M., Nielsen, D., Bundy, D. & Furtek, R. (1981). Use of water Raman emission to correct airborne laser fluorosensor data for effects of water optical attenuation. *Appl. Opt.*, **20**, 2889–906. *158–61*

61. Britz, S. J. & Briggs, W. R. (1976). Circadian rhythms of chloroplast orientation and photosynthetic capacity in *Ulva*. *Plant Physiol.*, **58**, 22–7. *352*

62. Brody, M. & Emerson, R. (1959). The effect of wavelength and intensity of light on the proportion of pigments in *Porphyridium cruentum*. *Amer. J. Bot.*, **46**, 433–40. *307, 317–8, 320, 338*

63. Bryant, D. A. (1981). The photoregulated expression of multiple phycocyanin genes. *Eur. J. Biochem.*, **119**, 425–9. *315, 317*

64. Brylinsky, M. (1980). Estimating the productivity of lakes and reservoirs. In *The functioning of freshwater ecosystems*, eds E. D. Le Cren & R. H. Lowe-McConnell, pp. 411–53. Cambridge: Cambridge University Press. *244, 290*

65. Buchwald, M. & Jencks, W. P. (1968). Properties of the crustacyanins and the yellow lobster shell pigments. *Biochemistry*, **7**, 844–59. *187*

66. Bukata, R. P. & Bruton, J. E. (1974). ERTS-1 digital classifications of the water regimes comprising Lake Ontario. *Proc. 2nd Canad. Symp. Rem. Sens.*, pp. 627–34. *154, 162, 165*

67. Bukata, R. P., Harris, G. P. & Bruton, J. E. (1974). The detection of suspended

solids and chlorophyll *a* utilizing multispectral ERTS-1 data. *Proc. 2nd Canad. Symp. Rem. Sens.*, pp. 551–64. *152, 154*

68. Bukata, R. P., Jerome, J. H., Bruton, J. E. & Jain, S. C. (1980). Nonzero subsurface irradiance reflectance at 670 nm from Lake Ontario water masses. *Appl. Opt.*, **19**, 2487–8. *147*

69. Burr, A. H. & Duncan, M. J. (1972). Portable spectroradiometer for underwater environments. *Limnol. Oceanogr.*, **17**, 466–75. *99*

70. Burt, W. V. (1958). Selective transmission of light in tropical Pacific waters. *Deep-Sea Res.*, **5**, 51–61. *57*

71. Bush, K. J. & Sweeney, B. M. (1972). The activity of ribulose diphosphate carboxylase in extracts of *Gonyaulax polyedra* in the day and night phases of the circadian rhythm of photosynthesis. *Plant Physiol.*, **50**, 446–51. *352*

72. Butler, W. L. (1978). Energy distribution in the photochemical apparatus of photosynthesis. *Ann. Rev. Plant Physiol.*, **29**, 345–78. *197, 251*

73. Calabrese, G. (1972). Research on red algal pigments. 2. Pigments of *Petroglossum nicaeense* (Duby) Schotter (Rhodophyceae, Gigartinales) and their seasonal variations at different light intensities. *Phycologia*, **11**, 141–6. *340*

74. Calabrese, G. & Felicini, G. P. (1973). Research on red algal pigments. 5. The effect of white and green light on the rate of photosynthesis and its relationship to pigment components in *Gracilaria compressa*. *Phycologia*, **12**, 195–9. *307*

75. Callame, B. & Debyser, J. (1954). Observations sur les mouvements des diatomées à la surface des sédiments marins de la zone intercotidale. *Vie milieu*, **5**, 242–9. *347*

76. Carpenter, D. J. (1981). Inland water quality assessment from Landsat data. *Proc. 15th Int. Conf. Rem. Sens. Environ.* (Ann Arbor), 483–490. *164*

77. Chan, A. T. (1978). Comparative physiological study of marine diatoms and dinoflagellates in relation to irradiance and cell size. *J. Phycol.*, **14**, 396–402. *305*

78. Clarke, G. L. (1939). The utilization of solar energy by aquatic organisms. In *Problems of lake biology*, ed. F. R. Poulton, pp. 27–38. American Association for the Advancement of Science, Publ. No. 10. Quoted by Talling (1982). *235*

79. Clarke, G. L. & Ewing, G. C. (1974). Remote spectroscopy of the sea for biological production studies. In *Optical aspects of oceanography*, ed. N. G. Jerlov & E. Steemann Nielsen, pp. 389–413. London: Academic Press. *164*

80. Clarke, G. L., Ewing, G. C. & Lorenzen, C. J. (1970). Spectra of backscattered light from the sea obtained from aircraft as a measure of chlorophyll concentration. *Science*, **167**, 1119–1121. *137, 139, 154–5*

81. Cox, C. & Munk, W. (1954). Measurement of the roughness of the sea surface from photographs of the sun's glitter. *J. Opt. Soc. Amer.*, **44**, 838–50. *41, 103*

82. Craigie, J. S. (1974). Storage Products. In *Algal physiology and biochemistry*, ed. W. D. P. Stewart, pp. 206–35. Oxford: Blackwell. *199*

83. Critchley, C. (1981). Studies on the mechanism of photoinhibition in higher plants. *Plant Physiol.*, **67**, 1161–5. *232*

84. Crossett, R. N., Drew, E. A. & Larkum, A. W. D. (1965). Chromatic adaptation in benthic marine algae. *Nature*, **207**, 547–8. *297, 299*

85. Cullen, J. J. & Horrigan, S. G. (1981). Effects of nitrate on the diurnal vertical migration, carbon to nitrogen ratio, and the photosynthetic capacity of the dinoflagellate, *Gymnodinium splendens*. *Mar. Biol.*, **62**, 81–9. *345*

86. Curran, R. J. (1972). Ocean colour determination through a scattering atmosphere. *Appl. Opt.*, **11**, 1857–66. *146*

87. Davis, M. A. & Dawes, C. J. (1981). Seasonal photosynthetic and respiratory responses of the intertidal red alga, *Bostrychia binderi* Harvey (Rhodophyta, Ceramiales) from a mangrove swamp and a salt marsh. *Phycologia*, **20**, 165–73. *340*

88. Day, H. R. & Felbeck, G. J. (1974). Production and analysis of a humic-acid-like exudate from the aquatic fungus *Aureobasidium pullulans*. *J. Amer. Water Works Ass.*, **66**, 484–8. *53*

89. Demers, S. & Legendre, L. (1981). Mélange vertical et capacité photosynthétique du phytoplancton estuarien (estuaire du Saint Laurent). *Mar. Biol.*, **64**, 243–50. *352–3*

90. Di Toro, D. M. (1978). Optics of turbid estuarine water: approximations and applications. *Water Res.*, **12**, 1059–68. *82–3*

91. Diakoff, S. & Scheibe, J. (1973). Action spectra for chromatic adaptation in *Tolypothrix tenuis*. *Plant Physiol.*, **51**, 382–5. *315–16*

92. Döhler, G., Bürstell, H. & Jilg-Winter, G. (1976). Pigment composition and photosynthetic $CO_2$ fixation of *Cyanidium caldarium* and *Porphyridium aerugineum*. *Biochem. Physiol. Pflanzen*, **170**, 103–10. *307, 335*

93. Dokulil, M. (1979). Optical properties, colour and turbidity. In *Neusiedlersee: the limnology of a shallow lake in Central Europe*, ed. H. Löffler, Monographiae biologicae, Vol. 37, pp. 151–67. The Hague: Junk. *58*

94. Doty, M. S., Gilbert, W. J. & Abbott, I. A. (1974). Hawaiian marine algae from seaward of the algal ridge. *Phycologia*, **13**, 345–57. *297*

95. Doty, M. S. & Oguri, M. (1957). Evidence for a photosynthetic daily periodicity. *Limnol. Oceanogr.*, **2**, 37–40. *352*

96. Drew, E. A. (1978). Factors affecting photosynthesis and its seasonal variation in the seagrasses *Cymodocea nodosa* (Ucria) Aschers., and *Posidonia oceanica* (L.) Delile in the Mediterranean. *J. exp. mar. Biol. Ecol.*, **31**, 173–94. *283*

97. Drew, E. A. (1979). Physiological aspects of primary production in seagrasses. *Aquat. Bot.*, **7**, 139–50. *227, 283, 350*

98. Dring, M. J. (1981). Chromatic adaptation of photosynthesis in benthic marine algae: an examination of its ecological significance using a theoretical model. *Limnol. Oceanogr.*, **26**, 271–84. *300–2*

99. Dubinsky, Z. & Berman, T. (1976). Light utilization efficiencies of phytoplankton in Lake Kinneret (Sea of Galilee). *Limnol. Oceanogr.*, **21**, 226–30. *242*

100. Dubinsky, Z. & Berman, T. (1979). Seasonal changes in the spectral composition of downwelling irradiance in Lake Kinneret (Israel). *Limnol. Oceanogr.*, **24**, 652–63. *112, 216*

101. Dubinsky, Z. & Berman, T. (1981). Light utilization by phytoplankton in Lake Kinneret (Israel). *Limnol. Oceanogr.*, **26**, 660–70. *237, 242, 245–6*

102. Duchrow, R. M. & Everhart, W. H. (1971). Turbidity measurement. *Amer. Fish. Soc. Trans.*, **100**, 682–90. *89*

103. Duntley, S. Q. (1963). Light in the sea. *J. Opt. Soc. Amer.*, **53**, 214–33. *122, 239*

104. Duntley, S. Q., Wilson, W. H. & Edgerton, C. F. (1974). Ocean colour analysis, part 1. San Diego: Scripps Inst. Oceanogr., Ref. 74–10, 41 pp. *115*

105. Durbin, E. G., Krawiec, R. W. & Smayda, T. J. (1975). Seasonal studies on the relative importance of different size fractions of phytoplankton in Narragansett Bay (U.S.A.). *Mar. Biol.*, **32**, 271–87. *341–2*

106. Dustan, P. (1979). Distribution of zooxanthellae and photosynthetic chloroplast pigments of the reef-building coral *Montastrea annularis* Ellis and Solander in relation to depth on a West Indian coral reef. *Bull. Mar. Sci.*, **29**, 79–95. *327–8*

107. Duysens, L. N. M. (1956). The flattening of the absorption spectrum of suspensions as compared to that of solutions. *Biochim. Biophys. Acta*, **19**, 1–12. *204, 206*

108. Egle, K. (1960). Menge und Verhältnis der Pigmente. In *Encyclopaedia of plant physiology*, ed. W. Ruhland, vol. 5.1, pp. 444–96. Berlin: Springer-Verlag. *181*

109. Emerson, R. (1958). Yield of photosynthesis from simultaneous illumination with pairs of wavelengths. *Science*, **127**, 1059–60. *251*

110. Emerson, R. & Lewis, C. M. (1942). The photosynthetic efficiency of phycocyanin in *Chroococcus* and the problem of carotenoid participation in photosynthesis. *J. gen. Physiol.* **25**, 579–95. *248, 250*

111. Emerson, R. & Lewis, C. M. (1943). The dependence of the quantum yield of *Chlorella* photosynthesis on wave length of light. *Amer. J. Bot.*, **30**, 165–78. *248–50*

112. Engelmann, T. W. (1883). Farbe und Assimilation. *Bot. Zeit.*, **41**, 1–13, 17–29. *292*

113. Engelmann, T. W. (1884). Untersuchungen über die quantitativen Beziehungen zwischen Absorption des Lichtes und Assimilation in Pflanzenzellen. *Bot. Zeit.*, **42**, 81–93, 97–108. *248*

114. Engelmann, T. W. & Gaidukov, N. I. (1902). Ueber experimentelle Erzeugung zweckmässiger Aenderungen der Färbung pflanzlicher Chromophylle durch farbiges Licht. *Arch. Anat. Physiol.*, *Lpz.: Physiol. Abt.*, **333–5**. Quoted by Rabinowitch (1945). *314*

115. Eppley, R. W. (1968). An incubation method for estimating the carbon content of phytoplankton in natural samples. *Limnol. Oceanogr.*, **13**, 547–82. *181*

116. Evans, G. T. & Taylor, F. R. J. (1980). Phytoplankton accumulation in Langmuir cells. *Limnol. Oceanogr.*, **25**, 840–5. *233*

117. Falkowski, P. G. & Dubinsky, Z. (1981). Light–shade adaptation of *Stylophora pistillata*, a hermatypic coral from the Gulf of Eilat. *Nature*, **289**, 172–4. *229, 306*

118. Falkowski, P. G. & Owens, T. G. (1978). Effects of light intensity on photosynthesis and dark respiration in six species of marine phytoplankton. *Mar. Biol.*, **45**, 289–95. *225*

119. Falkowski, P. G. & Owens, T. G. (1980). Light–shade adaptation: two strategies in marine phytoplankton. *Plant Physiol.*, **66**, 592–5. *305–6*

120. Faller, A. J. (1978). Experiments with controlled Langmuir circulations. *Science*, **201**, 618–20. *233*

121. Faust, M. A. & Gantt, E. (1973). Effect of light intensity and glycerol on the growth, pigment composition and ultrastructure of *Chroomonas* sp. *J. Phycol.*, **9**, 489–95. *306*

122. Fee, E. J. (1969). A numerical model for the estimation of photosynthetic production, integrated over time and depth, in natural waters. *Limnol. Oceanogr.*, **14**, 906–11. *286*

123. Fee, E. J. (1976). The vertical and seasonal distribution of chlorophyll in lakes of the Experimental Lakes Area, northwestern Ontario: implications for primary production estimates. *Limnol. Oceanogr.*, **21**, 767–83. *259, 280, 326–7*

124. Fleischhacker, P. & Senger, H. (1978). Adaptation of the photosynthetic apparatus of *Scenedesmus obliquus* to strong and weak light conditions. *Physiol. Plant*, **43**, 43–51. *305, 308–11*

125. Fogg, G. E., Stewart, W. D. P., Fay. P. & Walsby, A. E. (1973). *The blue-green algae*. London: Academic Press. *345, 347*

126. Fork, D. C. (1963). Observations on the function of chlorophyll *a* and accessory pigments in photosynthesis. In *Photosynthesis mechanisms in green plants*, pp. 352–61. Washington: National Academy of Science – National Research Council. *252–3, 301*

127. French, C. S. (1960). The chlorophylls *in vivo* and *in vitro*. In *Encyclopaedia of Plant Physiology*, vol. 5.1, ed. W. Ruhland, pp. 252–97. Berlin: Springer-Verlag. *182*

128. French, C. S., Brown, J. S. & Lawrence, M. C. (1972). Four universal forms of chlorophyll *a*. *Plant Physiol.*, **49**, 421–9. *187*

129. Friedman, E., Poole, L., Cherdak, A. & Houghton, W. (1980). Absorption coefficient instrument for turbid natural waters. *Appl. Opt.*, **19**, 1688–93. *49*

130. Fritsch, F. E. (1948). *The structure and reproduction of the algae*, vols 1 & 2. Cambridge: Cambridge University Press. *172*

131. Frost, B. W. (1980). Grazing. In *The physiological ecology of phytoplankton*, I. Morris, pp. 465–91. Oxford: Blackwell. *275*

132. Fujita, Y. & Hattori, A. (1960). Effect of chromatic lights on phycobilin formation in a blue-green alga, *Tolypothrix tenuis*. *Plant Cell Physiol.*, **1**, 293–303. *315*

133. Fujita, Y. & Hattori, A. (1962a). Photochemical interconversion between precursors of phycobilin chromoproteids in *Tolypothrix tenuis*. *Plant Cell Physiol.*, **3**, 209–20. *315–16*

134. Fujita, Y. & Hattori, A. (1962b). Changes in composition of cellular material during formation of phycobilin chromoproteids in a blue-green alga, *Tolypothrix tenuis*. *J. Biochem.*, **52**, 38–42. *315–16*

135. Gaidukov, N. (1902). Ueber den Einfluss farbigen Lichts auf die Faerbung lebender Oscillarien. *Abh. Preuss Akad. Wiss. Berlin*, No. 5. Quoted by Bogorad (1975). *314*

136. Ganf, G. G. (1974). Incident solar irradiance and underwater light penetration as factors controlling the chlorophyll *a* content of a shallow equatorial lake (Lake George, Uganda). *J. Ecol.*, **62**, 593–609. *216*

137. Ganf, G. G. (1975). Photosynthetic production and irradiance–photosynthesis

relationships of the phytoplankton from a shallow equatorial lake (Lake George, Uganda). *Oecologia*, **18**, 165–83. *226, 257, 280*

138. Gantt, E. (1975). Phycobilisomes: light-harvesting pigment complexes. *BioScience*, **25**, 781–8. *190*

139. Gantt, E. (1977). Recent contributions in phycobiliproteins and phycobilisomes. *Photochem. Photobiol.*, **26**, 685–9. *190*

140. Gates, D. M. (1962). *Energy exchange in the biosphere.* New York: Harper & Row. *31–2*

141. Gelin, C. (1975). Nutrients, biomass and primary productivity of nanoplankton in eutrophic Lake Vombsjön, Sweden. *Oikos*, **26**, 121–39. *216, 273*

142. Gershun, A. (1936). O fotometrii mutnykk sredin. *Tr. Gos. Okeanogr. Inst.*, **11**, 99. Quoted by Jerlov (1976). *22*

143. Gessner, F. (1937). Untersuchungen über Assimilation und Atmung submerser Wasserpflanzen. *Jahrb. Wiss. Bot.*, **85**, 267–328. *269*

144. Ghassemi, M. & Christman, R. F. (1968). Properties of the yellow organic acids of natural waters. *Limnol. Oceanogr.*, **13**, 583–97. *55*

145. Gilmartin, M. (1960). The ecological distribution of the deep water algae of Eniwetok atoll. *Ecology*, **41**, 210–21. *297*

146. Gjessing, E. T. (1965). Use of Sephadex gel for the estimation of molecular weight of humic substances in natural water. *Nature*, **208**, 1091–2. *55*

147. Glazer, A. N. (1981). Photosynthetic accessory proteins with bilin prosthetic groups. In *The biochemistry of plants*, vol. 8, ed. M. D. Hatch & N. K. Boardman, pp. 51–96. New York: Academic Press. *189, 190*

148. Glazer, A. N. & Cohen-Bazire, G. (1975). A comparison of Cryptophytan phycocyanins. *Arch. Mikrobiol.*, **104**, 29–32. *189*

149. Glenn, E. P. & Doty, M. S. (1981). Photosynthesis and respiration of the tropical red seaweeds, *Eucheuma striatum* (Tambalang and Elkhorn varieties) and *E. denticulatum. Aquat. Bot.* **10**, 353–64. *329, 352*

150. Goedheer, J. C. (1969). Energy transfer from carotenoids to chlorophyll in blue-green, red and green algae and greening bean leaves. *Biochim. Biophys. Acta*, **172**, 252–65. *253*

151. Goldman, C. R., Mason, D. T. & Wood, B. J. B. (1963). Light injury and inhibition in Antarctic freshwater plankton. *Limnol. Oceanogr.*, **8**, 313–22. *231*

152. Goldman, J. C. (1979a). Outdoor algal mass cultures. I. Applications. *Water Res.*, **13**, 1–19. *357*

153. Goldman, J. C. (1979b). Outdoor algal mass cultures. II. Photosynthetic yield limitations. *Water Res.*, **13**, 119–36. *357*

154. Gordon, H. R. (1978). Removal of atmospheric effects from satellite imagery of the oceans. *Appl. Opt.*, **17**, 1631–6. *146, 148*

155. Gordon, H. R. & Brown, O. B. (1973). Irradiance reflectivity of a flat ocean as a function of its optical properties. *Appl. Opt.*, **12**, 1549–51. *102*

156. Gordon, H. R., Brown, O. B. & Jacobs, M. M. (1975). Computed relationships between the inherent and apparent optical properties of a flat, homogeneous ocean. *Appl. Opt.*, **14**, 417–27. *102, 116*

157. Gordon, H. R. & Clark, D. K. (1980). Atmospheric effects in the remote sensing of phytoplankton pigments. *Boundary-layer Meteorol.*, **18**, 299–313. *137, 155–6*

158. Gordon, H. R. & Clark, D. K. (1981). Clear water radiances for atmospheric correction of coastal zone colour scanner imagery. *Appl. Opt.*, **20**, 4175–80. *148*

159. Gordon, H. R., Clark, D. K., Mueller, J. L. & Hovis, W. A. (1980). Phytoplankton pigments from the Nimbus-7 Coastal Zone Colour Scanner: comparison with surface measurements. *Science*, **210**, 63–6. *146, 148, 156, 166*

160. Gordon, H. R. & McLuney, W. R. (1975). Estimation of the depth of sunlight penetration in the sea for remote sensing. *Appl. Opt.*, **14**, 413–16. *135*

161. Gordon, J. I. (1969). San Diego: Scripps Inst. Oceanogr., Ref. 67-27. Quoted by Austin (1974a). *39–40*

162. Gorham, E. (1957). The chemical composition of lake waters in Halifax County, Nova Scotia. *Limnol. Oceanogr.*, **2**, 12–21. *58*

163. Gower, J. F. R. (1980). Observations of *in situ* fluorescence of chlorophyll *a* in Saanich Inlet. *Boundary-layer Meteorol.*, **18**, 235–45. *158*

164. Gugliemelli, L. A., Dutton, H. J., Jursinic, P. A. & Siegelman, H. W. (1981). Energy transfer in a light-harvesting carotenoid-chlorophyll *c*–chlorophyll *a*–protein of *Phaeodactylum tricornutum*. *Photochem. Photobiol.*, **33**, 903–7. *185*

165. Gurfink, A. M. (1976). Light field in the surface layers of the sea. *Oceanology*, **15**, 295–9. *41*

166. Hall, K. J. & Lee, G. F. (1974). Molecular size and spectral characterization of organic matter in a meromictic lake. *Water Res.*, **8**, 239–51. *55*

167. Halldal, P. (1974). Light and photosynthesis of different marine algal groups. In *Optical aspects of oceanography*, ed. N. G. Jerlov & E. Steemann Nielsen, pp. 345–60. London: Academic Press. *108, 296*

168. Hammer, U. T. (1980). Primary production: geographical variations. In *The functioning of freshwater ecosystems*, eds E. D. Le Cren & R. H. Lowe-McConnell, pp. 235–46. Cambridge: Cambridge University Press. *290, 356*

169. Hanisak, M. D. (1979). Growth patterns of *Codium fragile* ssp. *tomentosoides* in response to temperature, irradiance, salinity and nitrogen source. *Mar. Biol.*, **50**, 319–22. *282*

170. Harder, R. (1921). *Jahrb. Wiss. Botan.*, **60**, 531. Quoted by Rabinowitch (1951). *264*

171. Harder, R. (1923). Uber die Bedeutung von Lichtintensität und Wellenlänge fur die Assimilation Färbiger Algen. *Z. Bot.*, **15**, 305–55. *292*

172. Harding, L. W., Meeson, B. W., Prézelin, B. B. & Sweeney, B. M. (1981). Diel periodicity of photosynthesis in marine phytoplankton. *Mar. Biol.*, **61**, 95–105. *352*

173. Harding, L. W., Prézelin, B. B., Sweeney, B. M. & Cox, J. L. (1982a). Diel oscillations of the photosynthesis–irradiance (*P–I*) relationship in natural assemblages of phytoplankton. *Mar. Biol.*, **67**, 167–78. *352*

174. Harding, L. W., Prézelin, B. B., Sweeney, B. M. & Cox, J. L. (1982b). Primary production as influenced by diel periodicity of phytoplankton photosynthesis. *Mar. Biol.*, **67**, 179–86. *353–4*

175. Harper, M. A. (1969). Movement and migration of diatoms on sand grains. *Br. phycol. J.*, **4**, 97–103. *348*

176. Harris, G. P. (1980). The measurement of photosynthesis in natural populations of phytoplankton. In *The physiological ecology of phytoplankton*, ed. I. Morris, pp. 129–87. Oxford: Blackwell. *277*

177. Harris, G. P., Heaney, S. I. & Talling, J. F. (1979). Physiological and environmental constraints in the ecology of the planktonic dinoflagellate *Ceratium hirundinella. Freshwater Biol.*, **9**, 413–28. *112*

178. Harris, G. P. & Piccinin, B. B. (1977). Photosynthesis by natural phytoplankton populations. *Arch. Hydrobiol.*, **80**, 405–57. *231, 233–4, 350*

179. Hatcher, B. G., Chapman, A. R. O. & Mann, K. H. (1977). An annual carbon budget for the kelp *Laminaria longicruris. Mar. Biol.*, **44**, 85–96. *280–1, 283*

180. Haupt, W. (1973). Role of light in chloroplast movement. *BioScience*, **23**, 289–96. *350*

181. Haxo, F. T. & Blinks, L. R. (1950). Photosynthetic action spectra of marine algae. *J. Gen. Physiol.*, **33**, 389–422. *248–9*

182. Haxo, F. T., Kycia, J. H., Somers, G. F., Bennett, A. & Siegelman, H. W. (1976). Peridinin–chlorophyll *a* proteins of the dinoflagellate *Amphidinium carterae* (Plymouth 450). *Plant Physiol.*, **57**, 297–303. *188*

183. Heaney, S. I. & Talling, J. F. (1980). Dynamic aspects of dinoflagellate distribution patterns in a small productive lake. *J. Ecol.*, **68**, 75–94. *345*

184. Heckey, R. E. & Fee, E. J. (1981). Primary production and rates of algal growth in Lake Tanganyika. *Limnol. Oceanogr.*, **26**, 532–47. *112*

185. Hobson, L. A. (1981). Seasonal variations in maximum photosynthetic rates of phytoplankton in Saanich Inlet, Vancouver Island, British Columbia. *J. exp. mar. Biol. Ecol.*, **52**, 1–13. *342–3*

186. Hodkinson, J. R. & Greenleaves, J. I. (1963). Computations of light-scattering and extinction by spheres according to diffraction and geometrical optics and some comparisons with the Mie theory. *J. Opt. Soc. Amer.*, **53**, 577–88. *73–5*

187. Hoge, F. E. & Swift, R. N. (1981). Airborne simultaneous spectroscopic detection of laser-induced water Raman backscatter and fluorescence from chlorophyll *a* and other naturally occurring pigments. *Appl. Opt.*, **20**, 3197–205. *158, 161, 165*

188. Højerslev, N. K. (1973). Inherent and apparent optical properties of the western Mediterranean and the Hardangerfjord. *Univ. Copenhagen, Inst. Phys. Oceanogr. Rep.* **21**, 70 pp. *85*

189. Højerslev, N. (1975). A spectral light absorption meter for measurements in the sea. *Limnol. Oceanogr.*, **20**, 1024–34. *49, 98, 118–19*

190. Højerslev, N. (1979). On the origin of yellow substance in the marine environment. 17th General Assembly of I.A.P.S.O. (Canberra, 1979). Abstracts, p. 71. *55*

191. Holmes, R. W. (1970). The Secchi disk in turbid coastal waters. *Limnol. Oceanogr.*, **15**, 688–94. *96*

192. Hovis, W. (1978). The Coastal Zone Colour Scanner (CZCS) experiment. In *The Nimbus 7 users' guide*, ed. C. R. Madrid, pp. 19–31. Beltsville: Management & Technical Services Co. *141–2*

193. Hovis, W. A., Clark, D. K., Anderson, F., Austin, R. W., Wilson, W. H., Baker, E. T., Ball, D., Gordon, H. R., Mueller, J. L., El-Sayed, S. Z., Sturm, B., Wrigley, R. C. & Yentsch, C. S. (1980). Nimbus-7 Coastal Zone Colour

Scanner: system description and initial imagery. *Science*, **210**, 60–3.   *141, 166*

194. Hovis, W. A. & Leung, K. C. (1977). Remote sensing of ocean colour. *Opt. Eng.*, **16**, 158–66.   *139, 141*

195. Hutchinson, G. E. (1967). *A treatise on limnology*, vol. 2. New York: Wiley. *275*

196. Hutchinson, G. E. (1975). *A treatise on limnology*, vol. 3. New York: Wiley-Interscience.   *172, 265, 269*

197. Ikusima, I. (1970). Ecological studies on the productivity of aquatic plant communities. *Bot. Mag. (Tokyo)*, **83**, 330–41.   *227*

198. International Association for the Physical Sciences of the Ocean (1979). Sun Report (Report of the working group on symbols, units and nomenclature in physical oceanography). 99 pp. Paris.   *6*

199. Ivanoff, A., Jerlov, N. & Waterman, T. H. (1961). A comparative study of irradiance, beam transmittance and scattering in the sea near Bermuda. *Limnol. Oceanogr.*, **6**, 129–48.   *57*

200. Iverson, R. L. & Curl, H. (1973). Action spectrum of photosynthesis for *Skeletonema costatum* obtained with carbon-14. *Physiol. Plant.*, **28**, 498–502.   *248–50*

201. Jain, S. C. & Miller, J. R. (1976). Subsurface water parameters: optimization approach to their determination from remotely sensed water colour data. *Appl. Opt.*, **15**, 886–90.   *151*

202. Jamart, B. M., Winter, D. F., Banse, K., Anderson, G. C. & Lam, R. K. (1977). A theoretical study of phytoplankton growth and nutrient distribution in the Pacific Ocean off the northwestern U.S. coast. *Deep-Sea Res.*, **24**, 753–73.   *259*

203. James, H. R. & Birge, E. A. (1938). A laboratory study of the absorption of light by lake waters. *Trans. Wisc. Acad. Sci. Arts Lett.*, **31**, 1–154.   *53, 58*

204. Jassby, A. T. & Platt, T. (1976). Mathematical formulation of the relationship between photosynthesis and light for phytoplankton. *Limnol. Oceanogr.*, **21**, 540–7.   *225*

205. Jaworski, G. H. M., Talling, J. F. & Heaney, S. I. (1981). The influence of carbon dioxide depletion on growth and sinking rate of two planktonic diatoms in culture. *Br. phycol. J.*, **16**, 395–410.   *268*

206. Jeffrey, S. W. (1961). Paper chromatographic separation of chlorophylls and carotenoids in marine algae. *Biochem. J.*, **80**, 336–42.   *181*

207. Jeffrey, S. W. (1972). Preparation and some properties of crystalline chlorophyll $c_1$ and $c_2$ from marine algae. *Biochim. Biophys. Acta*, **279**, 15–33.   *181*

208. Jeffrey, S. W. (1976). The occurrence of chlorophyll $c_1$ and $c_2$ in algae. *J. Phycol.*, **12**, 349–54.   *181*

209. Jeffrey, S. W., Sielicki, M. & Haxo, F. T. (1975). Chloroplast pigment patterns in dinoflagellates. *J. Phycol.*, **11**, 374–84.   *180, 181, 183, 185*

210. Jeffrey, S. W. & Vesk, M. (1977). Effect of blue-green light on photosynthetic pigments and chloroplast structure in the marine diatom *Stephanopyxis turris*. *J. Phycol.*, **13**, 271–9.   *321–3*

211. Jeffrey, S. W. & Vesk, M. (1978). Chloroplast structural changes induced by white light in the marine diatom *Stephanopyxis turris*. *J. Phycol.*, **14**, 238–40.   *321–3*

212. Jerlov, N. G. (1951). Optical studies of ocean water. *Rep. Swedish Deep-Sea Exped.*, **3**, 1–59. *66–7*

213. Jerlov, N. G. (1961). Optical measurements in the eastern North Atlantic. *Medd. Oceanogr. Inst. Goteborg, Ser. B*, **8**, 40 pp. *79*

214. Jerlov, N. G. (1976). *Marine optics.* Amsterdam: Elsevier. *40, 53, 55, 57, 66–7, 72, 80, 85, 101, 105, 107, 115, 122, 296, 300–1*

215. Jerlov (Johnson), N. G. & Liljequist, G. (1938). On the angular distribution of submarine daylight and the total submarine illumination. *Sven. Hydrogr.-Biol. Komm. Skr., Ny Ser. Hydrogr.*, **14**, 15 pp. Quoted by Jerlov (1976). *122*

216. Jerlov, N. G. & Nygard, K. (1969). A quanta and energy meter for photosynthetic studies. *Univ. Copenhagen Inst. Phys. Oceanogr. Rep.*, **10**, 29 pp. *92, 94–5, 112*

217. Jewson, D. H. (1976). The interactions of components controlling net phytoplankton photosynthesis in a well-mixed lake (Lough Neagh, Northern Ireland). *Freshwater Biol.*, **6**, 551–76. *226, 244, 257, 273–4*

218. Jewson, D. H. (1977). Light penetration in relation to phytoplankton content of the euphotic zone of Lough Neagh, N. Ireland. *Oikos*, **28**, 74–83. *215, 216, 238*

219. Jewson, D. H. & Taylor, J. A. (1978). The influence of turbidity on net phytoplankton photosynthesis in some Irish lakes. *Freshwater Biol.*, **8**, 573–84. *58*

220. Jewson, D. H. & Wood, R. B. (1975). Some effects on integral photosynthesis of artificial circulation of phytoplankton through light gradients. *Ver. Internat. Verein. Limnol.*, **19**, 1037–44. *234*

221. Jitts, H. R. (1963). The simulated *in situ* measurement of oceanic primary production. *Aust. J. mar. freshwater Res.*, **14**, 139–47. *220*

222. Jitts, H. R., Morel, A. & Saijo, Y. (1976). The relation of oceanic primary production to available photosynthetic irradiance. *Aust. J. mar. freshwater Res.*, **27**, 441–54. *227*

223. Johnson, R. W. (1978). Mapping of chlorophyll *a* distributions in coastal zones. *Photogramm. Eng. Rem. Sens.*, **44**, 617–24. *149, 153, 156, 165*

224. Johnson, R. W. (1980). Remote sensing and spectral analysis of plumes from ocean dumping in the New York Bight Apex. *Remote Sens. Env.*, **9**, 197–209. *149, 153, 164*

225. Johnson, R. W. & Harriss, R. C. (1980). Remote sensing for water quality and biological measurements in coastal waters. *Photogramm. Eng. Rem. Sens.*, **46**, 77–85. *140*

226. Jones, D. & Wills, M. S. (1956). The attenuation of light in sea and estuarine waters in relation to the concentration of suspended solid matter. *J. mar. biol. Ass. U.K.*, **35**, 431–44. *89*

227. Jones, L. W. & Kok, B. (1966). Photoinhibition of chloroplast reactions. *Plant Physiol.*, **41**, 1037–43. *231–2*

228. Jones, L. W. & Myers, J. (1964). Enhancement in the blue-green alga, *Anacystis nidulans. Plant Physiol.*, **39**, 938–46. *253*

229. Jones, L. W. & Myers, J. (1965). Pigment variations in *Anacystis nidulans* induced by light of selected wavelengths. *J. Phycol.* **1**, 6–13. *320*

230. Jones, R. I. (1978). Adaptations to fluctuating irradiance by natural phytoplankton communities. *Limnol. Oceanogr.*, **23**, 920–6. *286, 343–4*

231. Jones, R. I. & Ilmavirta, V. (1978*a*). A diurnal study of the phytoplankton in the eutrophic lake Lovojärvi, Southern Finland. *Arch. Hydrobiol.*, **83**, 494–514. *277, 326*

232. Jones, R. I. & Ilmavirta, V. (1978*b*). Vertical and seasonal variation of phytoplankton photosynthesis in a brown-water lake with winter ice cover. *Freshwater Biol.*, **8**, 561–72. *343*

233. Jørgensen, E. G. (1968). The adaptation of plankton algae. II. Aspects of the temperature adaptation of *Skeletonema costatum*. *Physiol. Plant.*, **21**, 423–7. *272*

234. Jørgensen, E. G. (1969). The adaptation of plankton algae. IV. Light adaptation in different algal species. *Physiol. Plant*, **22**, 1307–15. *305, 311–12*

235. Jupp, B. P. & Spence, D. H. N. (1977). Limitations on macrophytes in a eutrophic lake, Loch Leven. *J. Ecol.*, **65**, 175–86, 431–6. *276*

236. Kageyama, A., Yokohama, Y. & Nisizawa, K. (1979). Diurnal rhythms of apparent photosynthesis of a brown alga, *Spatoglossum pacificum*. *Botanica Marina*, **22**, 199–201. *352*

237. Kain, J. M. (1979). A view of the genus *Laminaria*. *Oceanogr. mar. Biol. Ann. Rev.*, **17**, 101–61. *276, 356*

238. Kalff, J., Welch, H. E. & Holmgren, S. K. (1972). Pigment cycles in two high-arctic Canadian lakes. *Ver. internat. Verein Limnol.*, **18**, 250–6. *343*

239. Kalle, K. (1937). *Annln. Hydrogr. Berl.*, **65**, 276–82. Quoted by Kalle (1966). *53*

240. Kalle, K. (1961). What do we know about the 'Gelbstoff'? *Union Geod. Geophys. int. Monogr.*, **10**, 59–62. Quoted by Jerlov (1976). *57*

241. Kalle, K. (1966). The problem of the gelbstoff in the sea. *Oceanogr. mar. Biol. Ann. Rev.*, **4**, 91–104. *53, 56*

242. Karelin, A. K. & Pelevin, V. N. (1970). The FMPO-64 marine underwater irradiance meter and its application in hydro-optical studies. *Oceanology*, **10**, 282–5. *98*

243. Khorram, S. (1981*a*). Use of ocean colour scanner data in water quality mapping. *Photogramm. Eng. Rem. Sens.*, **47**, 667–76. *157, 165*

244. Khorram, S. (1981*b*). Water quality mapping from Landsat digital data. *Int. J. Rem. Sens.*, **2**, 145–53. *152*

245. Kiefer, D. A. (1973). Chlorophyll *a* and fluorescence in marine centric diatoms: responses of chloroplasts to light and nutrient stress. *Mar. Biol.*, **23**, 39–46. *350–2*

246. Kiefer, D. A., Olson, R. J. & Holm-Hansen, O. (1976). Another look at the nitrite and chlorophyll maxima in the central North Pacific. *Deep-Sea Res.*, **23**, 1199–208. *258, 325*

247. Killilea, S. D., O'Carra, P. & Murphy, R. F. (1980). Structures and apoprotein linkages of phycoerythrobilin and phycocyanobilin. *Biochem. J.*, **187**, 311–20. *191*

248. Kim, H., Fraser, R. S., Thompson, L. L. & Bahethi, O. (1980). A design study for an advanced ocean colour scanner system. *Boundary-layer Meteorol.*, **18**, 315–27. *139, 142, 148*

249. Kim, H., McClain, C. R. & Hart, W. D. (1979). Chlorophyll gradient map from high-altitude ocean-colour-scanner data. *Appl. Opt.*, **18**, 3715–16. *149, 157, 165*

250. King, R. J. & Schramm, W. (1976). Photosynthetic rates of benthic marine algae in relation to light intensity and seasonal variations. *Mar. Biol.*, **37**, 215–22.   *222–3, 225, 227–9, 231, 283, 339–40*

251. Kirk, J. T. O. (1975*a*). A theoretical analysis of the contribution of algal cells to the attenuation of light within natural waters. I. General treatment of suspensions of pigmented cells. *New Phytol.*, **75**, 11–20.   *204–9*

252. Kirk, J. T. O. (1975*b*). A theoretical analysis of the contribution of algal cells to the attenuation of light within natural waters. II. Spherical cells. *New Phytol.*, **75**, 21–36.   *204–9, 217*

253. Kirk, J. T. O. (1976*a*). A theoretical analysis of the contribution of algal cells to the attenuation of light within natural waters. III. Cylindrical and spheroidal cells. *New Phytol.*, **77**, 341–58.   *204–9, 215, 217*

254. Kirk, J. T. O. (1976*b*). Yellow substance (gelbstoff) and its contribution to the attenuation of photosynthetically active radiation in some inland and coastal southeastern Australian waters. *Aust. J. mar. freshwater Res.*, **27**, 61–71. *53–4, 56–60, 69*

255. Kirk, J. T. O. (1977*a*). Use of a quanta meter to measure attenuation and underwater reflectance of photosynthetically active radiation in some inland and coastal southeastern Australian waters. *Aust. J. mar. freshwater Res.*, **28**, 9–21.   *57–8, 111–2, 114–5, 118*

256. Kirk, J. T. O. (1977*b*). Thermal dissociation of fucoxanthin–protein binding in pigment complexes from chloroplasts of *Hormosira* (Phaeophyta). *Plant Sci. Lett.*, **9**, 373–80.   *185, 188*

256*a*. Kirk, J. T. O. (1979). Spectral distribution of photosynthetically active radiation in some south-eastern Australian waters. *Aust. J. mar. freshwater Res.*, **30**, 81–91.   *107–9*

257. Kirk, J. T. O. (1980*a*). Relationship between nephelometric turbidity and scattering coefficients in certain Australian waters. *Aust. J. mar. freshwater Res.*, **31**, 1–12.   *83, 118, 132*

258. Kirk, J. T. O. (1980*b*). Spectral absorption properties of natural waters: contribution of the soluble and particulate fractions to light absorption in some inland waters of southeastern Australia. *Aust. J. mar. freshwater Res.*, **31**, 287–96.   *48, 58, 60–1, 67–9*

259. Kirk, J. T. O. (1981*a*). A Monte Carlo study of the nature of the underwater light field in, and the relationships between optical properties of, turbid yellow waters. *Aust. J. mar. freshwater Res.*, **32**, 517–32.   *65–6, 82–3, 102, 116, 120, 124–33*

260. Kirk, J. T. O. (1981*b*). Estimation of the scattering coefficient of natural waters using underwater irradiance measurements. *Aust. J. mar. freshwater Res.*, **32**, 533–9.   *82–3, 86*

261. Kirk, J. T. O. (1981*c*). A Monte Carlo procedure for simulating the penetration of light into natural waters. *CSIRO Division of Plant Industry, Technical Paper*, No. 36, 18 pp.   *102, 120, 124–5*

262. Kirk, J. T. O. & Goodchild, D. J. (1972). Relationship of photosynthetic effectiveness of different kinds of light to chlorophyll content and chloroplast structure in greening wheat and in ivy leaves. *Aust. J. biol. Sci.*, **25**, 215–41.   *302*

263. Kirk, J. T. O. & Tilney-Bassett, R. A. E. (1978). *The plastids*, 2nd edn. Amsterdam: Elsevier.   *171, 180, 183, 188, 305*

264. Kirkman, H. & Reid, D. D. (1979). A study of the role of the seagrass *Posidonia australis* in the carbon budget of an estuary. *Aquat. Bot.*, **7**, 173–83. *281–2*

265. Klemas, V., Bartlett, D., Philpot, W., Rogers, R. & Reed, L. (1974). Coastal and estuarine studies with ERTS-1 and Skylab. *Remote Sens. Env.*, **3**, 153–74. *152, 162–3*

266. Klemas, V., Borchardt, J. F. & Treasure, W. M. (1973). Suspended sediments observations from ERTS-1. *Remote Sens. Env.*, **2**, 205–21. *151–2*

267. Klemer, A. R., Feuillade, J. & Feuillade, M. (1982). Cyanobacterial blooms: carbon and nitrogen limitation have opposite effects on the buoyancy of *Oscillatoria*. *Science*, **215**, 1629–31. *347*

268. Koblents-Mishke, O. I. (1965). Primary production in the Pacific. *Oceanology*, **5**, 104–16. *278–9, 289*

269. Koblents-Mishke, O. I. (1979). Photosynthesis of marine phytoplankton as a function of underwater irradiance. *Soviet Plant Physiol.*, **26**, 737–46. *232, 242, 244*

270. Kondratyev, K. Ya. (1954). *Radiant solar energy.* (In Russian) Leningrad. Quoted by Robinson (1966). *37*

271. Kopelevich, O. V. & Burenkov, V. I. (1971). The nephelometric method for determining the total scattering coefficient of light in sea water. *Izv. Atmos. Oceanic Phys.*, **7**, 835–40. *80, 85*

272. Kopelevich, O. V. & Burenkov, V. I. (1977). Relation beween the spectral values of the light absorption coefficients of sea water, phytoplanktonic pigments, and the yellow substance. *Oceanology*, **17**, 278–82. *55, 57*

273. Kopelevich, O. V. & Mezhericher, E. M. (1979). Improvement of the method of 'inversion' of the spectral values of the luminance coefficient of the sea. *Oceanology*, **19**, 621–4. *150*

274. Kowalik, W. S., Marsh, S. E. & Lyon, R. J. P. (1982). A relation between Landsat digital numbers, surface reflectance, and the cosine of the solar zenith angle. *Remote Sens. Env.*, **12**, 39–55. *146*

275. Kullenberg, G. (1968). Scattering of light by Sargasso Sea water. *Deep-Sea Res.*, **15**, 423–32. *78, 85, 89*

276. Kullenberg, G., Lundgren, B., Malmberg, S. A., Nygard, K. & Højerslev, N. K. (1970). Inherent optical properties of the Sargasso Sea. *Univ. Copenhagen Inst. Phys. Oceanogr. Rep.*, **11**, 18 pp. Quoted by Jerlov (1976). *57*

277. Langmuir, I. (1938). Surface motion of water induced by wind. *Science*, **87**, 1119–23. *232–3*

278. Larkum, A. W. D., Drew, E. A. & Crossett, R. N. (1967). The vertical distribution of attached marine algae in Malta. *J. Ecol.*, **55**, 361–71. *297–8, 301*

279. Larson, D. W. (1972). Temperature, transparency, and phytoplankton productivity in Crater Lake, Oregon. *Limnol. Oceanogr.*, **17**, 410–17. *260, 326*

280. Latimer, P. & Rabinowitch, E. (1959). Selective scattering of light by pigments *in vivo*. *Arch. Biochem. Biophys.*, **84**, 428–41. *212*

281. Le Cren, E. D. & Lowe-McConnell, R. H. (1980). (Eds) *The functioning of freshwater ecosystems.* Cambridge: Cambridge University Press. *254*

282. Leletkin, V. A., Zvalinskii, V. I. & Titlyanov, E. A. (1981). Photosynthesis of

zooxanthellae in corals of different depths. *Soviet Plant Physiol.*, **27**, 863–70.  *327*

283. Lévêque, C. Quoted in Le Cren & Lowe-McConnell (1980).  *280*

284. Levring, T. (1959). Submarine illumination and vertical distribution of algal vegetation. *Proc. 9th Int. Bot. Congr.*, pp. 189–193. Canada: University of Toronto Press.  *293*

285. Levring, T. (1966). Submarine light and algal shore zonation. In *Light as an ecological factor*, eds R. Bainbridge, G. C. Evans & O. Rackham, pp. 305–18. Oxford: Blackwell.  *298–9*

286. Levring, T. (1968). Photosynthesis of some marine algae in clear, tropical oceanic water. *Bot. Mar.*, **11**, 72–80.  *298–9*

287. Lewin, R. A. (1976). Prochlorophyta as a proposed new division of algae. *Nature*, **261**, 697–8.  *178*

288. Lewis, J. B. (1977). Processes of organic production on coral reefs. *Biol. Rev.*, **52**, 305–47.  *355*

289. Lewis, W. M. & Canfield, D. (1977). Dissolved organic carbon in some dark Venezuelan waters and a revised equation for spectrophotometric determination of dissolved organic carbon. *Arch. Hydrobiol.*, **79**, 441–5.  *58, 60*

290. Ley, A. C. & Butler, W. L. (1980). Effects of chromatic adaptation on the photochemical apparatus of photosynthesis in *Porphyridium cruentum*. *Plant Physiol.*, **65**, 714–22.  *318–20, 335, 338*

291. Lipkin, Y. (1977). Seagrass vegetation of Sinai and Israel. In *Seagrass ecosystems*, eds C. P. McRoy & C. Helfferich, pp. 263–93. New York: Marcel Dekker.  *304*

292. Lipkin, Y. (1979). Quantitative aspects of seagrass communities, particularly of those dominated by *Halophila stipulacea*, in Sinai (northern Red Sea). *Aquat. Bot.*, **7**, 119–28.  *333–4*

293. Lorenzen, C. J. (1968). Carbon/chlorophyll relationships in an upwelling area. *Limnol. Oceanogr.*, **13**, 202–4.  *181*

294. Love, R. J. R. & Robinson, G. G. C. (1977). The primary productivity of submerged macrophytes in West Blue Lake, Manitoba. *Can. J. Bot.*, **55**, 118–27.  *283–4*

295. Lüning, K. (1971). Seasonal growth of *Laminaria hyperborea* under recorded underwater light conditions near Heligoland. In *Fourth European Marine Biology Symposium*, ed. D. J. Crisp, pp. 347–61. Cambridge: Cambridge University Press.  *356*

296. Lüning, K. & Dring, M. J. (1979). Continuous underwater light measurement near Helgoland (North Sea) and its significance for characteristic light limits in the sublittoral region. *Helgoländer Wiss. Meeresunters*, **32**, 403–24.  *256, 282*

297. Lynch, M. & Shapiro, J. (1981). Predation, enrichment and phytoplankton community structure. *Limnol. Oceanogr.*, **26**, 86–102.  *276, 297–8*

298. Malone, T. C. (1977). Light-saturated photosynthesis by phytoplankton size fractions in the New York Bight, U.S.A. *Mar. Biol.*, **42**, 281–92.  *273*

299. Malone, T. C. & Neale, P. J. (1981). Parameters of light-dependent photosynthesis for phytoplankton size fractions in temperate estuarine and coastal environments. *Mar. Biol.*, **61**, 289–97.  *244*

300. Mann, K. H. (1972). Ecological energetics of the seaweed zone in a marine bay on the Atlantic coast of Canada. II. Productivity of the seaweeds. *Mar. Biol.*, **14**, 199–209. *356*

301. Mann, K. H. & Chapman, A. R. O. (1975). Primary production of marine macrophytes. In *Photosynthesis and productivity in different environments*, ed. J. P. Cooper, pp. 207–23. Cambridge: Cambridge University Press. *355–6*

302. Marcus, B. A. (1980). Relationship between light intensity and chlorophyll content in *Myriophyllum spicatum* L. in Canadice Lake (New York, U.S.A.). *Aquat. Bot.*, **9**, 169–72. *325*

303. Mariani Colombo, P., Orsenigo, M., Solazzi, A. & Tolomio, C. (1976). Sea depth effects on the algal photosynthetic apparatus. IV. Observations on the photosynthetic apparatus of *Halimeda tuna* (Siphonales) at sea depths between 7 and 16 m. *Mem. Biol. mar. Ocean.*, **6**, 197–208. *334*

304. Marra, J. (1978). Phytoplankton photosynthetic response to vertical movement in a mixed layer. *Mar. Biol.*, **46**, 203–8. *224, 230, 234*

305. Mathieson, A. C. (1979). Vertical distribution and longevity of subtidal seaweeds in northern New England, U.S.A. *Bot. Mar.*, **30**, 511–20. *294*

306. Mathieson, A. C. & Burns, R. L. (1971). Ecological studies of economic red algae. *J. exp. mar. Biol. Ecol.*, **7**, 197–206. *271*

307. Mathieson, A. C. & Norall, T. L. (1975). Physiological studies of subtidal red algae. *J. exp. mar. Biol. Ecol.*, **20**, 237–47. *328*

308. Mathis, P. & Paillotin, G. (1981). Primary processes of photosynthesis. In *The biochemistry of plants*, vol. 8, eds M. D. Hatch & N. K. Boardman, pp. 97–161. New York: Academic Press. *195*

309. McAllister, H. A., Norton, T. A. & Conway, E. (1967). A preliminary list of sublittoral marine algae from the west of Scotland. *Br. phycol. Bull*, **3**, 175–84. *294*

310. McLachlan, A. J. & McLachlan, S. M. (1975). The physical environment and bottom fauna of a bog lake. *Arch. Hydrobiol.*, **76**, 198–217. *58*

311. McRoy, C. P. & Helfferich, C. (1980). Applied aspects of seagrasses. In *Handbook of seagrass biology*, eds R. C. Phillips & C. P. McRoy, pp. 297–343. New York: Garland STPM Press. *277*

312. Medina, E. (1971). Effect of nitrogen supply and light intensity during growth on the photosynthetic capacity and carboxydismutase activity of leaves of *Atriplex patula* ssp. *hastata. Carnegie Inst. Wash. Yearbook*, **70**, 551–9. *308*

313. Megard, R. O., Combs, W. S., Smith, P. D. & Knoll, A. S. (1979). Attenuation of light and daily integral rates of photosynthesis attained by planktonic algae. *Limnol. Oceanogr.*, **24**, 1038–50. *112, 216, 237*

314. Melack, J. M. (1979). Photosynthesis and growth of *Spirulina platensis* (Cyanophyta) in an equatorial lake (Lake Simbi, Kenya). *Limnol. Oceanogr.*, **24**, 753–60. *112*

315. Mie, G. (1908). Beiträge zur Optik trüber Medien, speziell kolloidalen Metall-lösungen. *Ann. Phys.*, **25**, 377–445. *72–5*

316. Miller, J. R., Jain, S. C., O'Neill, N. T., McNeil, W. R. & Thomson, K. P. B. (1977). Interpretation of airborne spectral reflectance measurements over Georgian Bay. *Remote Sens. Env.* **6**, 183–200. *139, 151*

317. Mishkind, M. & Mauzerall, D. (1980). Kinetic evidence for a common photosynthetic step in diverse seaweeds. *Mar. Biol.*, **58**, 89–96. *305–6*

318. Mizusawa, M., Kageyama, A. & Yokohama, Y. (1978). Physiology of benthic algae in tide pools. *Jap. J. Phycol.*, **26**, 109–14.  *269*

319. Molinier, R. (1960). Etudes des biocénoses marines du Cap Corse. *Vegetatio*, **9**, 121–92, 219–312.  *297*

320. Monahan, E. C. & Pybus, M. J. (1978). Colour, ultraviolet absorbance and salinity of the surface waters off the west coast of Ireland. *Nature*, **274**, 782–4.  *53*

321. Monteith, J. L. (1973). *Principles of environmental physics*. London: Edward Arnold.  *28, 30–2, 36*

322. Moon, P. (1940). *J. Franklin Inst.*, **230**, 583. Quoted by Monteith (1973).  *28*

323. Moon, P. (1961). *The scientific basis of illuminating engineering*. New York: Dover.  *26*

324. Moon, R. E. & Dawes, C. J. (1976). Pigment changes and photosynthetic rates under selected wavelengths in the growing tips of *Euchaema isiforme* (C. Agardh) J. Agardh var *denudatum* Cheney during vegetative growth. *Br. phycol. J.*, **11**, 165–74.  *340*

325. Morel, A. (1966). Etude expérimentale de la diffusion de la lumière par l'eau, les solutions de chlorure de sodium et l'eau de mer optiquement pures. *J. Chim. Phys.*, **10**, 1359–66.  *80*

326. Morel, A. (1973). Diffusion de la lumière par les eaux de mer. Résultats expérimentaux et approche théorique. In *Optics of the sea*, AGARD Lect. Ser. 61, pp. 3.1–1 to 3.1–76. Neuilly-sur-Seine: NATO.  *84, 85, 89*

327. Morel, A. (1974). Optical properties of pure water and pure seawater. In *Optical aspects of oceanography*, eds N. G. Jerlov & E. Steemann Nielsen, pp. 1–24. London: Academic Press.  *83–5*

328. Morel, A. (1978). Available, usable and stored radiant energy in relation to marine photosynthesis. *Deep-Sea Res.*, **25**, 673–88.  *235, 240, 242, 244, 246*

329. Morel, A. (1980). In-water and remote measurement of ocean colour. *Boundary-layer Meteorol.*, **18**, 177–201.  *137, 150, 155*

330. Morel, A. (1982). Optical properties and radiant energy in the waters of the Guinea dome and of the Mauritanian upwelling area: relation to primary production. *Rapp. P. V. Reun. Cons. int. explor. Mer*, **180**, 94–107.  *85–6, 105, 107*

331. Morel, A. & Bricaud, A. (1981). Theoretical results concerning light absorption in a discrete medium, and application to specific absorption of phytoplankton. *Deep-Sea Res.*, **28**, 1375–93.  *204, 206, 209*

332. Morel, A. & Prieur, L. (1975). Analyse spectrale des coefficients d'attenuation diffuse, de reflection diffuse, d'absorption et de retrodiffusion pour diverses régions marines. Rep. No. 17, Laboratoire d'Oceanographie Physique, Villefranche-sur-Mer. 157 pp. Quoted by Jerlov (1976).  *68*

333. Morel, A. & Prieur, L. (1977). Analysis of variations in ocean colour. *Limnol. Oceanogr.*, **22**, 709–22.  *64, 89, 115, 116, 118, 154, 158*

334. Morel, A. & Smith, R. C. (1974). Relation between total quanta and total energy for aquatic photosynthesis. *Limnol. Oceanogr.*, **19**, 591–600.  *5, 238*

335. Morris, I. (1980). (Ed.) *The physiological ecology of phytoplankton*. Oxford: Blackwell.  *254*

336. Morris, I. & Farrell, K. (1971). Photosynthetic rates, gross patterns of carbon dioxide assimilation and activities of ribulose diphosphate carboxylase in

marine algae grown at different temperatures. *Physiol. Plant.*, **25**, 372–7. *272*

337. Munday, J. C. & Alföldi, T. T. (1979). Landsat test of diffuse reflectance models for aquatic suspended solids measurement. *Remote Sens. Env.*, **8**, 169–83. *145–6, 152–3*

338. Myers, J. & Graham, J.-R. (1963). Enhancement in *Chlorella. Plant Physiol.*, **38**, 105–16. *251*

339. Myers, J. & Graham, J.-R. (1971). The photosynthetic unit in *Chlorella* measured by repetitive short flashes. *Plant Physiol.*, **48**, 282–6. *305*

340. Nakamura, K., Ogawa, T. & Shibata, K. (1976). Chlorophyll and peptide compositions in the two photosystems of marine green algae. *Biochim. Biophys. Acta*, **423**, 227–36. *181*

341. Neville, R. A. & Gower, J. F. R. (1977). Passive remote sensing of phytoplankton via chlorophyll *a* fluorescence. *J. geophys. Res.*, **82**, 3487–93. *137, 139, 158*

342. Norton, T. A. (1968). Underwater observations on the vertical distribution of algae at St. Mary's, Isles of Scilly. *Br. phycol. Bull.*, **3**, 585–8. *294*

343. Norton, T. A., Hiscock, K. & Kitching, J. A. (1977). The ecology of Lough Ine. XX. The Laminaria forest at Carrigathorna. *J. Ecol.*, **65**, 919–41. *295–6*

344. Norton, T. A., McAllister, H. A., Conway, E. & Irvine, L. M. (1969). The marine algae of the Hebridean island of Colonsay. *Br. phycol. J.*, **4**, 125–36. *294*

345. Nultsch, W. & Pfau, J. (1979). Occurrence and biological role of light-induced chromatophore displacements in seaweeds. *Mar. Biol.*, **51**, 77–82. *349–50*

346. O'Carra, C. & O'h Eocha, C. (1976). Algal biliproteins and phycobilins. In *Chemistry and biochemistry of plant pigments*, vol. 1, 2nd edn., ed. T. W. Goodwin, pp. 328–76. London: Academic Press. *189, 190*

347. Oertel, G. F. & Dunstan, W. M. (1981). Suspended-sediment distribution and certain aspects of phytoplankton production off Georgia, U.S.A. *Marine Geol.*, **40**, 171–97. *162, 261*

348. Ohad, I., Schneider, H. A. W., Gendel, S. & Bogorad, L. (1980). Light-induced changes in allophycocyanin. *Plant Physiol.*, **65**, 6–12. *317*

349. O'h Eocha, C. (1965). Phycobilins. In *Chemistry and biochemistry of plant pigments*, 1st edn., ed. T. W. Goodwin, pp. 175–96. London: Academic Press. *190*

350. O'h Eocha, C. (1966). Biliproteins. In *Biochemistry of chloroplasts*, vol. 1, ed. T. W. Goodwin, pp. 407–21. London: Academic Press. *189*

351. O'Kelly, C. J. (1982). Chloroplast pigments in selected marine Chaetophoraceae and Chaetosiphonaceae (Chlorophyta): the occurrence and significance of siphonaxanthin. *Bot. Mar.*, **25**, 133–7. *183, 298*

352. Oltmanns, F. (1892). Uber die Kultur und Lebensbedingungen der Meeresalgen. *Jb. Wiss. Bot.*, **23**, 349–440. *292*

353. Ortner, P. B., Wiebe, P. H. & Cox, J. L. (1980). Relationships between oceanic epizooplankton distributions and the seasonal deep chlorophyll maximum in the northwestern Atlantic Ocean. *J. Mar. Res.*, **3**, 507–31. *258, 335*

354. Paasche, E. (1964). A tracer study of the inorganic carbon uptake during coccolith formation and photosynthesis in the coccolithophorid *Coccolithus huxleyi. Physiol. Plant.*, **3** (suppl.), 1–82. *265*

355. Paerl, H. W. & Ustach, J. F. (1982). Blue-green algal scums: an explanation for their occurrence during freshwater blooms. *Limnol. Oceanogr.*, **27**, 212–17. *347*

356. Palmer, K. F. & Williams, D. (1974). Optical properties of water in the near infrared. *J. Opt. Soc. Amer.*, **64**, 1107–10. *50–1*

357. Pearse, J. F. & Hines, A. H. (1979). Expansion of a Central California kelp forest following the mass mortality of sea urchins. *Mar. Biol.*, **51**, 83–91. *277*

358. Pelevin, V. N. (1978). Estimation of the concentration of suspension and chlorophyll in the sea from the spectrum of outgoing radiation measured from a helicopter. *Oceanology*, **18**, 278–82. *150*

359. Pelevin, V. N. & Rutkovskaya, V. A. (1977). On the optical classification of ocean waters from the spectral attenuation of solar radiation. *Oceanology*, **17**, 28–32. *66–7, 302*

360. Perez-Bermudez, P., Garcia-Carrascosa, M., Cornejo, M. J. & Segura, J. (1981). Water-depth effects in photosynthetic pigment content of the benthic algae *Dictyota dichotoma* and *Udotea petiolata*. *Aquat. Bot.*, **11**, 373–7. *324–5, 339*

361. Perkins, E. J. (1960). The diurnal rhythm of the littoral diatoms of the river Eden estuary, Fife. *J. Ecol.*, **48**, 725–8. *348*

362. Perry, M. J., Talbot, M. C. & Alberte, R. S. (1981). Photoadaptation in marine phytoplankton: response of the photosynthetic unit. *Mar. Biol.*, **62**, 91–101. *305–6, 310*

363. Petzold, T. L. (1972). *Volume scattering functions for selected ocean waters*. San Diego: Scripps Inst. Oceanogr., Ref. 72–78, 79 pp. *78–9, 82, 85, 87–8, 127*

364. Phillips, G. L., Eminson, D. & Moss, B. (1978). A mechanism to account for macrophyte decline in progressively eutrophicated freshwaters. *Aquat. Bot.*, **4**, 103–26. *276*

365. Plass, G. N., Humphreys, T. J. & Kattawar, G. W. (1981). Ocean–atmosphere interface: its influence on radiation. *Appl. Opt.*, **20**, 917–31. *102*

366. Plass, G. N. & Kattawar, G. W. (1972). Monte Carlo calculations of radiative transfer in the Earth's atmosphere–ocean system. I. Flux in the atmosphere and ocean. *J. phys. Oceanogr.*, **2**, 139–45. *102*

367. Plass, G. N., Kattawar, G. W. & Catchings, F. E. (1973). Matrix operator theory of radiative transfer. *Appl. Opt.*, **12**, 314–29. *102*

368. Platt, T. (1969). The concept of energy efficiency in primary production. *Limnol. Oceanogr.*, **14**, 653–9. *245–6*

369. Platt, T., Gallegos, C. L. & Harrison, W. G. (1980). Photoinhibition of photosynthesis in natural assemblages of marine phytoplankton. *J. Mar. Res.*, **38**, 687–701. *225*

370. Platt, T. & Jassby, A. D. (1976). The relationship between photosynthesis and light for natural assemblages of coastal marine phytoplankton. *J. Phycol.*, **12**, 421–30. *227, 243, 244, 273, 341*

371. Platt, T. & Subba Rao, D. V. (1975). Primary production of marine microphytes. In *Photosynthesis and productivity in different environments*, ed. J. P. Cooper, pp. 249–80. Cambridge: Cambridge University Press. *289*

372. Poole, H. H. (1945). The angular distribution of submarine daylight in deep water. *Sci. Proc. R. Dublin Soc.*, **24**, 29–42. *122*

373. Poole, H. H. & Atkins, W. R. G. (1929). Photoelectric measurements of submarine illumination throughout the year. *J. Mar. Biol. Ass. U.K.*, **16**, 297–324. *95*

374. Preisendorfer, R. W. (1957). Exact reflectance under a cardioidal luminance distribution. *Q. J. Roy. Meteorol. Soc.*, **83**, 540. Quoted by Jerlov (1976). *40*

375. Preisendorfer, R. W. (1959). Theoretical proof of the existence of characteristic diffuse light in natural waters. *J. mar. Res.*, **18**, 1–9. *122, 124*

376. Preisendorfer, R. W. (1961). Application of radiative transfer theory to light measurements in the sea. *Union Geod. Geophys. Inst. Monogr.* **10**, 11–30. *12, 18, 20, 23, 82, 85, 119, 122, 131*

377. Preisendorfer, R. W. (1976). *Hydrologic optics*, vol. 1. Washington: U.S. Department of Commerce. *3*

378. Prézelin, B. B. (1976). The role of peridinin–chlorophyll *a* proteins in the photosynthetic light adaptation of the marine dinoflagellate, *Glenodinium* sp. *Planta*, **130**, 225–33. *307, 309–11*

379. Prézelin, B. B. & Alberte, R. S. (1978). Photosynthetic characteristics and organization of chlorophyll in marine dinoflagellates. *Proc. Natl. Acad. Sci., Wash.*, **75**, 1801–4. *189*

380. Prézelin, B. B. & Haxo, F. T. (1976). Purification and characterization of peridinin–chlorophyll *a* proteins from the marine dinoflagellates *Glenodinium* sp. and *Gonyaulax polyedra*. *Planta*, **128**, 133–41. *188*

381. Prézelin, B. B., Ley, A. C. & Haxo, F. T. (1976). Effects of growth irradiance on the photosynthetic action spectra of the marine dinoflagellate, *Glenodinium* sp. *Planta*, **130**, 251–6. *336*

382. Prézelin, B. B., Meeson, B. W. & Sweeney, B. M. (1977). Characterization of photosynthetic rhythms in marine dinoflagellates. I. Pigmentation, photosynthetic capacity and respiration. *Plant Physiol.*, **60**, 384–7. *352*

383. Prézelin, B. B. & Sweeney, B. M. (1977). Characterization of photosynthetic rhythms in marine dinoflagellates. II. Photosynthesis-irradiance curves and *in vivo* chlorophyll fluorescence. *Plant Physiol.*, **60**, 388–92. *352–3*

384. Prieur, L. (1976). Transfer radiatif dans les eaux der mer. D.Sc. thesis, Univ. Pierre et Marie Curie, Paris, 243 pp. *101–2, 116*

385. Prieur, L. & Morel, A. (1971). Etude théorique du régime asymptotique: relations entre caractéristiques optiques et coefficient d'extinction relatif à la pénétration de la lumière du jour. *Cah. Oceanogr.*, **23**, 35–47. *122*

386. Prieur, L. & Morel, A. (1973). *Apercu sur les théories du transfert radiatif applicables a la propagation dans la mer*. AGARD lect. Ser. No. 61, pp. 1.3–1 to 1.3–25. Neuilly-sur-Seine: NATO. *101*

387. Prieur, L. & Sathyendranath, S. (1981). An optical classification of coastal and oceanic waters based on the specific spectral absorption curves of phytoplankton pigments, dissolved organic matter, and other particulate materials. *Limnol. Oceanogr.*, **26**, 671–89. *68*

388. Rabinowitch, E. I. (1945). *Photosynthesis*, vol. I. New York: Interscience. *292*

389. Rabinowitch, E. I. (1951). *Photosynthesis*, Vol. II.1. New York: Interscience. *264*

390. Radmer, R. & Kok, B. (1977). Photosynthesis: limited yields, unlimited dreams. *BioScience*, **27**, 599–605. *239*

391. Ramus, J., Beale, S. I. & Mauzerall, D. (1976). Correlation of changes in pigment content with photosynthetic capacity of seaweeds as a function of depth. *Mar. Biol.*, **37**, 231–8. *228–9, 302, 328*

392. Ramus, J., Beale, S. I., Mauzerall, D. & Howard, K. L. (1976). Changes in photosynthetic pigment concentration in seaweeds as a function of depth. *Mar. Biol.*, **37**, 223–9. *323–4*

393. Ramus, J., Lemons, F. & Zimmerman, C. (1977). Adaptation of light-harvesting pigments to downwelling light and the consequent photosynthetic performance of the eulittoral rockweeds *Ascophyllum nodosum* and *Fucus vesiculosus*. *Mar. Biol.*, **42**, 293–303. *324, 329*

394. Raschke, E. (1972). Multiple-scattering calculation of the transfer of solar radiation in an atmosphere–ocean system. *Beitr. Phys. Atmos.*, **45**, 1–19. *102*

395. Raven, J. A. (1970). Endogenous inorganic carbon sources in plant photosynthesis. *Biol. Rev.*, **45**, 167–221. *265, 268*

396. Raymont, J. E. (1980). *Plankton and productivity in the oceans. vol. 1. Phytoplankton*, 2nd edn. Oxford: Pergamon. *172, 254, 275, 277, 289–90*

397. Reynolds, C. S. & Walsby, A. E. (1975). Water blooms. *Biol. Rev.*, **50**, 437–81. *345*

398. Rhee, C. & Briggs, W. R. (1977). Some responses of *Chondrus crispus* to light. I. Pigmentation changes in the natural habitat. *Bot. Gaz.*, **138**, 123–8. *307, 325, 339, 340*

399. Ried, A., Hessenberg, B., Metzler, H. & Ziegler, R. (1977). Distribution of excitation energy amongst Photosystem I and Photosystem II in red algae. *Biochim. Biophys. Acta*, **459**, 175–86. *253*

400. Riley, G. A. (1942). The relationship of vertical turbulence and spring diatom flowerings. *J. Mar. Res.*, **5**, 67–87. *278*

401. Ritchie, J. C., Schiebe, F. R. & McHenry, J. R. (1976). Remote sensing of suspended sediments in surface waters. *Photogramm. Eng. Rem. Sens.*, **42**, 1539–45. *151–2*

402. Robinson, N. (1966). *Solar radiation.* Amsterdam: Elsevier. *26–7, 31*

403. Rodhe, W. (1969). Crystallization of eutrophication concepts in Northern Europe. In *Eutrophication: causes, consequences, correctives*, pp. 50–64. Washington: National Academy of Science. *290*

404. Roemer, S. C. & Hoagland, K. D. (1979). Seasonal attenuation of quantum irradiance (400–700 nm) in three Nebraska reservoirs. *Hydrobiologia*, **63**, 81–92. *112*

405. Round, F. E. & Happey, C. M. (1965). Persistent vertical-migration rhythms in benthic microflora. IV. A diurnal rhythm of the epipelic diatom association in non-tidal flowing water. *Br. phycol. Bull.*, **2**, 465–71. *347*

406. Round, F. E. & Palmer, J. D. (1966). Persistent vertical-migration rhythms in benthic microflora. II. Field and laboratory studies on diatoms from the banks of the River Avon. *J. Mar. Biol. Ass. U.K.*, **46**, 191–214. *348*

407. Rouse, L. J. & Coleman, J. M. (1976). Circulation observations in the Louisiana Bight using Landsat imagery. *Remote Sens. Env.*, **5**, 55–66. *162*

408. Ryther, J. H. (1956). Photosynthesis in the ocean as a function of light intensity. *Limnol. Oceanogr.*, **1**, 61–70. *278*

409. Ryther, J. H. (1969). Photosynthesis and fish production in the sea. *Science*, **166**, 72–6. *275*

410. Ryther, J. H. & Menzel, D. (1959). Light adaptation by marine phytoplankton. *Limnol. Oceanogr.*, **4**, 492–7. *221, 223, 330*

411. Sakamoto, M. (1966). Primary production by phytoplankton community in some Japanese lakes and its dependence on lake depth. *Arch. Hydrobiol.*, **62**, 1–28. *255*

412. San Pietro, A. (1971, 1972, 1980) (Ed.). *Methods in enzymology*, vols. 23, 24, 69. New York: Academic Press. *219*

413. Sasaki, T., Watanabe, S., Oshiba, G., Okami, N. & Kajihara, M. (1962). On the instrument for measuring angular distribution of underwater radiance. *Bull. Jap. Soc. Sci. Fish*, **28**, 489–96. *101*

414. Sauberer, F. (1945). Beiträge zur Kenntnis der optischen Eigenschaften der Kärntner Seen. *Arch. Hydrobiol.*, **41**, 259–314. *53*

415. Scheibe, J. (1972). Photoreversible pigment: occurrence in a blue-green alga. *Science*, **176**, 1037–9. *316–7*

416. Scherz, J. P., van Domelen, J. F., Holtje, K. & Johnson, W. (1974). Lake eutrophication as indicated by ERTS satellite imagery. In *Symposium on remote sensing and photo interpretation*. (Canada), pp. 247–58. International Society for Photogrammetry. *144, 162*

417. Schnitzer, M. (1978). Humic substances: chemistry and reactions. In *Soil organic matter*, eds. M. Schnitzer & S. U. Khan, pp. 1–64. Amsterdam: Elsevier. *52*

418. Scott., B. D. (1978). Phytoplankton distribution and light attenuation in Port Hacking estuary. *Aust. J. mar. freshwater Res.*, **29**, 31–44. *112*

419. Sears, J. R. & Cooper, R. A. (1978). Descriptive ecology of offshore, deep-water, benthic algae in the temperate western North Atlantic Ocean. *Mar. Biol.*, **44**, 309–14. *304*

420. Senger, H. & Fleischhacker, P. (1978). Adaptation of the photosynthetic apparatus of *Scenedesmus obliquus* to strong and weak light conditions. *Physiol. Plant.*, **43**, 35–42. *308–11*

421. Shepherd, S. A. & Sprigg, R. C. (1976). Substrate sediments and subtidal ecology of Gulf St. Vincent and Investigator Strait. In *Natural History of the Adelaide region*, eds C. R. Twidale, M. J. Tyler & B. P. Webb, pp. 161–74. Adelaide: Royal Society of South Australia. *299, 304*

422. Shepherd, S. A. & Womersley, H. B. S. (1970). The sublittoral ecology of West Island, South Australia: environmental features and algal ecology. *Trans. R. Soc. South Aust.*, **94**, 105–38. *299–300*

423. Shepherd, S. A. & Womersley, H. B. S. (1971). Pearson Island Expedition 1969. 7. The sub-tidal ecology of benthic algae. *Trans. R. Soc. South Aust.*, **94**, 155–67. *299–300*

424. Shepherd, S. A. & Womersley, H. B. S. (1976). The subtidal algal and seagrass ecology of St. Francis Island, South Australia. *Trans. R. Soc. South Aust.*, **100**, 177–91. *299–300*

425. Shibata, K. (1959). Spectrophotometry of translucent biological materials – opal glass transmission method. *Methods of biochemical Analysis*, **7**, 77–109. *47*

426. Shimura, S. & Fujita, Y. (1975). Changes in the activity of fucoxanthin-excited photosynthesis in the marine diatom *Phaeodactylum tricornutum* grown under different culture conditions. *Mar. Biol.*, **33**, 185–94. *311*

427. Shimura, S. & Ichimura, S. (1973). Selective transmission of light in the ocean waters and its relation to phytoplankton photosynthesis. *J. Oceanogr. Soc. Japan.*, **29**, 257–66. Quoted by Shimura & Fujita (1975). *331*

428. Short, N. M. (1976). *Mission to Earth: Landsat views the world.* Washington: NASA. *140*

429. Shulenberger, E. (1978). The deep chlorophyll maximum and mesoscale environmental heterogeneity in the western half of the North Pacific central gyre. *Deep-Sea Res.*, **25**, 1193–208. *258*

430. Sieburth, J. McN. & Jensen, A. (1968). Studies on algal substances in the sea. I. Gelbstoff (humic material) in terrestrial and marine waters. *J. exp. mar. Biol. Ecol.*, **2**, 174–89. *53*

431. Sieburth, J. McN. & Jensen, A. (1969). Studies on algal substances in the sea. II. The formation of gelbstoff (humic material) by exudates of Phaeophyta. *J. exp. mar. Biol. Ecol.*, **3**, 278–89. *54*

432. Simenstad, C. A., Estes, J. A. & Kenyon, K. W. (1978). Aleuts, sea otters, and alternative stable-state communities. *Science*, **200**, 403–11. *277*

433. Slater, P. N. (1980). *Remote sensing: optics and optical systems.* Reading: Addison-Wesley. *137, 140*

434. Smith, E. L. (1936). Photosynthesis in relation to light and carbon dioxide. *Proc. Natl Acad. Sci., Wash.*, **22**, 504–11. *225, 285*

435. Smith, E. L. (1938). Limiting factors in photosynthesis: light and carbon dioxide. *J. gen. Physiol.*, **22**, 21–35. *264*

436. Smith, F. A. & Walker, N. A. (1980). Photosynthesis by aquatic plants: effects of unstirred layers in relation to assimilation of $CO_2$ and $HCO_3^-$ and to carbon isotopic discrimination. *New Phytol.*, **86**, 245–59. *268*

437. Smith, R. C. (1968). The optical characterization of natural waters by means of an 'extinction coefficient'. *Limnol. Oceanogr.*, **13**, 423–9. *94*

438. Smith, R. C. (1969). An underwater spectral irradiance collector. *J. mar. Res.*, **27**, 341–51. *92*

439. Smith, R. C., Austin, R. W. & Tyler, J. E. (1970). An oceanographic radiance distribution camera system. *Appl. Opt.*, **9**, 2015–22. *101*

440. Smith, R. C. & Baker, K. S. (1978*a*). The bio-optical state of ocean waters and remote sensing. *Limnol. Oceanogr.*, **23**, 247–59. *216*

441. Smith, R. C. & Baker, K. S. (1978*b*). Optical classification of natural waters. *Limnol. Oceanogr.*, **23**, 260–7. *67, 70*

442. Smith, R. C. & Baker, K. S. (1980). Biologically effective dose transmitted by culture bottles in $^{14}C$ productivity measurements. *Limnol. Oceanogr.*, **25**, 364–6. *232*

443. Smith, R. C. & Baker, K. S. (1981). Optical properties of the clearest natural waters (200–800 nm). *Appl. Opt.*, **20**, 177–84. *50–1*

444. Smith, R. C. & Baker, K. S. (1982). Oceanic chlorophyll concentrations as determined by satellite (Nimbus-7 Coastal Zone Colour Scanner). *Mar. Biol.*, **66**, 269–79. *166*

445. Smith, R. C., Baker, K. S., Holm-Hansen, O. & Olson, R. (1980). Photoinhibition of photosynthesis in natural waters. *Photochem. Photobiol.*, **31**, 585–92. *232*

446. Smith, R. C., Ensminger, R. L., Austin, R. W., Bailey, J. D. & Edwards, G. D. (1979). Ultraviolet submersible spectroradiometer. *Soc. Photo-Opt. Instr. Eng.*, **208**, 127–40. *99*

447. Smith, R. C. & Tyler, J. E. (1967). Optical properties of clear natural water. *J. Opt. Soc. Amer.*, **57**, 589–95. *68*

448. Smith, R. C. & Wilson, W. H. (1980). Ship and satellite bio-optical research in the California Bight. In *Oceanography from space*, ed. J. F. R. Gower. New York: Plenum Press. *146*

449. Smith, W. O. (1977). The respiration of photosynthetic carbon in eutrophic areas of the ocean. *J. mar. Res.* **35**, 557–65. *225*

450. Solazzi, A. & Tolomio, C. (1976). Sea depth effects on the algal photosynthetic apparatus. I. Chlorophyll pigments of *Halimeda tuna* Lam. (Chlorophyceae, Siphonales). *Mem. Biol. mar. Ocean*, **6**, 21–7. *324*

451. Sournia, A. (1974). Circadian periodicities in natural populations of marine phytoplankton. *Adv. mar. Biol.*, **12**, 325–89. *277*

452. Sournia, A. (1976). Primary production of sands in the lagoon of an atoll and the role of Foraminiferan symbionts. *Mar. Biol.*, **37**, 29–32. *255*

453. Spence, D. H. N. (1976). Light and plant response in fresh water. In *Light as an ecological factor*, eds G. C. Evans, R. Bainbridge & O. Rackham, pp. 93–133. Oxford: Blackwell. *348*

454. Spence, D. H. N., Campbell, R. M. & Chrystal, J. (1971). Spectral intensity in some Scottish freshwater lochs. *Freshwater Biol.*, **1**, 321–37. *112*

455. Spence, D. H. N., Campbell, R. M. & Chrystal, J. (1973). Specific leaf areas and zonation of freshwater macrophytes. *J. Ecol.*, **61**, 317–28. *333–4*

456. Spence, D. H. N. & Chrystal, J. (1970). Photosynthesis and zonation of freshwater macrophytes. I. Depth distribution and shade tolerance. II. Adaptability of species of deep and shallow water. *New Phytol.*, **69**, 205–15, 217–27. *313–14*

457. Spencer, J. W. (1971). Fourier series representation of the position of the sun. *Search*, **2**, 172. *33*

458. Spilhaus, A. F. (1968). Observations of light scattering in seawater. *Limnol. Oceanogr.*, **13**, 418–22. *79*

459. Spinrad, R. W., Zaneveld, J. R. V. & Pak, H. (1978). Volume scattering function of suspended particulate matter at near-forward angles: a comparison of experimental and theoretical values. *Appl. Opt.*, **17**, 1125–30. *79*

460. Spitzer, D. & Wernand, M. R. (1979*a*). Photon scalar irradiance meter. *Appl. Opt.*, **18**, 1698–1700. *98*

461. Spitzer, D. & Wernand, M. R. (1979*b*). *Irradiance and absorption spectra measurements in the tropical East Atlantic.* 17th General Assembly of IAPSO (Canberra, 1979). Abstracts p. 73. San Diego: International Association for the Physical Sciences of the Ocean. *49*

462. Stadelmann, P., Moore, J. E. & Pickett, E. (1974). Primary production in relation to temperature structure, biomass concentration, and light conditions at an inshore and offshore station in Lake Ontario. *J. Fish Res. Bd Canada*, **31**, 1215–32. *112*

463. Steane, M. S. (1979). Light penetration and stratification in Tasmanian lakes and reservoirs. Honours thesis, Botany Dept., University of Tasmania. *59, 113, 158*

464. Steele, J. H. & Baird, I. E. (1965). The chlorophyll *a* content of particulate organic matter in the northern North Sea. *Limnol. Oceanogr.*, **10**, 261–7. *181*

465. Steele, J. H. & Yentsch, C. S. (1960). The vertical distribution of chlorophyll. *J. Mar. Biol. Ass. U.K.*, **39**, 217–26. *259*

466. Steemann Nielsen, E. (1952). The use of radioactive carbon ($^{14}$C) for measuring organic production in the sea. *J. Cons. perm. int. Explor. Mer*, **18**, 117–40. *220*

467. Steemann Nielsen, E. (1974). Light and primary production. In *Optical aspects of oceanography*, eds N. G. Jerlov & E. Steemann Nielsen, pp. 361–88. London: Academic Press. *255*

468. Steemann Nielsen, E. (1975). *Marine photosynthesis.* Amsterdam: Elsevier. *225, 265–6, 271–3, 279, 311, 313, 330, 335*

469. Steemann Nielsen, E. & Hansen, V. K. (1961). Influence of surface illumination on plankton photosynthesis in Danish waters (56° N) throughout the year. *Physiol. Plant.*, **14**, 595–613. *227, 340–1*

470. Steemann Nielsen, E. & Jørgensen, E. G. (1968). The adaptation of plankton algae. *Physiol. Plant.*, **21**, 401–13. *271–2*

471. Sternberg, R. W., Baker, E. T., McManus, D. A., Smith, S. & Morrison, D. R. (1974). An integrating nephelometer for measuring particle concentrations in the deep sea. *Deep-Sea Res.*, **21**, 887–92. *89*

472. Stokes, A. N. (1975). Proof of a law for calculating absorption of light by cellular suspensions. *Arch. Biochem. Biophys.*, **167**, 393–4. *204*

473. Strong, A. E. (1974). Remote sensing of algal blooms by aircraft and satellite in Lake Erie and Utah Lake. *Remote Sens. Env.*, **3**, 99–107. *154*

474. Strong, A. E. (1978). Chemical whitings and chlorophyll distributions in the Great Lakes as viewed by Landsat. *Remote Sens. Env.*, **7**, 61–72. *164, 165*

475. Stuermer, D. H. & Harvey, G. R. (1974). Humic substances from seawater. *Nature*, **250**, 480–1. *55*

476. Stuermer, D. H. & Harvey, G. R. (1978). Structural studies on marine humus. *Mar. Chem.*, **6**, 55–70. *56*

477. Stuermer, D. H. & Payne, J. R. (1976). Investigation of seawater and terrestrial humic substances with carbon-13 and proton nuclear magnetic resonance. *Geochim. Cosmochim. Acta*, **40**, 1109–14. *56*

478. Sverdrup, H. U. (1953). On conditions for the vernal blooming of phytoplankton. *J. Cons. int. Explor. Mer*, **18**, 287–95. *257–8*

479. Swift, E. & Taylor, W. R. (1967). Bioluminescence and chloroplast movement in the dinoflagellate *Pyrocystis lunula. J. Phycol.*, **3**, 77–81. *350*

480. Sydor, M. (1980). Remote sensing of particulate concentrations in water. *Appl. Opt.*, **19**, 2794–800. *145, 153*

481. Taguchi, S. (1976). Relationship between photosynthesis and cell size of marine diatoms. *J. Phycol.*, **12**, 185–9. *243*

482. Talling, J. F. (1957a). Photosynthetic characteristics of some freshwater plankton diatoms in relation to underwater radiation. *New Phytol.*, **56**, 29–50. *226, 257, 273*

483. Talling, J. F. (1957b). The phytoplankton population as a compound photosynthetic system. *New Phytol.*, **56**, 133–49. *215, 256, 285–9*

484. Talling, J. F. (1960). Self-shading effects in natural populations of a planktonic diatom. *Wett. Leben*, **12**, 235–42. *216, 238*

485. Talling, J. F. (1970). Generalized and specialized features of phytoplankton as a form of photosynthetic cover. In *Prediction and measurement of photosynthetic productivity*, pp. 431–45. Wageningen: Pudoc. *58, 68*

486. Talling, J. F. (1971). The underwater light climate as a controlling factor in the production ecology of freshwater phytoplankton. *Mitt. int. Verein. Limnol.*, **19**, 214–43. *216, 257, 277*

487. Talling, J. F. (1976). The depletion of carbon dioxide from lake water by phytoplankton. *J. Ecol.*, **64**, 79–121. *265, 267–8*

488. Talling, J. F. (1979). Factor interactions and implications for the prediction of lake metabolism. *Arch. Hydrobiol. Beih. ergebn. Limnol.*, **13**, 96–109. *267–8*

489. Talling, J. F. (1982). Utilization of solar radiation by phytoplankton. In *Trends in photobiology*, eds C. Helene, M. Charlier, T. Montenay-Garestier & G. Laustriat, pp. 619–31. New York: Plenum Press. *235*

490. Talling, J. F., Wood, R. B., Prosser, M. V. & Baxter, R. M. (1973). The upper limit of photosynthetic productivity by phytoplankton: evidence from Ethiopian soda lakes. *Freshwater Biol.*, **3**, 53–76. *244–5*

491. Tam, C. K. N. & Patel, A. C. (1979). Optical absorption coefficients of water. *Nature*, **280**, 302–4. *51*

492. Tamiya, H. (1957). Mass culture of algae. *Ann. Rev. Plant Physiol.*, **8**, 309–34. *356*

493. Tanada, T. (1951). The photosynthetic efficiency of carotenoid pigments in *Navicula minima*. *Amer. J. Bot.*, **38**, 276–83. *250*

494. Tandeau de Marsac, N. (1977). Occurrence and nature of chromatic adaptation in cyanobacteria. *J. Bacteriol.*, **130**, 82–91. *314–5, 317*

495. Taylor, W. R. (1959). Distribution in depth of marine algae in the Caribbean and adjacent seas. *Proc. 9th Int. Botan. Congr.*, pp. 193–7. Canada: University of Toronto Press. *296*

496. Taylor, W. R. (1964). Light and photosynthesis in intertidal benthic diatoms. *Helgoländer Wiss. Meeresunters*, **10**, 29–37. *227*

497. Techtum Instrument (Umea, Sweden). *QSM-2500 quanta spectrometer*. *99*

498. Thayer, G. W., Wolfe, D. A. & Williams, R. B. (1975). The impact of man on seagrass ecosystems. *Amer. Scientist*, **63**, 288–95. *356*

499. Thekaekara, M. P. & Drummond, A. J. (1971). Standard values for the solar constant and its spectral components. *Nature*, **229**, 6–9. *25*

500. Thornber, J. P., Markwell, J. P. & Reinman, S. (1979). Plant chlorophyll–protein complexes: recent advances. *Photochem. Photobiol.*, **29**, 1205–16. *188*

501. Thorne, S. W., Newcomb, E. H. & Osmond, C. B. (1977). Identification of chlorophyll $b$ in extracts of prokaryotic algae by fluorescence spectroscopy. *Proc. Natl Acad. Sci., Wash.*, **74**, 575–8. *178, 181*

502. Thurman, E. M. Quoted by Josephson, J. (1982) in Humic substances. *Environ. Sci. Technol.*, **16**, 20A–24A. *52*

503. Tilzer, M. M. (1973). Diurnal periodicity in the phytoplankton assemblage of a high mountain lake. *Limnol. Oceanogr.*, **18**, 15–30. *277, 343–4*

504. Tilzer, M. M. & Goldman, C. R. (1978). Importance of mixing, thermal stratification and light adaptation for phytoplankton productivity in Lake Tahoe (California–Nevada). *Ecology*, **59**, 810–21. *260, 326, 332–3*

505. Tilzer, M. M., Goldman, C. R. & De Amezaga, E. (1975). The efficiency of photosynthetic light energy utilization by lake phytoplankton. *Verh. int. verein. Limnol.*, **19**, 800–7. *245*

506. Tilzer, M. M. & Schwarz, K. (1976). Seasonal and vertical patterns of

phytoplankton light adaptation in a high mountain lake. *Arch. Hydrobiol.*, **77**, 488–504. *343*

507. Timofeeva, V. A. (1971). Optical characteristics of turbid media of the sea-water type. *Izv. atmos. oceanic Phys.*, **7**, 863–5. *87–8*

508. Timofeeva, V. A. (1974). Optics of turbid water. In *Optical aspects of oceanography*, eds N. G. Jerlov & E. Steemann Nielsen, pp. 177–219. London: Academic Press. *122*

509. Timofeeva, V. A. & Gorobets, F. I. (1967). On the relationship between the attenuation coefficients of collimated and diffuse light fluxes. *Izv. atmos. oceanic Phys.*, **3**, 166–9. *77*

510. Titus, J. E. & Adams, M. S. (1979). Coexistence and the comparative light relations of the submersed macrophytes *Myriophyllum spicatum* L. and *Vallisneria americana* Michx. *Oecologia*, **40**, 273–86. *227*

511. Tyler, J. E. (1960). Radiance distribution as a function of depth in an underwater environment. *Bull. Scripps Inst. Oceanogr.*, **7**, 363–411. *101, 117, 119, 122–3*

512. Tyler, J. E. (1968). The Secchi disc. *Limnol. Oceanogr.*, **13**, 1–6. *96*

513. Tyler, J. E. (1975). The *in situ* quantum efficiency of natural phytoplankton populations. *Limnol. Oceanogr.*, **20**, 976–80. *112*

514. Tyler, J. E. (1978). Optical properties of water. In *Handbook of optics*, eds W. G. Driscoll & W. Vaughan, pp. 15-1 to 15-38. New York: McGraw-Hill. *126*

515. Tyler, J. E. & Richardson, W. H. (1958). Nephelometer for the measurement of volume scattering function *in situ*. *J. Opt. Soc. Amer.*, **48**, 354–7. *79*

516. Tyler, J. E. & Smith, R. C. (1967). Spectroradiometric characteristics of natural light under water. *J. Opt. Soc. Amer.*, **57**, 595–601. *158*

517. Tyler, J. E. & Smith, R. C. (1970). *Measurements of spectral irradiance underwater*. New York: Gordon & Breach. *29, 99, 106–9, 112, 117–8*

518. Van, T. K., Haller, W. T. & Bowes, G. (1976). Comparison of the photosynthetic characteristics of three submersed aquatic plants. *Plant Physiol.*, **58**, 761–8. *222–3, 226, 263*

519. van de Hulst, H. C. (1957). *Light scattering by small particles*. New York: Wiley. *75–6*

520. Van Wijk, W. R. & Scholte Ubing, D. W. (1963). In *Physics of plant environment*, ed. W. R. Van Wijk. Amsterdam: North Holland. Quoted by Monteith (1973). *36*

521. Vesk, M. & Jeffrey, S. W. (1977). Effect of blue-green light on photosynthetic pigments and chloroplast structure in unicellular marine algae from six classes. *J. Phycol.*, **13**, 280–8. *321*

522. Vierling, E. & Alberte, R. S. (1980). Functional organization and plasticity of the photosynthetic unit of the cyanobacterium *Anacystis nidulans*. *Physiol. Plant.*, **50**, 93–8. *305–6*

523. Viollier, M., Deschamps, P. Y. & Lecomte, P. (1978). Airborne remote sensing of chlorophyll content under cloudy skies as applied to the tropical waters in the Gulf of Guinea. *Remote Sens. Env.*, **7**, 235–48. *139, 150, 164*

524. Viollier, M., Tanré, D. & Deschamps, P. Y. (1980). An algorithm for remote sensing of water colour from space. *Boundary-layer Meteorol.*, **18**, 247–67. *148, 149, 150, 157*

525. Virgin, H. I. (1964). Some effects of light on chloroplasts and plant protoplasm.

In *Photophysiology*, ed. A. C. Giese, pp. 273–303. New York: Academic Press. *348*

526. Vollenweider, R. A. (1969) (Ed.). *A manual on methods for measuring primary production in aquatic environments.* (IBP Handbook No. 12). Oxford: Blackwell. *219*

527. Vollenweider, R. A. (1970). Models for calculating integral photosynthesis and some implications regarding structural properties of the community metabolism of aquatic systems. In *Prediction and measurement of photosynthetic productivity*, pp. 455–72. Wageningen: Pudoc. *286*

528. Waaland, J. R., Waaland, S. D. & Bates, G. (1974). Chloroplast structure and pigment composition in the red alga *Griffithsia pacifica*: regulation by light intensity. *J. Phycol.*, **10**, 193–9. *306*

529. Walker, G. A. H., Buchholz, V. L., Camp, D., Isherwood, B., Glaspey, J., Coutts, R., Condal, A. & Gower, J. (1974). A compact multichannel spectrometer for field use. *Rev. sci. Instrum.*, **45**, 1349–52. *139*

530. Wallen, D. G. & Geen, G. H. (1971*a*). Light quality in relation to growth, photosynthetic rates and carbon metabolism in two species of marine algae. *Mar. Biol.*, **10**, 34–43. *320–1*

531. Wallen, D. G. & Geen, G. H. (1971*b*). Light quality and concentrations of protein, RNA, DNA and photosynthetic pigments in two species of marine plankton algae. *Mar. Biol.*, **10**, 44–51. *320–1*

532. Wallentinus, I. (1978). Productivity studies on Baltic macroalgae. *Bot. Mar.*, **21**, 365–80. *228–9*

533. Walsby, A. E. (1975). Gas vesicles. *Ann. Rev. Plant Physiol.*, **26**, 427–39. *345*

534. Walsby, A. E. & Booker, M. J. (1976). The physiology of water-bloom formation by planktonic blue-green algae. *Br. phycol. J.*, **11**, 200. *347*

535. Walsh, J. J. (1981). A carbon budget for overfishing off Peru. *Nature*, **290**, 300–4. *289*

536. Weinberg, S. (1975). Ecologie des Octocoralliaires communs du substrat dur dans la région de Banyuls-sur-Mer. *Bijdr. Dierk*, **45**, 50–70. *282*

537. Welschmeyer, N. A. & Lorenzen, C. J. (1981). Chlorophyll-specific photosynthesis and quantum efficiency at subsaturating light intensities. *J. Phycol.*, **17**, 283–93. *243*

538. Westlake, D. F. (1967). Some effects of low-velocity currents on the metabolism of aquatic macrophytes. *J. exp. Bot.*, **18**, 187–205. *269*

539. Westlake, D. F. (1980*a*). Photosynthesis: macrophytes. In *The functioning of freshwater ecosystems*, eds E. D. Le Cren & R. H. Lowe-McConnell, pp. 177–82. Cambridge: Cambridge University Press. *284, 356*

540. Westlake, D. F. (1980*b*). Biomass changes: macrophytes. In *The functioning of freshwater ecosystems*, eds E. D. Le Cren & R. H. Lowe-McConnell, pp. 203–6. Cambridge: Cambridge University Press. *276, 284*

540a. Westlake, D. F. (1980*c*). Effects of macrophytes. In *The functioning of freshwater ecosystems*, eds E. D. Le Cren & R. H. Lowe-McConnell, pp. 161–2. Cambridge: Cambridge University Press. *218*

541. Whatley, J. M. (1977). The fine structure of *Prochloron. New Phytol.*, **79**, 309–13. *178*

542. Wheeler, J. R. (1976). Fractionation by molecular weight of organic substances in Georgia coastal water. *Limnol. Oceanogr.*, **21**, 846–52. *55, 58*

543. Wheeler, W. N. (1980*a*). Pigment content and photosynthetic rate of the fronds of *Macrocystis pyrifera*. *Mar. Biol.*, **56**, 97–102.  *324, 329*

544. Wheeler, W. N. (1980*b*). Effect of boundary layer transport on the fixation of carbon by the giant kelp, *Macrocystis pyrifera*. *Mar. Biol.*, **56**, 103–10.  *269*

545. Whitney, L. V. (1941). The angular distribution of characteristic diffuse light in natural waters. *J. mar. Res.*, **4**, 122–31.  *122*

546. Wiginton, J. R. & McMillan, C. (1979). Chlorophyll composition under controlled light conditions as related to the distribution of seagrasses in Texas and the U.S. Virgin Islands. *Aquat. Bot.*, **6**, 171–84.  *324–5*

547. Wild, A., Ke, B. & Shaw, E. R. (1973). The effect of light intensity during growth of *Sinapis alba* on the electron-transport components. *Z. Pflanzenphysiol.*, **69**, 344–50.  *308*

548. Williams, R. B. & Murdoch, M. B. (1966). Phytoplankton production and chlorophyll concentration in the Beaufort Channel, North Carolina. *Limnol. Oceanogr.*, **11**, 73–82.  *274, 279*

549. Womersley, H. B. S. (1981). Marine ecology and zonation of temperate coasts. In *Marine botany: an Australasian perspective*, eds M. N. Clayton & R. J. King, pp. 211–40. Melbourne: Longman Cheshire.  *299*

550. Yamada, T., Ikawa, T. & Nisizawa, K. (1979). Circadian rhythm of the enzymes participating in the $CO_2$-photoassimilation of a brown alga, *Spatoglossum pacificum*. *Bot. Mar.*, **22**, 203–9.  *353*

551. Yentsch, C. S. (1960). The influence of phytoplankton pigments on the colour of sea water. *Deep-Sea Res.*, **7**, 1–9.  *63–4*

552. Yentsch, C. S. (1962). Measurement of visible light absorption by particulate matter in the ocean. *Limnol. Oceanogr.*, **7**, 207–17.  *62*

553. Yentsch, C. S. (1974). Some aspects of the environmental physiology of marine phytoplankton: a second look. *Oceanogr. mar. Biol. Ann. Rev.*, **12**, 41–75.  *272–3*

554. Yentsch, C. S. & Lee, R. W. (1966). A study of photosynthetic light reactions, and a new interpretation of sun and shade phytoplankton. *J. mar. Res.*, **24**, 319–37.  *330*

555. Yentsch, C. S. & Scagel, R. F. (1958). Diurnal study of phytoplankton pigments. An *in situ* study in East sound, Washington. *J. mar. Res.*, **17**, 567–83.  *354*

556. Yocum, C. S. & Blinks, L. R. (1958). Light-induced efficiency and pigment alterations in red algae. *J. gen. Physiol.*, **41**, 1113–17.  *318–9, 338*

557. Yoder, J. A., Atkinson, L. P., Lee, T. N., Kim, H. H. & McLain, C. R. (1981). Role of Gulf Stream frontal eddies in forming phytoplankton patches on the outer southeastern shelf. *Limnol. Oceanogr.*, **26**, 1103–10.  *165*

558. Yokohama, Y. (1973). A comparative study on photosynthesis–temperature relationships and their seasonal changes in marine benthic algae. *Int. Rev. ges. Hydrobiol.*, **58**, 463–72.  *269–70*

559. Yokohama, Y., Kageyama, A., Ikawa, T. & Shimura, S. (1977). A carotenoid characteristic of Chlorophycean seaweeds living in deep coastal waters. *Bot. Mar.*, **20**, 433–6.  *183, 187, 298*

560. Yokohama, Y. & Misonou, T. (1980). Chlorophyll *a*:*b* ratios in marine benthic green algae. *Jap. J. Phycol.*, **28**, 219–23.  *181*

561. Yu, M., Glazer, A. N., Spencer, K. G. & West, J. A. (1981). Phycoerythrins of the red alga *Callithamnion*, **68**, 482–8.  *192, 307*

# INDEX TO ORGANISMS

389

# SUBJECT INDEX

............................................................

absorbance, 45–8, 201–7, 212–3; and
action spectra, 248
absorptance, 12–14, 201–2, 209–12; and
action spectra, 247–53; of plant tissue,
209–10; of spherical cells, 206
absorption coefficient, 12–15, 18–20,
21–22, 45–70, 102; due to different
components, 49–66; due to
phytoplankton, 203, 209, 213–4;
influence on field, 122–33; measurement,
45–9, 63–4, 82, 98; and reflectance,
115–8, 149–50, 153; specific, 204–5,
213–4; total, 48–9, 65–6, 201
absorption cross-section, 75; and action
spectra, 247–8, 253; of photosystems,
319; of phytoplankton, 204–5, 208–9,
211–13
absorption of light (general), 42–70; and
chloroplast movement, 350–2;
components responsible for, 49–66,
68–70; measurement, 45–69; mechanism,
42–45; in photosynthetic system, 193–4;
rate in aquatic plants, 209–14
absorption meter, 48–9
absorption spectra: of aquatic angiosperm,
210; of gilvin, (yellow substances,
dissolved humics) 51–60; of multicellular
algae, 249, 252; of phytoplankton, 62–4,
65–6, 208, 212, 249, 321, 337; total,
65–6; of tripton, 60–2; of water, 49–51
action spectra (photosynthesis), 247–53,
300–1, 318–20
adaptation: criteria, 330, 332; depth,
323–9; intensity, 304–14; morphogenetic,
333–4; rapid, 344–54; seasonal, 339–44;
spectral quality, 293–304, 314–23;
temperature, 269–74, 283
adaptochrome, 316
aerosol scattering, 146–9
algal distribution, 255–6; Australian,

299–300; and chromatic adaptation,
291–304; northern temperate, 293–6;
tropical, 296–9
algal mass culture, 356–7
allophycocyanin, 189–92, 194, 317
alloxanthin, 183–4
Angstrom exponent, 148
angular distribution (light field), 121–9 (*see
also* radiance distribution)
apparent optical properties, 18–20
asymptotic radiance distribution, 121–9
atmosphere, effects of, 26–36; aerosol, 137,
144–9; in remote sensing, 136–7, 144–9,
152, 155–7; Rayleigh, 137, 146–8
atmospheric transmittance, 146–8
ATP, 171, 195–9, 261–2, 270, 274
attenuance, 13–14
attenuation coefficient (beam), 13–14, 22,
82, 89, 96, 161; measurement, 76–7
attenuation length, 20
average cosines, 10–11, 19, 22, 82, 100,
103, 124–8, 131
azimuth angle, 6, 101, 103, 124, 127, 143

backscattering coefficient, 17–20; ratio to
absorption coefficient, 115–18, 149–50,
153
Beer's Law, 204
benthic algae, 174, 255–6, 277, 280–4, 355;
diatoms, 227, 280–4, 355; distribution,
291–304
bicarbonate utilization, 264–9
biliproteins, 45, 174, 176, 178–9, 186,
189–92, 203, 217, 296; absorption
spectra, 189–90; action spectra, 252–3;
and chromatic adaptation, 293–304,
314–20, 336–9; chromophores, 191–2;
distribution, 191; energy transfer, 194,
197; ratio to chlorophyll, 306–7, 317–20,
337–8, 344; structure, 191–2

# INDEX TO WATER BODIES

*General references to waters of a particular type are listed in the subject index.*